MIND AT LARGE:
KNOWING IN THE TECHNOLOGICAL AGE

RESEARCH IN PHILOSOPHY & TECHNOLOGY, Supplement 2
Series Editor: Frederick Ferré, University of Georgia
Supplement Editor: Carl Mitcham, Polytechnic University, New York

Series Editor: Frederick Ferré, *University of Georgia*

Supplement Editor: Carl Mitcham, *Philosophy & Technology Studies Center, Polytechnic University, New York*

Editorial Board for Research in Philosophy & Technology

George Allan
Dickinson College

Hans Lenk
University of Karlsruhe

Ian G. Barbour
Carlton College

W. Davis Lewis
Auburn University

John Crompton
Vanderbilt University

Carl Mitcham
Polytechnic University

Don Ihde
State University of New York Stony Brook

Friedrich Rapp
University of Dortmund

Nicholas Rescher
University of Pittsburgh

Joseph Kockelmans
Pennsylvania State University

Egbert Schuurman
Free University, Amsterdam Technical University, Delft

Erazin Kohak
Boston University

Melvin Kranzberg
Georgia Institute of Technology

Stephen Toulmin
University of Chicago

Walther Zimmerli
Technical University Braunschweig

MIND AT LARGE:
KNOWING IN THE TECHNOLOGICAL AGE

by **PAUL LEVINSON**
Fairleigh Dickinson University

JAI PRESS INC.

Greenwich, Connecticut   London, England

Copyright © 1988 JAI PRESS INC.
55 Old Post Road, No. 2
Greenwich, Connecticut 06830

JAI PRESS LTD.
3 Henrietta Street
London WC2E 8LU
England

All rights reserved. No part of this publication may be reproduced, stored on a retrieval system, or transmitted in any form or by any means, electronic, mechanical, photocopying, filming, recording or otherwise without prior permission in writing from the publisher.

ISBN 0-89232-816-9

Manufactured in the United States of America

For Marshall McLuhan

CONTENTS

SERIES EDITOR'S INTRODUCTION	
Frederick Ferré	xi
PREFACE	
Paul Levinson	xiii
ACKNOWLEDGMENTS	xvii

Chapter 1. MIND AS A PRODUCT OF EVOLUTION — 1

Evolutionary Epistemology	2
Evolution, Reality, and Truth	3
The Provocation of Innate Knowledge	6
Evolution and the Uncertain A Priori	7
Biological Embodiments of Knowledge	9
The Technological Preemption	11
Notes	12

Chapter 2. RATIONALITY AND EVOLUTION — 17

The Biological Basis of Reason	20
Are the Roots of Reason Rational?	22
Prerational vs. Irrational: Self-Trancendence and the Bootstrap Emergence of Reason	25
On the Necessary Nonrationality of First Causes	31
Inescapable Logic	35
Notes	37

Chapter 3. FALLIBILISM AND OPTIMISM — 45

Evolutionary Origins and Limitless Knowledge	47
Radical Fallibilism: The Evolutionary Advantages of Error	50
Bringing Ideas to Life	56
Notes	57

viii / *Mind at Large*

Chapter 4. **TECHNOLOGY: THE EMBODIMENT OF HUMAN IDEAS AND THE UNNOTICED PHILOSOPHIC REVOLUTION**   63

    Unideated Material and Immaterial Ideas   64
    The Missing Link in Kant's Interactionism   66
    Marx on Mind in Material   73
    A Technological Adjustment of Popper's "Three Worlds"   76
    Productive Knowledge   81
    Notes   82

Chapter 5. **TECHNOLOGY AS AN AGENT OF COGNITIVE EVOLUTION**   89

    The Unintended Technological Library of Successful Knowledge   90
    The Biological Antiquity of Technology and the Pursuit of Knowledge   94
    Telescopes, Microscopes, and Human Equation to the Cosmos   96
    Computers and Tractable Immensity   101
    Meta-Cognitive Technologies   106
    Notes   109

Chapter 6. **THE DOUBLE ENTENDRE OF COMMUNICATIONS MEDIA**   117

    The Agony and the Ecstasy of Abstraction   119
    The Book in History   128
    Electronic Facilitations of Print   135
    The Conquest of Immediacy and Eternity   138
    The Immaculate Conception of Photography   139
    Electronic Mercury   141
    Return to Sender   146
    Notes   148

Chapter 7. **SOCRATIC TECHNOLOGY: MEDIA AS MIRRORS OF THE MIND**   163

    Escaping the Cell of Subjectivity   164
    Lights, Camera, Cognition!   166
    Artificial Intelligence and Real Life   172
    "Anthropotropic" Media   183
    Notes   186

Chapter 8. **TECHNOLOGY AND THE GLOBALIZATION**
OF INTELLECTUAL CIRCLES   **197**

    A Realistic Sociology of Knowledge   200
    Communities of Representations   203
    Flesh and the Growth of Knowledge   209
    Beyond Planet   213
    Notes   214

Chapter 9. **TECHNOLOGY AND THE EVOLUTION**
OF THE COSMOS   **221**

    The Burden of Rational Technology   224
    Turning the Universe Inside Out   228
    The Face in Our Future   231
    Notes   233

BIBLIOGRAPHY   239

BIOGRAPHICAL SKETCH OF THE AUTHOR   253

INDEX   255

Series Editor's Introduction

With this Supplementary Volume, *Research in Philosophy & Technology* makes two transitions. First, a new General Editor and Editorial Board are making their initial appearance (although Supplement Editor Mitcham has done the work). Second, an original book-length, single-author work is being published. We hope our readers will enjoy Paul Levinson's style and provocative arguments as much as we have. If others are working on longer manuscripts on our themes, this venture should be an encouraging precedent. *Research in Philosophy & Technology* is eager to serve our growing field in whatever ways are most useful.

Current changes are not by any means total: Carl Mitcham, our capable Supplement Editor, is known to all in our field for his leadership in philosophy and technology and for his important contributions to this publication. Our publisher remains the same and has reiterated firm commitment to our enterprise. Several members of the previous Editorial Board remain on the new Board to help both with continuity and with new ideas for the future. Outgoing Editor Paul Durbin, whose wisdom and energy have been prime movers of this project through all of its previous existence, has generously made the transition easy and deserves boundless thanks for all he has done and continues to do for our field.

One new development that our readers should anticipate, as the regular annual series continues, is our plan to provide a series of volumes highlighting specific themes over the next few years. The next regular volume will concentrate on Technology and Ethics. Future volumes will include both invited and submitted papers on themes such as Technology and Religion, Technology and Politics, Technology and Economic Theory, Technology and Aesthetics, Technology and Phenomenology, Technology and Language, Technology and History, and Environmental Technology.

Specific volume numbers and deadlines for submission of papers to these theme volumes will be announced.

In addition to the featured thematic sections of these announced volumes we shall continue to publish important ongoing discussion papers, regardless of topic. In addition, *Research in Philosophy & Technology* will continue its tradition of providing its readers bibliographical information on works in philosophy of technology, inclusively conceived.

The inclusiveness of our conception of philosophy of technology should be stressed. The whole philosophical gamut of questions is—or should be—raised by comprehensive and critical reflection on technology. To deal with them the whole range of philosophical methods is needed. This requires an editorial policy that is respectful of a healthy pluralism in philosophical approaches. We shall look for good philosophy in all the articles we publish, but "good philosophy" will not become a code phrase for trimming submissions to anyone's party line.

We shall also be careful not to become identified with either "pro"-technology or "anti"-technology points of view. The issues are far too complex for such capsule thinking. Well-argued defenses and attacks will be equally welcome, but most of our articles will be neither attacks nor defenses but philosophical examinations of conceptual issues raised by the insistent presence of technology in human history and in our own time.

Frederick Ferré
Department of Philosophy
The University of Georgia
Athens, Georgia

Preface

Mind at Large studies human knowledge as a technological product. It examines the ways that the human brain works through technology to produce accurate knowledge of the world—knowledge that serves as the basis for subsequent technological transformations of the world to human specifications. This growth of knowledge and human reshaping of the environment is linked to the evolution of organisms and human life.

By "mind" I intend nothing exclusive of the human brain. Rather, I refer to the unique kilogram of material in the human skull capable of thinking, and therein transcending in a variety of ways the living material of organisms and the nonliving material of the cosmos from which it arose. The most enduring of these ways are technological, by which I mean the embodiment of ideas produced by our brains in external material forms, or, from the perspective of the external world, the remaking of the external material to human dictates.

Trees in the wild are examples of nontechnological pieces of material (in this case, living material). The notions of somehow using these trees for shelter or transportation or communication are ideas situated at first in the substrate of the human brain. Huts—the fusion of the material tree with the human idea—are resultant technology, as are canoes and paper. The perspective of this book is that huts, canoes, and paper make all the difference.

The book begins with a discussion of how our cognitive capacities or modes of knowledge generation are consequences of our natural selection as a species no less than our modes of respiration and digestion. This evolutionary approach tells us that our theories grow in an imperfect trial-and-error manner akin to the growth of living organisms, with external reality acting as an editor or regulator of the many ideas we propose to it. Chapter 2 tackles the perennial philosophic problem of how can we be rational, and offers an evolutionary account of the emergence and continued operation of our rationality. The ubiquity of error that accompanies trial-and-error growth is considered in Chapter 3, and is found to be a necessary ingredient in evolution of both organisms and ideas. Today's mistakes are more than just castoffs; they often become tomorrow's truths and workable technologies.

Chapter 4 shifts the focus to technology proper, and considers the impact that the embodiment or fulfillment of human ideas in material technology has on both the human creators and the recreated world. Anticipations of this "techno-evolutionary" perspective are traced in the philosophies of Kant, Marx, Dewey, Karl Popper, and others, each of which is found, however, to lack a crucial piece of the technological picture.

The intended and unintended contributions of technology to the pursuit of knowledge are the subject of the next four chapters. Chapter 5 starts with the observation that every working technology, regardless of its purpose, embodies some minimally accurate knowledge of external reality—accurate enough, that is, to allow the technology to function. The sum total of material culture on our planet, then, comprises a sort of grand unintended Library of Congress of minimally accurate knowledge, ideas that have survived a punishing confrontation with reality, there for the reading if we choose.

Technologies also have been deliberately employed in the generation, criticism, and dissemination of knowledge—the three stages of knowledge growth that correspond to the generation, elimination, and propagation of organisms in the natural world. Knowledge originates in the impact of the external world upon our cognitive capacities, and technologies assist us in all aspects of this generation process. Telescopes and microscopes bring our eyes into contact with the far reaches and minute corners of the cosmos, making our perception more equal to these immensities of experience; and the computer, in the first of many services to knowledge considered in this book, predigests the huge quantities of experience supplied by telescopes and microscopes, rendering them more amenable to human comprehension. The heightening of human perception and cognitive power by such technologies reduces the disproportion of the human being to the universe that so vexed Pascal.

Chapter 6 examines the role of technology in the criticism and dissemination of knowledge, a role performed by special types of technology known as communications media. All technologies embody knowledge; some, however, not only embody knowledge but also carry or represent an additional cache of ideas. A book, for example, is both an embodiment of an idea about how to communicate (as opposed to, say, speaking or making a film) and a carrier of usually quite different ideas in the words on its pages. These words may convey romance, or scientific hypotheses, or in the case of the book you now are reading, philosophic and historical accounts of the evolution of knowledge. In all cases the idea embodied and the ideas represented (communicated) are distinct, although they may overlap in a book whose subject is the book as a communications medium.

We join the history of communications media at the highly abstract processes of speech, writing, and print that were the only media of earlier civilizations, and proceed to the more recent replicative devices of photography and its motion, color, and video offspring. Electronics integrate this gamut of media, permitting the instant or delayed communication of voices, print, and pictures.

The impact of dominant systems of media on larger social-cognitive institutions is explored, with special attention to the relationship of the alphabet and the rise of monotheism, and the printing press as a spur to both the Protestant Reformation and the Age of Discovery. Just as the previous chapter on knowledge generation may be viewed as a technological extension of the interactionism of Kant and Piaget, so Chapter 6 extends to cognition a study of media pioneered by McLuhan.

An unintended consequence of a different kind, generally unexplored in connection with communications media, is the subject of Chapter 7. In transmitting information, media also externalize elements of our perceptual and cognitive processes that produced this information; thus, these media allow us to examine elements of our cognition that heretofore were locked inside our skulls. The motion picture is one such example, projecting on the screen the Kantian faculties of time and space organization that we use to make sense of our real, nonreel, world. Computers are another, although here several fundamental differences between current artificial intelligences and intelligence of the human sort are examined. The gulf between the computer and the human brain amounts to the computer not being alive, with all the connotations and characteristics that being alive entails.

Whatever the enabling technology, the growth of knowledge has always come down to people-to-people, or the exchange of ideas from one person to another. Chapter 8 looks at how technologies have created and transformed the places in which this essential human transfer occurs. One arm of this transformation is transportation technology, which now allows people to travel from country to country with something approaching the convenience of walking from hall to hall on a large university campus. The other arm is long distance communication, in particular the telephone's recreation of the ambience of a Victorian parlor on a global scale. The transmission of texts via computer to and from people's studies may be even more significant to the growth of knowledge, providing for instant communication of highly abstract, detailed descriptions at the stroke of a finger, sent and received in the comfort of home. Twenty-three hundred years after Socrates decried writing as destructive of immediate dialogue, descendants of technology made possible by writing have given the written word all the facility of the spoken word and more. Indeed, the ascendency of intellect made possible by such easy and powerful communication of ideas via personal computer may be as profound a revolution for the human psyche as the catapult of emotionality by Freudian psychology at the turn of the century.

The book concludes with a consideration of what this escalating whirlwind of human knowledge holds for the larger cosmos in which our species and planet are situated. As far as we know, until the arrival of human brains the evolution of the cosmos proceeded in a blind, unplanned, naturally selective manner. Now with each idea that we bring into material being through technology, we are in effect increasing the planned or intended content of the universe, shifting the ratio ever more slightly towards a humanly proposed rather than blindly evolved existence. Because we and our scheming brains are themselves products of natural selection, the flow of planned technologies from our brains can be seen as a case of the unplanned universe turning itself inside out.

The deliberate application of technology not only to transforming our surroundings but also to generating and disseminating knowledge that serves as the basis of the transformations greatly accelerates this process. Very soon now our embodied theories about space travel will give us permanent residence in the byways beyond Earth. And with this residence comes the responsibility of knowing that if we destroy ourselves, if our rationality fails, we will take a much larger portion of the cosmos along with us. For better or for worse, we may well be the custodians of the cosmos, and the handwriting is on the wall: through the agency of technology, the future of the universe lies increasingly in the human mind.

Paul Levinson
New York City
November 1986

Acknowledgments

I first and foremost thank Carl Mitcham, without whose encouragement and good offices this book would not have been published. Mitcham is the rare twentieth century editor who assesses ideas and manuscripts not on the basis of how they compare with his own scholarly theses; his support for my work, whether it agreed with his views or not, was a constant source of strength for me throughout the writing of this book.

I also thank the following people for helpful comments on all or parts of the manuscript in various stages of completion: Susan O'Connell of the American Association for the Advancement of Science; Peter Skagestad of Trinity College; Aharon Kantorovich of the Hebrew University of Jerusalem; Walter Orr Roberts, Edward Glenn, Jeff Stamps, and Christopher Wright of the Western Behavioral Sciences Institute computer teleconferencing network; Mario Bunge of McGill University; and Sylvia Engdahl of Connected Education. My students at The New School for Social Research, The Western Behavioral Sciences Institute, and Fairleigh Dickinson University also provided valuable forums for discussion of some of the ideas in this volume.

Thanks also to the staff of the Fairleigh Dickinson University Library, Peter Haratonik of the New School's MA in Media Studies Program, and Joseph Zuckerman for some last minute assistance in tracking down citations.

My family, immediate and extended, also was crucial to my commencement and completion of this book. I began the volume in the summer of 1982 at our cottage on Cape Cod Bay, and the keen interest in some of the themes by my nephew Zachary Thacher (9 years old at the time) was a valued stimulant to the project. My wife, Tina Vozick, provided essential assistance with everything from the minutist details to the general concept of the volume. The book was further made possible by the family responsibilities she shouldered to give me time to devote to this book. Our son Simon, born in 1983, and our newborn daughter Molly have been humanizing influences in all aspects of my life, and I hope a bit of this shows in the following pages.

Chapter 1

Mind as a Product of Evolution

All inquiries must start with some foundation or assumptions. In this age of scepticism, acknowledgment of foundations is not very fashionable, but even the most devastating nihilists must make use of logic if their nihilistic arguments are to have any force. The only hope irrationalists have of convincing us is to argue for the logical sense of irrationality: to appeal to us on rational grounds. And of course, all appeals and arguments must be presented through some communicative system or language.

Logic, rationality, and language, then, are necessary tools of any intellectual activity. And when we see sceptics making use of these tools, we have at that instant refuted their claim that no foundations or assumptions are needed in our understanding of this world. Actually, the sceptics have defeated these claims on their own.

This book, however, rests on an assumption that goes beyond the prerequisites of logic and language. In investigating the nature and operation of human cognition, I accept the Darwinian theory of evolution, and in particular its relevance to human beings. Seen in an evolutionary context, human activities such as thought and technology are closely related, and traditional philosophic problems such as "how do we know" become more tractable.

In assuming the accuracy of Darwinian theory and using it as a foundation for an inquiry into the relationship of mentality and technology, I do not endorse any particular interpretation or genre of Darwinian theory. Our inquiry need take no stand, for example, on the current controversy regarding whether evolution proceeds in gradual steps or in sudden spurts. To be sure, each of these views has implications as to whether knowledge develops gradually or in revolutionary leaps, but both are fully compatible with the more fundamental view of knowledge as a biological, technological activity.[1] For the purposes of this book, I would not even mind if there were no such things as genes, though we have no reason to doubt their existence. Nor would I want to suggest that any aspect or variant of Darwinian theory is completely correct. Indeed, I hold to the contrary, along with Karl Popper and other fallibilist philosophers, that all scientific theories are imperfect conjectures, which, if we have luck and persevere, we may improve through criticism of all sorts including empirical testing. The assertion of the general accuracy of Darwinism does not exempt this theory from such a critical process.

So what, then, am I asserting? First, human beings, along with both our organic attributes such as mind and our inorganic creations such as ideas and technologies, are the products of developmental, historical, and continuing interchanges with the environment. This claim assumes the accuracy of evolutionary theory as opposed to, say, creationism as an explanation of our existence. I then claim that the relationship between us and our environment is one in which we generate organic and inorganic possibilities, and the environment eliminates "unfit" possibilities, leaving the remaining ones standing as at least temporary actualities. This assumption asserts the accuracy of a Darwinian as opposed to a Lamarckian conception of evolution.

Notwithstanding the recent revival of creationism and Lamarckian accounts of evolution[2] (the two to be sure are in no manner comparable, as creationism is simply unscientific and uncriticizable dogma), I am aware that the assumption of a generally Darwinian approach to evolution is nothing extraordinary or even provocative. Nevertheless, I call attention to the Darwinian underpinning of this book because in the unlikely event that Darwinian theory should prove fundamentally wrong—if, say, creationism were indeed the correct explanation of the genesis of human life—then much of what follows in this book would likely be wrong as well.[3]

EVOLUTIONARY EPISTEMOLOGY

The linking of mind to evolution opens up many possibilities for our understanding of mental operations. For example, Donald T. Campbell notes that just as biological evolution proceeds in a three-step process of generating new organs and/or organisms, eliminating unfit characteristics and/or species, and propagating those that survive, so the growth of knowledge proceeds in a similar process of generating or creating new ideas, eliminating via criticism and testing those ideas found to be incorrect, and disseminating through education and mass media those ideas that have not yet been disproven.[4] These and other evolutionary metaphors will be taken up further in the course of the book.

But the heart of what Campbell calls "evolutionary epistemology" is the direct, nonmetaphoric assertion that the mind is literally a product of biological evolution—that just as we have lungs and breathe the way we do because our respiratory system works in the Earth's atmosphere, so we have brains and think the way we do because our cognitive system is well adapted to our environment, that is, our cognition told us what we needed to know in order to survive as a species. Note that the connection between respiration and cognition is quite real in that the failure of either system would have resulted in our physical nonexistence. As George Gaylord Simpson has bluntly described the cognition-related area of perception, "The monkey who did not have a realistic perception of the tree branch he jumped for was soon a dead monkey—and therefore did not become one of our ancestors."[5]

The evolutionary view of cognition, then, provides an initial answer to the old philosophic problem of how we come to know so much about the world: we know so much because our ways of knowing are evolutionary adaptations to our environment, honed through eons of exposure so as to provide us a means of understanding our environment accurate enough to have enabled our survival to date. Of course, evolutionary adaptation is not necessarily perfect: just as we make mistakes with our hands, so we make mistakes with our minds. More soberingly, we use our fingers to pull the triggers of guns that kill us, and use our evolutionarily adapted mentalities to invent far more monstrous weapons such as atom bombs. Evolutionary adaptation thus by no means guarantees continuing survival even if the environment that is adapted to remains the same; in the case of our high-powered and flexible mental capacities, which not only understand but also alter the environment via technology, we incur special dangers of misuse.

Nonetheless, an optimism of sorts is a consequence of evolutionary epistemology. In obtaining a natural answer to the question of how we know so much about the world, we undermine the sceptic's claim that we have no such knowledge, which in turn was a response to the ancient view that our knowledge was a gift from God or a mysterious reflection of some unfathomable, eternal verities. Mind evolutionarily construed offers an alternative to the despair of nihilism on the one hand, and the naiveté of blind faith and the fantasy of extreme metaphysics (which provoked nihilism) on the other.

EVOLUTION, REALITY, AND TRUTH

An evolutionary account of cognition also provides an alternative to two rival claims within the tradition of rational philosophy, namely the "subjectivist" or "idealist" view that nothing of the world exists independent of our conception and perception of it, and the "empiricist" claim that nothing exists in our minds independent of the impressions the world makes upon us. But in order for the environment to have some shaping effect upon our cognition—in order for evolution to have some impact upon mind—there must be an environment or world that exists independently of the mind's conception of it, and some way of passing knowledge from generation to generation that goes beyond individual perception, that is, through inheritance or genetics. Evolutionary epistemology thus posits both the existence of an external world which is real and independent of the mind (as against the arguments of subjectivists), and a significant aspect of mentality that is inherited or genetically derived, and thus is innate or a priori to our first individual perceptions of the world (contrary to the claims of the extreme empiricists).[6]

Since no one other than Eastern mystics, some 20th century physicists, and a few philosophers throughout the ages has ever seriously denied the existence of external reality,[7] evolutionary support for "the real" might hardly seem noteworthy. But if the notion of reality has not often been seriously rejected, the weight of much 20th century thought has been brought to bear against the related concept of objective, independent truth. Evolution's insistence on reality thus derives much of its epistemological significance from the direct support that such insistence lends to the related notion of truth.

The oppression of truth as a regulative mechanism has resulted primarily from many strains of cultural relativism that contend or imply that truth (if it is to be admitted as a useful or meaningful construct at all) is a consequence of language, or mode of communication, or social institution or environment, or historical circumstance, or of some combination.[8] In each of these cases, because truth is depicted as a product of some central human activity, and thus stands completely inside the sphere of this activity, truth cannot serve as a standard against which to compare this activity with other activities, or even to evaluate ideas outside of their performance in the given activity. Truth thus becomes trivialized as an operational or instrumental measure of how well ideas and actions work in specific, nonequatable realms of human life, or as a simple shorthand for "that which works," regardless of why. Frequently lost in this shorthand is a quest for deeper explanation and meaning: the curative power of spider webs applied to wounds is plain, functional truth, and the antibiotic effect of spores attached to webs may be irrelevant to a culture that stops at this functional level (and considers spiders magical and sacred). Yet the antibiotic action of these molds is the source of the spider web's power, and is the aspect that can be applied with greatest success beyond the realm of the spider web.

Dilutions of the notion of truth are actually a partial consequence of the untenability of a very high-powered mathematical definition of truth prevalent in the early part of this century. Seeking to clarify and elevate the concept of truth beyond the commonsense notion of correspondence to facts or reality, Bertrand Russell and others had defined the truth of a theory or system in terms of its internal consistency, an approach that allows elegant, formulaic "proofs" of truth. But when Kurt Gödel and others later demonstrated on mathematical grounds that no system could ever be both complete and consistent—which, according to Russell's criterion of truth, meant that no system could ever be true—the general notion of truth and not merely the specific consistency approach to truth fell victim.[9] If truth was impossible in the clean, lean lines of mathematics, how could it possibly exist in the messier realm of the real world?

This messy real world, and the commonsense concept of truth as correspondence to the world, is precisely the world and the truth to which evolutionary theory speaks. Remember that evolution presupposes an external, independent real world, which acts as a selecting force on organisms and their attributes, including the human mind. Since to act as a selecting force is also to act as a regulative standard, we might then say that evolution is the truth of the environment or reality clashing with and eliminating unfit possibilities proposed by life. Alternately, evolution is the reality of truth expressing itself organically through natural selection. Note that this goes well beyond the useful analogy that just as reality selects organisms for survival in physical evolution, so truth selects ideas for survival in the development of theories. The relationship of reality and truth in the perspective of evolution is the far more powerful one of equation or identity, in which reality and truth are merely different faces (physical and epistemological, respectively) of the same selecting agent.

The truth of evolution suffers from neither the triviality of relativistic truth nor the dogmatism of presumed "absolute" truth, for although the selecting environment/reality/truth is very much intertwined with the organisms it acts upon (indeed the environment is largely comprised of these organisms), the vast majority of the environment is clearly separate and external to any one organism or attribute. Thus, environment is not at all a relative product of the organism in the way that, say, truth is held to be a completely relative product of language by a linguistic theory of truth. But although environmental truth is not relative to the organism in this sense, it is changeable and to some extent influenced by the actions of the organism and thus in no sense eternal or absolute. Indeed, the way in which humans change the truth of their environment through technology will be a primary theme of this book.

Conceived of in terms of evolutionary reality, then, truth is objective without being absolute and alterable without being relative. Although the old commonsense view of truth as correspondence to reality has already been reasserted by some philosophers (and by Alfred Tarski in particular)[10] on grounds independent of evolution, evolutionary epistemology enriches and gives flesh and blood to this notion of truth by situating it in a living world.

THE PROVOCATION OF INNATE KNOWLEDGE

A few comments are in order about the other consequence of an evolutionary account of cognition: that significant aspects of mentality must be innate or in place prior to individual experience in the world.

Aside from a perennial philosophic appeal, the issue of whether, how much, and what part of our mental capacity is innate has given rise to a huge amount of lively and sometimes acrimonious debate in psychology, political science, and social theory.[11] Certainly the genetic justification used by the Nazis gives cause for serious concern about the misapplication of innatist theories. This concern has been fanned more recently by the claims of Arthur Jensen and others that humans differ quantitatively in their cognitive capacities along innate, racial lines.[12] On the other hand, as Noam Chomsky for one often has pointed out, focus on the extent to which aspects of human mentality exist prior to individual experience with the environment can provide powerful arguments for an egalitarian point of view;[13] most genetically figured traits (such as eyes) vary insignificantly across our species, usually differing in superficial packaging (for example, eye color) rather than in performance. It thus seems fair to say that notwithstanding the intensity of debate about the social consequences of innatist theories, pursuit of the possibilities of evolutionarily determined or innate knowledge does not necessarily entail commitment to either elitist or egalitarian political structures.

Moreover, the focus of this book will be on the ways that human species-wide attributes, and not individual or group variations, operate in production of knowledge and technology, and therein change the world. Since the proximate agents of knowledge and technology production are individuals and groups, the assumpion here will be that the knowledge capacities discussed (for example, Kant's capacities of space and time organization) are operative in all but the most severely damaged human individuals.

Nonetheless, evolutionary epistemology ought not on moral or political grounds be dissuaded from investigating the possible biological basis of individual cognitive differences—for example, the cognition of genius—that may exist, or from searching for differences in cognitive styles that may be tied to genetically structured groups. Thus, just as eye and skin color are genetically associated with different human populations, so we someday may identify subtly different "colors" or "flavors" of cognitive modes that vary along genetically determined group lines, with no more pejorative implication than we place (or should place) on eye and skin color. Awareness of such hypothetical subtleties could be of great help in tasks such as design of research teams, and fear of the abuse of such possible cognitive discoveries by bigots and lunatics should not be allowed to preclude their investigation.

Nor, sadly, if the testimony of history is any authority, would preclusion of any line of research, genetic or otherwise, save us from social abuses. Stalin the environmentalist was arguably as deplorable as Hitler the geneticist (and, conversely, Jefferson the environmentalist was as fervently committed to freedom as the innatist Rousseau). About the only proof we can ever muster against such despoilers is truth. And truth too must be the sole goal of science and philosophy, regardless of perceived social consequences.[14] Understanding ourselves and our universe as we really are and not as we want these to be is our only hope of becoming what we want to be.

EVOLUTION AND THE UNCERTAIN A PRIORI

What is the nature of those aspects of mentality that an evolutionary account of cognition suggests are innate? In the history of philosophy, conceptions of a priori knowledge have in general progressed from notions of numerous clear and distinct innate ideas matching discrete units of external experience (in the Cartesian mode, for example) to more recent views of innate cognitive abilities as a smaller number of generalized structures or capacities that process or make sense out of a wide variety of external experiences. This more recent "process" approach is in evidence both in the genetic epistemology of Piaget and in the innate rules of language use postulated by Chomsky.[15] Such a view of innate knowledge as a limited number of broadly applicable processes is also very much in keeping with what we would expect of mentality as an evolutionary adaptation: just as our mouths are used for a variety of tasks including eating, drinking, breathing, speaking, and nonverbal demonstrations of affection, so should any individual innate cognitive capacity serve a variety of purposes. Evolution is not generous with the number of radically different patterns it permits, but makes the most out of those that it does allow.

The traditional philosophy most appropriate to evolutionary epistemology was propounded by Immanuel Kant, who died five years before the birth of Darwin in 1809 but nonetheless offered a view of human mental operations that fits almost presciently into evolutionary theory. Kant explained cognition as the action of a handful of innate intellectual categories ("structures" in contemporary terminology) upon the raw experience provided by our organs of sensation. Although Kant offered no reason other than the logical necessity of his own analysis why such innate capacities should exist (the only other explanation available in Kant's day would have been that they were gifts from God), the situation of the mind in an evolutionary framework explains Kant's structures as species-wide adaptations to the environment which are genetically transmitted to every individual.

Konrad Lorenz, who has done much to illuminate the link between Kant and evolutionary theory, cautions against a complete equation of Kantian and evolutionary epistemology on the grounds that Kant intended "a priori" to mean absolutely prior to any and all possible experience, whereas in evolution innate structures are subsequent or a posteriori to the selecting action of the environment upon the species, and in this sense are not prior to the species' experience with the environment. (Lorenz agrees that genetically derived or innate knowledge is absolutely prior to all experience for any individual.)[16] This raises a most interesting question about the ultimate origin of knowledge in the universe, whose consideration in Darwinian terms may yet uphold Kant. For if the environment selects from the possibilities first proposed, as it were, by the organism (rather than directly imposing characteristics on the organism, as in the Lamarckian scheme), then even though a cognitive faculty might come to exist as a result of the environment's selecting activity, that selecting activity could take place only after the appearance of some type of cognitive characteristic for the environment to act upon. In other words, natural selection requires preexisting characteristics from which the environment can select. While such characteristics are themselves no doubt the product of previous species experience with the environment, the natural selection model is quite clear in assigning the initial generative function to the organism rather than the environment. Kant's insistence on the absolute priority of the human mind in its sojourn in the world thus seems not incompatible with a naturally selective epistemology after all.[17]

Kant's epistemology may be more in need of evolutionary improvement in its claim of certainty for a priori operations, and in its neglect of the vast majority of times that our innate expectations run into environmental brick walls. In life, nothing is certain and success is rarely the rule; indeed the evolutionary process itself can be defined in large part as the elimination of previously fit characteristics now rendered unfit by the shifting sands of fortune and environment.

Although Charles Sanders Peirce wrote about both fallibilism and evolutionary theory, Karl Popper more than any other modern philosopher has emphasized the uncertainty and innateness of our cognitive processes in a systematic epistemology that finally brings Kant fully into our evolutionary century.[18] Popper views knowledge as generated by imperfect probes sent out by speculative faculties to meet the realities of the environment, and he therein has constructed a model of human knowing both analogous to and continuous with the hit-and-miss nature of biological evolution. Moreover, whereas Kant dwelt upon the way our innate faculties condition and shape our experience, Popper considers the most significant aspect of knowledge growth to be the disappointment of expectations or the refutation of conjectures. This "falsificationist" view of learning and science, in which incorrect ideas are sought out through criticism and empirical confrontation and then provisionally stripped from the body of conjectural knowledge,[19] also corresponds beautifully and is of a piece with the indirect, trial-and-error process of natural selection.

Thus refined through Popper's fallibilism and falsificationism, Kant's epistemology becomes an apt basis for an evolutionary epistemology. Popper's improvement, however, is no small adjustment of Kant's philosophy. Unlike Kant's formless environment waiting to be structured by human intellect, Popper posits in his falsificationism an environment of intrinsic structure and organization sufficient to confound many of our best conjectures about it. Such refinements make Popper's contribution to evolutionary epistemology as important as Kant's, and warrant Campbell's description of Popper as the "modern father" of a "natural selection epistemology" (although Campbell may be co-deserving of this designation himself).[20]

BIOLOGICAL EMBODIMENTS OF KNOWLEDGE

Thus far we have listed some of the lessons that evolution holds for what we might vainly but accurately term the upper end of the biological knowledge continuum, that is, the implications of evolution for the operation of human mentality, and in particular the abstract human mental realm of philosophic debate. The evolutionary approach also has something to say about knowledge throughout all levels of organic existence, and these living (as distinct from solely human) manifestations of information are essential to the picture of technology that we will detail in this book.

When applied to the organic world as a whole, evolutionary epistemology suggests that evolution can be profitably seen not only as the development of living organisms, but also as the evolution of physical embodiments and transmitters of knowledge about the external environment. The material and structures of all organisms—cilia, leaves, claws, hearts, eyes, and brains—are physical expressions or embodiments of information or knowledge encoded in DNA about how to construct such structures from surrounding proteins. The very survival of these structures suggests that the knowledge they embody is in some sense an accurate reflection of some small or large corner of the world (or not grossly inaccurate, apropos Simpson's monkeys).

Chromosomes and reproductive material occupy a most fascinating and unique position in this schema, for they carry not only the knowledge that is embodied in other living structures, but also the knowledge necessary for the replication of their own transmission or carrier abilities. DNA—the CP/M or MS-DOS of the living world[21]—thus does a double task in biological knowledge transactions, much like the social task of communications media such as books, which not only convey the information on their pages, but also are themselves embodiments of ideas about how to communicate, that is, how to encode information in a book, as opposed to, say, a recording.

Environment-sensitive organs such as eyes and ears (or membranes in less-developed organisms) are also of special interest in that they not only embody knowledge of the mechanisms of the external world (in the case of eyes, information about the nature of light), but also are vehicles for discovery of additional knowledge of external reality. These knowledge-acquisition organs find their technological equivalents in telescopes and microscopes.

Most astounding of all is the kilogram or so of material that comprises the thinking human brain, for this embodiment of DNA instructions has the ability to inquire into the workings and meaning of these instructions. Indeed, through its amanuensis, the hand, and the technologies they construct, the brain can change the shape (conventional technology) and sometimes the structure (as in atom smashing and element building) of the external world, including the instructions themselves (gene splicing) and, to complete the cycle, even the possible sources of thought themselves (as in artificial intelligence).

The biological history of knowledge thus brings us to the technological embodiment and pursuit of knowledge that is the hallmark of human life and the subject of this book. Recognition of organic knowledge embodiment also underscores the symbiotic relationship of information and material in the living and thinking worlds: Whether in the generative and disseminative cores of DNA and the brain, or in physical embodiments of cells and tissue or tools and machines, knowledge goes nowhere without some physical accompaniment. Similarly, cells, wings, and machines are what they are because of their animation by information.

THE TECHNOLOGICAL PREEMPTION

An evolutionary account of cognition thus provides us with:

1. an explanation of the human knowledge process as a biological adaptation to the environment, an explanation that a) asserts against the sceptic that we do indeed have the capacity to know and do in fact possess knowledge, and b) suggests an origin of knowledge that is more accessible and investigatible than the obscure sources posited by religions and purely metaphysical arguments;
2. sibling notions of reality and truth as objective, independent, regulative mechanisms on the growth of life and knowledge;
3. what might be termed a Kantian/Popperian view of innate knowledge, that is, a view that knowledge commences from the action of innate, generalized, imperfect mental capacities that to some extent condition and shape our experience with the world, but operate more significantly as probes that test and challenge the environment; and
4. a view of all organic structures as physical embodiments of DNA-encoded information, suggesting a continuity between embodiments of naturally selected knowledge in the structures of organisms and embodiments of humanly produced knowledge in technologies (bird nests and beaver dams would be an intermediate example).

Evolutionary epistemology thus gives us an orientation, a basis, for attempting to understand how we understand and for explaining how we explain. All epistemology, philosophy, and science aims at explanation. But to paraphrase Marx, what characterizes us as human beings is not only our ability to explain ourselves and our world, but our capacity to change them.[22] Change is a property of evolution and indeed of all existence, and as such is not at all a peculiarly human capacity, but the ability to deliberately change things in accordance with preconceived specifications is (as far as we know) a uniquely human attribute.

The agent through which we give our ideas material expression, and thereby alter the material of which we and the world are composed, is technology. Through technology we embody and extend our ideas, inject our minds into the world and disperse our theories to the far corners of the universe, and therein begin to mold the universe to human design. We may know all there is to know about salt water, chemical reactions, electricity, and simple mechanics; we may even have a theory about the desalinization of water; but until we embody such knowledge and theory in a desalinization plant, the water of the world flows on unchanged.

With technology, then, humans change from products of evolution to the producers of evolution or change, from comprehenders of the existing world to creators of new worlds. The story of technology is actually a story of three protagonists—evolution, mind, and technology—and how the second, a product or consequence of the first, gives rise to and expresses itself through the third, so as to in turn command the first. This book considers some of the many facets of this saga.

Before proceeding with the story, we must consider two issues whose disposition will provide important guidelines for our analysis of technology, but which can be addressed wholly within the realm of the initial two principals, evolution and mind. The first issue concerns the possibility of deliberate, rationally planned activity (technological or otherwise), which amounts to an examination of rationality itself. Cast in evolutionary terms, an investigation of rationality leads to a consideration of the relationship between the nonpurposive, "blind" process of evolution, and the very purposeful, directive procedures of human rationality—in other words, the question of how the human mind, evolutionarily derived, can come to behave in a purposeful, and in this sense, counter-evolutionary, fashion. The second issue, not unrelated to the first, is one of value, of whether the evolutionary account of mentality in which technology is situated is a hopeful or despairing account of human destiny; this problem requires a discussion of evolution and optimism.

We begin with the question of rationality.

NOTES

1. See Stephen J. Gould and Niles Eldridge, "Punctuated Equilibria: The Tempo and Mode of Evolution Reconsidered," *Paleobiology,* 1977, pp. 115–151; Niles Eldridge and Joel Cracraft, *Phylogenic Patterns and the Evolutionary Process,* New York: Columbia University Press, 1980; and Stephen J. Gould, "Darwinism and the Expansion of Evolutionary Theory," *Science,* April 23, 1982, pp. 380–387, for presentations of Gould and Eldridge's "punctuated equilibrium" thesis, or the view that natural selection proceeds in uneven leaps

rather than gradual implements. The emphasis in P-E theory on species rather than individuals as the locus of natural selection suggests a preeminence of groups as opposed to individuals in production and evolution of knowledge—a preeminence not entirely supported by the techno-evolutionary epistemology developed in the present book (see "Metacognitive Technologies" in chapter 4 and see chapter 8). A more drastic challenge to the general Darwinian model comes from Motoo Kimura's "neutral molecular" theory ("The Neutral Theory of Molecular Evolution," *Scientific American,* November 1979, pp. 98–126), which asserts that not only generation but also survival of new organisms and traits are results of processes wholly at the internal molecular biological level, and thus not of the external environment as held by natural selection. G. Ledyard Stebbins and Francisco J. Ayala argue in "Is a New Evolutionary Synthesis Necessary?" *Science,* August 28, 1981, pp. 967–971, and "The Evolution of Darwinism," *Scientific American,* July 1985, pp. 72–82, however, that the evidence of internal molecular clocks as well as punctuated equilibria can be accommodated in the neo-Darwinian or "new synthesis" natural selection model of evolution in place since the 1930s and 1940s. (Gould, "Darwinism," reasonably counsels that attempts to include "neutral" mechanisms as special cases of a larger, naturally selective process run the risk of co-opting the neutralist claims not by evidence but by redefinition of Darwinism; he nonetheless concludes that current molecular, P-E, and other related critiques of Darwinism are compatible if not literally with Darwin's original theory at least with "the fundamental feature of Darwin's vision—direction of evolution by selection.") All discussion of Darwinian theory in the present volume, unless otherwise noted, thus refers to the neo-Darwinian or synthesis model (natural selection, genetics, and population biology) developed by Ernst Mayr, Theodosius Dobzhansky, George Gaylord Simpson, Julian Huxley, G. Ledyard Stebbins, and others, and refined more recently by studies in paleontology and molecular biology.

2. See E. J. Steele, *Somatic Selection and Adaptive Evolution,* Toronto: Williams & Wallace, 1979 for a report of evidence in support of a Lamarckian mechanism (transmission of RNA material from somatic cells to sex cells via a viral mechanism), and Roger Lewin, "Lamarck Will Not Lie Down," *Science,* July 17, 1981, pp. 316–321 for a report of failure to reproduce these results in subsequent, independent experimentation.

3. The exception would be a creationist model whch posited a divine agency that as a first act set the universe in motion along Darwinian lines, and as a second and last act willed itself out of existence (or otherwise disappeared). Such self-effacing divinity, however, would not only be superfluous to natural selection theory, but in its relegation of the divine from a living to a one-time historical entity would deprive creationism of any real impact. See also note 20 in chapter 2.

4. Donald T. Campbell, "Unjustified Variation and Selective Retention in Scientific Discovery," in F. J. Ayala and T. Dobzhansky, eds., *Studies in the Philosophy of Biology,* Berkeley: University of California Press, 1974, pp. 139–161. Campbell defined this area in his "Evolutionary Epistemology," in P. A. Schilpp, ed., *The Philosophy of Karl Popper,* La Salle, Ill: Open Court, 1974, pp. 413–463, which concludes with a listing of more than 150 earlier works that consider mental operations as naturally selective activities (usually in passing rather than in Campbell's comprehensive manner). See Campbell's "Evolutionary Epistemology: Partial Supplementary Bibliography" (distributed at the First International Convocation of the Open Society and Its Friends, New York City, November 1982, and updated regularly since) for more recent works in the area.

5. George Gaylord Simpson, "Biology and the Nature of Science," *Science,* January 11, 1963, p. 84. See also Bertrand Russell's observation that "not only intellect, but all of our [mental] faculties . . . have developed under the stress of practical utility. Intuition is seen at its best where it is directly useful, for example in regard to other people's characters and dispositions. . . . [T]he savage deceived by false friendship is likely to pay for this mistake with his life" (*Mysticism and Logic,* London: Allen & Unwin, 1914/1917, p. 15).

14 / Mind at Large

6. See my "What Technology Can Teach Philosophy" in P. Levinson, ed., *In Pursuit of Truth: Essays on the Philosophy of Karl Popper,* Atlantic Highlands, N.J.: Humanities, 1982, pp. 157–175 for further discussion of idealist versus empiricist traditions, and the evolutionary alternative to this debate.

7. Fritjof Capra's *The Tao of Physics,* New York: Bantam, 1976, exemplifies the confluence of these three sources of reality denial—Far Eastern mysticism, some 20th century physics, occasional philosophers—in its linking of the Buddhist derogation of physical reality with Werner Heisenberg's quantum mechanical rendering of subatomic reality as a function of the observer (and/or the observing instrument). Heisenberg's view was not shared completely by other architects of the quantum mechanical model such as Niels Bohr. See Karl R. Popper, *Quantum Theory and the Schism in Physics,* London: Hutchinson's, 1982, Preface, for discussion of the variety of interpretations of quantum mechanic phenomena. See also Sal Restivo, "Parallels and Paradoxes in Modern Physics and Eastern Mysticism, Parts I & II," *Social Studies of Science,* 1978, pp. 143–181, and 1982, pp. 37–71 for a sociological analysis of the appeal of quantum mechanical–mystical convergence.

8. Wittgenstein's exaltation of language, Kuhn's examination of historical paradigms, and McLuhan's focus on communications media are examples that have refined to the point of replacing the classic notion of truth as correspondence to objective, physical reality.

Wittgenstein distinguishes between truth and "truthfulness," and holds the latter—a measure of the importance or value of information to a particular task—to be far more significant than the former. "The importance of a true confession does not reside in its being a correct confession," he says in *Philosophical Investigations,* trans. G. E. M. Anscombe, Oxford: Blackwell, 1972, p. 222, "it resides rather in the special consequences which can be drawn from a confession . . ." (that is, regardless of whether the confession is objectively true or false, the impact it will have upon the confessor's social group).

Kuhn argues that paradigms or collective scientific perceptions and expectations are constitutive not only of scientific knowledge but in a significant sense "of nature as well" (*The Structure of Scientific Revolutions,* Chicago, Ill.: University of Chicago Press, 1962/1970, p. 110 and ch. 10).

McLuhan is the least explicitly epistemological of this group—and, as we shall see in this book, highly illuminating on the role of technology in the growth of objective knowledge—but his view that media "configure the awareness and experience of each one of us" (*Understanding Media,* New York: Mentor, 1964, p. 35) has a similar effect of distancing the human knower from the comprehension of external reality.

See Peter Munz, *Our Knowledge of the Growth of Knowledge,* London: Routledge & Kegan Paul, 1985, for discussions of the cultural "circularity" of Wittgenstein and Kuhn.

9. Russell's consistency definition of truth is presented in Alfred N. Whitehead and Bertrand Russell, *Principia Mathematica,* 3 vols., Cambridge: Cambridge University Press, 1910, 1912, 1913; Gödel's deflating theorem is in his 1931 paper, "Über Formal Unentscheidel Satze der *Principia Mathematica* und Verwaandter System I," translated and reprinted in Gödel, *On Formally Undecidable Propositions,* New York: Basic Books, 1962. See Douglas Hofstadter, *Gödel, Escher, Bach,* New York: Basic Books, 1979, pp. 18–24, for an accessible discussion of the impact of Russell and Gödel on conceptions of truth.

10. See Tarski, *Logic, Semantics, Metamathematics, Papers from 1923 to 1938,* trans. J. H. Woodger, New York: Oxford University Press, 1956. Popper, who has written extensively about the evolutionary necessity to see truth as a correspondence to reality (and correspondence to reality as a regulative mechanism on the growth of organisms and knowledge), recalls his "intense joy and relief when I learned in 1935" about Tarski's correspondence theory of truth, *Objective Knowledge,* New York: Oxford University Press, 1972/1979, p. 322.

11. See, for example, Gould's *The Mismeasure of Man,* New York: Norton, 1981, for a partisan summary and argument against the existence of an innate capacity "intelligence."
12. Arthur Jensen, *Bias in Mental Testing,* London: Methuen, 1980.
13. Chomsky, *Reflections on Language,* New York: Pantheon, 1975, pp. 128–134. See also the series of articles by Harry Bracken cited on these pages by Chomsky, and my review of *Reflections* in the *Media Ecology Review,* May 1976, pp. 24–26.
14. I agree with Bertrand Russell, writing of his disappointment with the Soviet government in his *The Practice and Theory of Bolshevism,* London: Allen and Unwin, 1920/1949, p. 9, that "If a more just economic system were only attainable by closing men's minds against free inquiry . . . I should consider the price too high."
15. See M. Piatelli-Palmarini, ed., *Language and Learning: The Debate Between Jean Piaget and Noam Chomsky,* Cambridge: Harvard University Press, 1980, for an account of the many similarities and few differences in the work of these two cognitive structuralists. Piaget's emphasis on environmental interaction makes him an heir of Kant, whereas Chomsky's detailed focus on the explicit innate rules of language puts him more in the Cartesian tradition. (Chomsky, however, sees the importance of the environmental role as elicitor of language behavior—what I would call the pouring of the hot water of the environment on the structured powder of instant coffee to produce a cup of coffee.) See the extended discussion of Kant and his relevance for a technological evolutionary epistemology in chapter 4 of the present volume.
16. Lorenz's discussion of Kantian epistemology and evolutionary theory is presented primarily in "Kant's Doctrine of the A Priori in the Light of Contemporary Biology" (1941), trans. Donald T. Campbell in R. I. Evans, ed., *Konrad Lorenz: The Man and His Ideas,* New York: Harcourt Brace Jovanovich, 1975, pp. 181–217; and *Behind the Mirror,* trans. R. Taylor, New York: Harcourt Brace Jovanovich, 1973/1977, pp. 8–9, ff. See also my "Evolutionary Epistemology Without Limits," *Knowledge,* June 1982, pp. 465–502; and chapter 4, note 5 in the present volume.
17. Additional evidence for the a priori character of the "initiating" organism in natural selection lies in the different organic responses that a given environmental pressure may elicit. R. C. Lewontin, for example, points out at least three distinct organic responses to the problem of cold: fur, body fat (as in seals), and brains (in humans who wear fur and eat fat) ("Organism and Environment," in H. Plotkin, ed., *Learning, Development, and Culture,* New York: Wiley, 1982, p. 159). Since the selection pressure (cold) is similar in all three cases, the source of the difference in responses must reside in the a priori structure of the organism.
18. Popper's development of fallibilism dates to his *Logik der Forschung,* 1935, translated and revised for English publication as *The Logic of Scientific Discovery,* London: Hutchinson's, 1959. More recent work includes *Conjectures and Refutations,* London: Routledge and Kegan Paul, 1962/1968, and *Objective Knowledge.* Peirce's work in the area appears in "The Scientific Attitude and Fallibilism," 1896–1899, reprinted in Justus Buchler, ed., *Philosophical Writings of Peirce,* New York: Dover, 1955, pp. 42–59. See Eugene Freeman and Henryk Skolimowski, "The Search for Objectivity in Peirce and Popper," in Schilpp, ed., *The Philosophy of Karl Popper,* pp. 464–519 for a comparison of aspects of the two philosophies.
    Popper has developed most facets of fallibilism far more thoroughly than did Peirce. One consequence of fallibilism especially relevant to technological possibilities, and discussed more by Peirce than by Popper, however, is recognition of the folly of negative dogmatism— the view that we can know with certainty what we can never do or accomplish. See chapter 3 in this volume, and my "Evolutionary Epistemology Without Limits."
19. The elimination of error is provisional because the identification of error may itself be

in error. See also the discussion in chapter 3 about the possibility of an error in one environment functioning accurately in another environment.

20. Campbell, "Evolutionary Epistemology," p. 450. Popper returns the compliment in his "Replies to My Critics," Schilpp, p. 1059, writing of Campbell's essay that "there is scarcely anything in the whole of modern epistemology to compare with it; certainly not in my own work." As W. W. Bartley, III points out in his "Biology and Evolutionary Epistemology," *Philosophia*, September–December 1976, p. 468, the emergence of a naturally selective evolutionary epistemology in the past two decades thus may best be viewed as an effort in which both Popper and Campbell have played complementary founding roles.

21. CP/M and MS-DOS are personal computer operating systems that direct the activity of various computer tasks (such as word processing) and have the capacity to generate copies of themselves.

22. Marx's 11th thesis on Feuerbach (1845): "The philosophers have only interpreted the world, in various ways; the point, however, is to change it."

Chapter 2

Rationality and Evolution

The central question in the history of serious thinking about rationality boils down to this: does such a process exist? Like philosophical doubts about the existence of reality, truth, beauty, and most other things our common sense takes for granted, such questions seem to have a hollow ring. Of course rationality exists: we have a clear intuitive sense that tells us when a thought or action is rational and when it is not. If, for example, I had left a sheet of paper in my typewriter overnight, it would be rational for me to assume that the paper would be there the next morning, and irrational for me to think that it might not, unless I had some reason (in other words, some rational cause) to think that someone had access to my typewriter overnight. In that case, any thought I might have about my paper being missing the next morning would no longer be irrational; indeed, it would be just as rational as my original expectation of finding the paper in place.

In our daily lives, we all have a definite idea of what it means to be rational, and no doubt we would insist quite strenuously that we have the ability to think and act rationally. However, when we come to the philosopher's task of pinpointing precisely what makes a thought or action rational—or of discerning general principles upon which rationality operates—then the situation gets a bit more difficult.

We can see that, in a very direct way, rationality is related to the use of logic, and that the commission of logical fallacies or errors in reason would render a train of thought irrational. One such case is the error of self-justification, or of using an argument in its own defense. Let us say we wished to claim (as some psychologists do) that people have difficulty understanding arguments or principles that in some way threaten their sense of intelligence or personality; let us further say that we present this claim to a person who has trouble understanding it. Most of us would agree that it would not be rational to cite this person's lack of understanding as support for our original claim: to do so would assume the correctness of our original claim on evidence that extends no further than the claim itself. With no evidence other than our original hypothesis that our friend's lack of understanding was a psychological defense, we have merely reiterated, not furthered, supported, or strengthened, our original claim. (Note that the irrationality of citing our friend's lack of understanding in support of our hypothesis does not mean that the hypothesis is untrue, nor does it deny that the mechanism suggested by the hypothesis was in fact operating in our friend's behavior. This book might be criticized as nonsense on the grounds that its author was born in the Bronx, and we would presumably disregard such criticism as irrational. Nonetheless, this book might indeed be nonsense.) Rational support, then, must come from evidence or arguments which are independent of (though of course are logically deducible from) the original claim.

Now in our investigation of the nature of rationality, we might well ask "Why is it rational to be rational?" or perhaps "Why is rationality desirable?" Obviously, we could provide a large number of seemingly satisfactory answers, such as "Rationality is desirable because it enables us to distinguish truth from falsity, and the identification of truths in the long run increases human happiness," or "Rationality helps us understand the world as it really is, and a life based on reality is less likely to result in disappointment than a life based on illusion." We might even undertake to demonstrate the existence of rationality by arraying a series of historical successes in which rationality apparently was at work; or we might appeal to reason by asking what the world would be like with no rationality, and, as the world is in fact not like that, conclude that therefore rationality does in fact exist. While the specifics of such arguments all seem capable of answering our questions about rationality, however, the arguments all suffer from a fatal flaw in structure as answers to questions concerning rationality: these answers are themselves all constructed in a rational way or form, and thus they assume the validity of the very mode of discourse—rationality—under question. To employ a rational argument to demonstrate the desirability of rationality is to assume before the argument that rationality is desirable, and to rationally argue for the existence of rationality is to already assume the existence of the very rational process whose existence is at issue. And so it comes to pass that rationality, which operates by attempting to identify and disqualify all manner of irrationality in discourse, including the commission of fallacies such as arguments used in their own defense, appears to disqualify itself by relying on the fallacy of self-justification.

This problem and its variants explain why such an intuitively obvious process as rationality has been such a contested issue in the history of philosophy. But the epistemological challenge to rationality has manifestly practical consequences, for as Bertrand Russell, Karl Popper, and others have painstakingly pointed out, the abandonment of rationality and the institutionalization of irrational doctrines do indeed often result in an increase of real human suffering.[1] Moreover, the undermining of rationality would be of enormous import even were it not for the dire political ramifications, since the untenability of rationality means that science and all human activities that use rational procedures would be called into question.

The examination of rationality and (one hopes) its defense is thus a task that sooner or later pages every thinking person. This chapter attempts to make some contribution to this task by considering rationality from the standpoint of evolutionary theory in three interrelated ways. First, temporarily disregarding the logical problems of rationality just described, we will consider the biological necessity of rationality—that is, the ways in which an evolutionary account of cognition would lead us to conclude that the human organism must have a rational component in order to survive; or, the extent to which evolutionary epistemology insists upon an innately derived rationality. Second, the logical problem of rationality's self-justification will be considered with the aid of evolutionary metaphors, specifically the evolutionary process of "self-transcendence." This evolutionary approach to the question of rationality's own justification will suggest a solution that is superior, I hope, to both the admission of Popper and many others of a regrettable though unavoidable "irrational" faith in rationality as the basis of all rationality,[2] and the attempt by Popper's student W. W. Bartley, III to avoid such an admission by claiming that a rationalist can logically hold everything, including initial faith in rationality, open to criticism and potential abandonment.[3] Third, the tensions between the blind, nondirective process of evolution and the foresighted, purposeful procedures of rationality will be considered, with rationality proffered as the evolution of evolution, or the emergence, through evolution, of a human process that can now attempt to direct, and to some extent transcend, the evolutionary matrix from which it arose.

Since these discussions will be rationally conducted, a certain amount of the self-referential problem of rationality—or the assumption of the efficacy of rationality in consideration of rationality's efficacy—will be unavoidably present. This situation need not be construed, however, as necessarily a victory for the irrationalist, since the irrationalist can call attention to this problem only by appealing to our sense of what is rational (in this case, our sense that use of a process under question in its own defense violates rationality), and thus enlist the very process the irrationalist wishes to discredit. Consequently, there seems to be a stand-off between the rationalist and the irrationalist on this particular aspect of the problem.[4]

In any case, consideration of rationality in the light of evolution should provide some fresh approaches toward the solution of this perennial, seemingly inexhaustible problem. In so doing, evolutionary epistemology reverses the usual relationship between philosophy and science in which the former illuminates and guides the latter: here the counsel of a scientific theory is offered in the treatment of an age-old philosophic question. Advantages which may result from this evolutionary application will assist assessments of technology and other human activities predicated on the operation of rationality.

THE BIOLOGICAL BASIS OF REASON

In the watershed essay entitled "Evolutionary Epistemology," Donald T. Campbell[5] argues that all types of knowing and perceiving are substitutes for touching, vicarious means of bumping into the world. These stand-ins for physical contact are evolutionarily advantageous. Starting at the origins of life, the one-celled organism or simple amoeba gains most of its knowledge about the environment by coming into direct physical contact with it. Such a mode of knowing is both highly accurate and dangerous—accurate in that whatever the amoeba bumps into is almost certainly there, dangerous in that should the amoeba bump into something hostile, the amoeba has little opportunity to profit by this knowledge and avoid the danger. Amoebas are very often obliged to die along with their highly accurate knowledge.

Further up the evolutionary scale, organisms develop indirect ways of knowing the environment, processes such as hearing, smell, and vision which provide representations of the world without the need of direct physical contact. Such modes of perception are a good deal less accurate than touching—our eyes are more likely to fool us than our hands (and amoebas presumably do not suffer from "tactile" illusions)—but obviously also a good deal safer. The errors introduced by representative rather than direct contact with the world are apparently more than offset by the increase in time and maneuverability that such knowing from afar gives its possessors, if we accept as evidence the abundance of organisms that utilize such indirect perceptual methods.

In human beings, this hierarchy of vicariousness is crowned by the processes of thinking, imagining, abstracting, and other mental faculties that in effect provide representations of representations. Such extremely indirect modes of knowing intensify both aspects of the inaccuracy/safety ratio, offering us second-generation pictures of the world that have greater capacity for error, but which immensely increase our ability to respond to and eventually control the world that the pictures imperfectly represent. The biological value of such distanced knowing, of seeing with the mind's eye, is again attested to by our success as a species thus far. (Communications technologies, unaddressed by Campbell, are the next step in this evolutionary process, for they allow us some of the distance of thinking, and virtually all the accuracy of direct sensory perception. The remote probes on Mars and other planets, for example, provide us with images nearly as direct as those that would be provided by our own unaided eyes, but at a distance that could formerly be spanned only by thought. These and other evolutionary aspects of communications media will be considered in detail in chapters 6 and 7 of this book.)

To be human, then, is to find oneself at the center of many possible levels of cognitive modality, ranging from the concrete, generally unmistakable knowing of touch,[6] through increasingly vicarious modes of perception and knowledge, culminating (pretechnologically) in abstract mental operations. Moreover, since some of our mental operations are in and of themselves capable of generating an almost limitless number of pictures of reality—as in the case of our faculty of imagination—the human organism is at every instant bathed in a plethora of cognitive possibilities and competing representations of real and imagined experience. Extrapolating from Campbell's model, we thus can see that humans must, as a biological necessity, possess some way of mediating and assessing these teeming possibilities, or of deciding which is actually the best representation of reality.

Such is precisely the power of rationality, which adjudicates among the host of sensations, impressions, ideas, and theories that we may have, separating them according to truth and falsity, usefulness and inapplicability, and sorting these products of our vicarious experience into whatever further categories we may deem appropriate. So fundamental is this rational power and its ability to help us distinguish reality from fantasy, that it allows us to cry at a sad movie—that is, get involved in a fictional plot and the motion picture illusion to the extent that we cry real tears—and yet know throughout this experience with perfect assuredness that what we are seeing on the screen is fiction and illusion.[7]

Indeed, without this ability to sort and assess the reality content of our experiences we would be, quite literally, crazy—lost in a psychotic's world of numerous, equally tempting and often contradictory, possible realities. Without rationality, the biological grant of multiple cognitive avenues would become a confusing superhighway cloverleaf that would kill us. Just as we would starve amid an abundance of food without the power to digest it, so we would wither in the plenty of cognitive experience without the rational means to make sense of it.

The amoeba has no such problem, as it has only one available mode of cognitive experience. Neither do other organisms less multiply endowed than humans—lacking, that is, our powers of imagination and abstraction. But in the human being, the cacophony of a nearly infinite array of cognitive possibilities demands a mediating rational faculty; whatever the origins of rationality, its continued presence was dictated by the extreme need of the emerging human mental environment in which rationality arrived. To accept Campbell's hierarchy of vicariousness as a description of human cognition, then, is to identify at least one clear evolutionary reason for the intuitively obvious biological necessity of rationality.

In sum, an evolutionary account of cognition posits not only the existence of the philosophically questioned entities of reality, truth, and innate knowledge as seen in the past chapter, but also the existence and necessary operation of the philosophically problematic faculty of rationality as a specific component of our innate mentality.

ARE THE ROOTS OF REASON RATIONAL?

A demonstrated biological need for rationality does not in itself solve the philosophical problem of rationality's self-justification, for the need is demonstrated by the very use of reason under question. Consequently, we turn now to the much more difficult challenge of giving a purely logical account of rationality's existence.

The ball-and-chain of rationality, as we have seen, is its reliance on rationally constructed arguments in its own evaluation, that is, employment of the mode of discourse whose existence, necessity, and value is at issue. This circularity can be heard clanging in all attempts to account for rationality, and is telling, since rationality itself apparently insists that self-certifying procedures are irrational or subverting of rational argument.

Those who have sought to uphold rationality have addressed this problem in a variety of ways. Karl Popper, arguing along lines similar to Bertrand Russell and George Santayana, has suggested that the only reasonable course for a rationalist under these trying circumstances is to admit that rationality begins with an "irrational" faith in reason, that is, a choice that cannot be supported without the irrationality of self-justification.[8] The difference between the rationalist and the irrationalist, then, is that after this initial concession to irrationality, the rationalist may live a life totally in accordance with rational principles, whereas the irrationalist continues to indulge in all manner of illogical and inconsistent gambits.

Approaches such as Popper's do seem reasonable, but their admission of even a one-time irrational commitment to rationality has been seized upon by irrationalists as arguments in irrationality's favor, and thus these approaches have been found crucially inadequate by some supporters of the rational mode. Popper's student, W. W. Bartley, III, in particular, views the stakes in the explication of rationality as so high as to deserve something better than a concession, even a minimal one such as Popper's, to irrational faith in reason.[9] Bartley wants to posit a rationality free of concessions and commitments, and most especially free of a commitment, not rationally supportable, to itself.

Bartley proposes to erect such a commitment-free rationality by equating rationality, totally, with an openness to criticism, or a willingness to abandon any position or principle, including rationality itself, in the face of effective criticism. Since such a critical procedure would itself be open to criticism and potential abandonment, Bartley thinks that a "pancritical" rationalism avoids the self-defeating pitfalls of an irrational commitment to itself. In other words, if such a system works—if rationality is indeed openness to criticism, and openness to criticism is itself open to potential rejection—then rationality may flourish without recourse to an initial, indispensable, self-serving, and thus irrational commitment to its own method.

Does such a system work? The crux of the pancritical attempt to avoid irrational commitment is that the critical method itself may be criticized and thus conceivably dispensed with. However, if we were to encounter a criticism of the critical method so deadly that we were induced to abandon it, would we not be abandoning the critical method only because we were, paradoxically, accepting its value or at least the value of the criticism that was in this case brought to bear against the critical method itself? Is not the criticism of criticism no less self-defeating than the rational justification of rationality in that, should such criticism be correct, it would invalidate the very critical method used in mounting such criticism? Bartley's claim that the pancritical rationalist can be argued out of pancritical rationalism (and therefore is not committed to it) thus is impossible for two reasons. First, to be argued out of criticism presupposes the worth of the critical method used in the argument, and second, the success of such an argument destroys the very mode of argument used to achieve such success.

Moreover, when we inquire as to what would make such a criticism of criticism count—indeed, when we ask what makes any criticism count—

we find that separating worthwhile from worthless criticism pitches us back into a dependence on some traditional sense of rationality. Consider again the type of criticism described earlier in this chapter that states that the arguments presented in this book are incorrect because its author was born in the Bronx, and contrast this with a possible criticism that the arguments in this book are incorrect because its author misrepresents Bartley's thesis, and so on. Presumably everyone reading these words would reject the first type of criticism out of hand, without considering whether or not the criticism were factually correct (that is, whether I was indeed born in the Bronx or in Borneo); but presumably everyone would at least be willing to entertain the second type of criticism, and would indeed be interested in the factuality of the claim, which could be investigated by a firsthand reading of Bartley's work. Quite clearly, then, there are standards quite independent of content to which criticism must adhere in order for it to be taken seriously. In the above instance concerning my birth in the Bronx, the criticism is worthless because it is irrelevant, and the judgement of relevance is a rational assessment made according to rational principles that transcend any specific criticism and thus serve as independent regulative mechanisms of the entire critical process. Thus, even were pancritical rationalism tenable as a self-maintaining system (which, for reasons indicated in the preceding two paragraphs, it is not), it would nonetheless fail to meet Bartley's expectation as a replacement for traditional rationality, since rather than superseding or even substituting for rationality, the critical method presupposes and depends on it.[10]

The notion of holding everything open to criticism, including criticism itself, is thus a noble but impossible dream, punctured by the complexity of presuppositions and self-referents that the practice of criticism entails. But does the untenability of pancritical rationalism mean that we should abandon Bartley's concern that an uncriticizable, "irrational" faith in reason gives the irrationalist too much? Does the recognition that the critical method presupposes both itself and rationality mean that we can do no better than describe the initial source of rationality as an irrational leap of faith?

In the next section, I will suggest that while Popper, Russell, et al. are correct in their view that the initial acceptance of rationality is rationally insupportable and thus occurs outside the realm of rationality, Bartley is right in wanting to dispute the irrationality of this initial step. In other words, I will argue that the ultimate roots of rationality are neither rational nor irrational, but something that is logically (and, via analogy, biologically) very much different. The designation of the origins of rationality as irrational is far worse than a semantic error, and is as out of touch with evolutionary reality as the designation of an embryo as stupid, or a grain of sand as dead.

PRERATIONAL VERSUS IRRATIONAL: SELF-TRANSCENDENCE AND THE BOOTSTRAP EMERGENCE OF REASON

The practice of rationality and rational criticism seems predicated on indispensable and hence ultimately uncriticizable first principles which, because they themselves are of a rational nature, seem to situate the origins of rationality in the irrational quagmire of self-justification. But is this initial leap into rationality really irrational?

We can begin to answer this question by examining the commonsense use of the term irrational. When we say that someone is behaving or thinking irrationally, we usually mean that someone is through conscious or unconscious choice (or perhaps through misunderstanding or accident) pursuing a nonrational course in a situation in which rational activity is possible. Irrationality is thus the rejection or denial of rationality. The key prerequisite of irrationality, in this colloquial perception, is the presence of a rational alternative: in order for rejection or negation of rationality to take place, rationality must already be a possibility, that is, one must have something to reject or negate.

Furthermore, when we move from irrationality in everyday parlance to the exaltation of irrationality as a philosophy, we find that the contemplation and rejection of rationality is precisely what gives irrationality its bite. Irrationalists such as Nietzsche and Feyerabend are rarely the voices of uncivilized, primal intuition, whatever their public images. To the contrary, they speak from a posture of having already seen the vast expanse of rationality; and with the wisdom of this experience they purport to offer an avenue somewhere above and beyond the rational plane.[11] Irrationality in practice and in philosophy is thus intrinsically postrational, and bears the same relationship to rationality as death does to life. In both cases, the negation is powerful and noteworthy directly in proportion to the power of the negated entity. Death and irrationality owe their existence and their impact to life and reason.

Is this derivative power of negation, this atmosphere of informed rejection, characteristic of the irrational leap of faith that Popper and many others confess begets rationality? How could this be, when the initial leap into rationality is initial only insofar as rationality does not exist prior to the leap? The absence of rationality at the origins of reason—the inability of reason to rationally account for itself—understandably may seem like a failure of rationality to those for whom reason is second nature. But absence of a process before its existence can only be an absence, not a failure, denial, or rejection of the process; the lack of rationality at the roots of reason thus bears as much resemblance to irrationality as a stone that was never alive bears to a dead organism, or a tree that was never aware or awake bears to an unconscious human. The recognition that rationality ultimately cannot explain its operation wholly within the precincts of rationality poses problems indeed for the rationalist, but need not, as so many rationalists have feared, turn the field over to the smirking irrationalist.

The biological analogies help clarify this situation further. Let us consider the three-way relationship of inanimate matter, living matter, and dead matter; or to use an example with less pejorative connotations for irrationality, consider the relationship of the level of consciousness of a tree (or even of a human zygote), a person's awakeness, and a person's sleep. Clearly there are substantial similarities between the first and third members of the trios as compared to the second: neither inanimate nor dead matter is alive, and neither zygotes nor sleepers are awake. But there are equally clear distinctions between the first and third members, and the obliteration of such distinctions with lazy terminology palpably worsens our picture of reality. The most important of these distinctions seems to be that the first members of the triads precede and may lead to the second in an evolutionary or developmental sense, whereas the third members follow and in effect cancel the second members. (On occasion death may safeguard or even generate life in a self-sacrificial way, and of course sleep regularly precedes wakefulness. But my point here is that life and death, awakeness and sleep, are in most instances mutually exclusive opposites which interact on a same temporal, developmental plane, whereas inanimate matter is always prior to, and sometimes generative of, both phases of the life and death cycle, and the zygote is a life form that neither wakes nor sleeps but grows into an organism that does both.) In view of the generative possibilities of inanimate matter vis-à-vis life and zygotes vis-à-vis wakefulness, we might further refine the notion of inanimate material by calling it preliving rather than nonliving (certainly not dead), and more clearly convey the status of the zygote's awakeness by calling it preawake rather than unawake (certainly not asleep).

The application of these biological realities to the problem of rationality suggests the following: not only is the lack of rationality at the origins of rationality not irrational, this nonrationality of the ultimate presuppositions of reason is actually a prerational condition, or one which in some way fosters the emergence of rationality. The error of viewing the roots of reason as irrational thus goes beyond the confusion of absence and denial to the far more serious confusion of generation and rejection. Irrationality is to the prerational constituents of reason not only as death is to inanimate material, but, more tellingly, as death is to the special configurations of inanimate matter that give rise to life.

What can the biological analogies tell us about the specific ways in which the prerational engenders the rational, that is, how rationality emerges from its nonrational beginnings? Are there structural similarities in the emergence of life from nonlife and of reason from lack of reason? The historical origin of life from inanimate material is yet to be explained in scientific detail, though several general processes seem evident. A cardinal characteristic of every generative situation is its apparent ability to change or transform itself "autocatalytically" (without identifiable external cause) into a process with quite different properties. Manfred Eigen, for example, has described how the accidental arrival of a new polymer with the capacity to produce not only itself but also a second protein which acted as a catalyst to the production of the first would greatly increase the survival of the primary polymer at the expense of other existing and mutant polymers that only reproduced and did not generate the catalytic second substance as a by-product. The eventual result would be an environment dominated by the special type of primary polymer—a self-transformation from the original environment of numerous types of polymers, and a possible step in the emergence of fully living organisms.[12] Whatever the specific mechanisms of self-transcendence, such a more–from–less process seems to have been at work in the way void developed into matter, matter gave rise to life, and life produced intelligent life. All of these are transformations with a peculiar, counterintuitive set of properties: 1) the second or resultant system has characteristics which are radically divergent or even opposite from the first; 2) the first system is nonetheless not obliterated in the transformation, and continues as a sort of substratum to the second system; 3) the impetus or stimulus for the transformation apparently resides wholly within the first system. Erich Jantsch has gone so far as to anthropomorphically describe such progressive instability as altruistic.[13] (The notion of God or a divine agent, incidentally, merely postpones rather than eliminates the need for some principle of self-transcendence, since God presumably willed Him/Her/Itself into existence—that is, came to be through some ultimately primordial act of self-transcendence.)

Taking care not to adopt a physical metaphor to an abstract task too glibly, may we not say that the emergence of rationality from nonrational roots occurs through a process of self-transcendence akin to the rise of matter from void, and life from nonlife? The "decomposition" of nonreason into reason seems to show all three of the properties of self-transformation: 1) the result, reason, is substantially different or even opposite from nonreason; 2) nonreason nonetheless continues as an available option in the world of reason (that is, one may choose to be irrational); and 3) the transformation happens entirely within the initial realm of nonreason. (We need not postulate a reasonable cause for this emergence, that is, the rationalist's attempt to find a rational basis for the emergence of reason is fruitless because the nonrational state is transformed entirely by virtue of its own nonrational conditions.) The last point is key to understanding why the search for a rational basis of reason has been and always will be unrewarding and unnecessary: just as the origins of life cannot possibly be explained as a product of life, so the origins of reason cannot be rationally accounted for. Instead, we must search for Eigen's polymers of the prerational state—the peculiar characteristics that make certain types of nonrational conditions prerational, or that prefigure and stimulate the emergence of rationality.

To better grasp how such an abstract self-transcendence might operate, let us briefly shift from the realm of biology to the realm of language[14] and consider the interesting sentence cited by Douglas Hofstadter in his discussion of self-reference: "To understand this sentence, change one pig."[15] On one level of scrutiny, the sentence is incomprehensible. (This level might correspond to the paradoxical inability of rationality to account for its own operation: the nonrationality of the prerational condition.) But this sentence is not incomprehensible in the same way as a sentence such as "To kiss this sentence, fly two mules." (This sentence might correspond to the nonrational state of the goldfish—it is merely incomprehensible in the same sense that the goldfish is nonrational not prerational.) The difference in the sentence about the pig is that it is progressive or self-transcending: after studying the sentence, we discover that we can render it comprehensible by following its own (disguised) suggestion. We have but to change the word "pig" to the word "word" and we have a sentence that reads, "To understand this sentence, change one word." While still rather pointless as a literal command, this sentence is no longer patently incomprehensible. How were we able to wrest comprehensibility out of the initial confusion? The peculiar structure of the original sentence served to guide us to the higher level of meaning—indeed, the original sentence is self-obliterating or self-subsuming, in that once we attain the minimal meaning by changing the absurd word ("pig"), the original absurdity is no longer apparent.

Of course, sentences cannot be literally self-transcending in the manner of Eigen's polymers: the leap from meaninglessness to minimal meaning in any sentence requires the comprehension capacity of a human being which, from the perspective of the sentence, is in part an outside agent. Nevertheless, this situation is not wholly inapplicable to the emergence of rationality which, after all, occurs in an environment of human consciousness and possible encouragement, and therefore does not arise entirely on its own. We thus gain at least two insights about the prerational condition from the study of self-referential sentences. First, the prerational state is to the more general, nonrational state as the "pig" sentence is to the "mule"; that is, the self-improving structure of the pig example provides a linguistic analogy to special nonrational circumstances needed for the emergence of reason. Second, the actual characteristics of this emergence likely fit somewhere between Eigen's polymers and Hofstadter's pigs; in other words, the cognitive emergence of rationality is more humanly directed than the development of self-encouraging polymers, but less so than the changing of "pig" to "word."

The question of human, conscious encouragement of rationality calls attention to the importance of momentum in self-transcending processes. At the polymer level, the production of catalysts, which in turn encourage the production of the primary polymers, which in turn produce catalysts, results in an escalating situation of more and more polymers and catalysts: once the first catalyst is produced, the results of the transformation increasingly render more origins into results. (Eigen calls this process the "hypercycle.") Similarly, once we have made the leap from "pig" to "word," our recognition that the sentence makes more sense with "word" reinforces our transformation of the pig. The emergence of rationality is in the same way not only a self-transcending but also a self-escalating process. Once rationality emerges in an individual, he or she can logically see the value of reason and is thereby encouraged to be rational. Further, rational individuals are moved to persuade nonrational (and for that matter, irrational) individuals of the value of the rational life, and thus the emergence of reason is self-escalating on a group-wide basis as well. (Self-escalation, however, offers no assurance of complete success, as the large number of nonliving molecules and nonrational behaviors attests.)

The incomplete emergence of rationality in individuals and groups (as opposed to the emergence of rationality as a species-wide potential or property) underscores the fact that rationality, like all innate or genetically derived human cognitive functions, needs to surmount two crises of self-transcendence in order to operate. First, rationality must emerge in a phylogenic evolutionary sense—that is, become possible as a capacity of our entire species. Few deny that human beings are in some sense capable of rationality, and even the most extreme irrationalists seldom dispute what has been recognized since Aristotle's time as the human capacity for reason.[16] The second process of self-transcendence is more problematic, and occurs in an ontogenic or developmental way as individuals actualize their capacities for rationality. Just as Kant's innate categories cannot operate without external experience, and Chomsky's innate rules of grammar need activation by general external linguistic stimuli (as the example of linguistically incompetent feral children tragically shows), so rationality is an evolutionarily provided faculty which must be discovered anew by every human individual. The infant or very young child who has never entertained or used a rational argument is thus prerational in a developmental sense which is analogous to the phylogenic prerationality of our hominid ancestors. Moreover, since these ancestors no longer exist, we may further refine our understanding of the prerational state by locating it solely in human beings with an unactualized capacity for reason. Prerationality thus functions as a cognitive stage much like the various early stages outlined by Piaget, with the difference that the prerational stage seems less operationally defined than Piaget's stages (that is, general, commonsense rationality entails a far greater variety of functions than the cognitive abilities investigated by Piaget), and emerges less evenly and with greater complications in human individuals.[17] Indeed, this is where debates about the use and value of reason usually focus: although a demonstration of rationality's logical impossibility would invalidate rationality on all levels—and indicate that even the general human capacity for rationality is an illusion—irrationalist claims that rationality is self-defeating have by and large been aimed at presumed overextensions of rationality by scientific and philosophic communities.[18]

The intent of this section has been to show not that rationality is never misapplied—for surely scientific development of germ warfare and nuclear weapons are such misapplications—but that the logical problem of rationality's inability to account for its origins, which has been cited by irrationalists (for whatever motive) as an argument against the use of rationality by individuals (that is, developmental rather than evolutionary emergence of rationality), may be clarified and partially resolved by the application of evolutionary analogies. In particular, we've seen that the roots of reason are logically not irrational but nonrational, since in common sense and professional philosophic usage irrationality entails the denial or rejection of rationality, and rationality by definition cannot exist before its origin. We've further seen that the roots of reason are not merely nonrational but also prerational. In other words, they have a capacity for self-transcendence similar to that described in Eigen's hypercycle and found in certain self-referential sentences, an ability to give rise on their own accord to a system with radically divergent, often opposing qualities, much as matter has given rise to life, and life to intelligence.

Since this analysis is itself rationally constructed, it is vulnerable to criticisms of rationality as a general activity, and thus cannot answer every objection to the rational process: in the event that a criticism of rationality were well-founded, it would unavoidably count against this analysis by virtue of the latter's rationality. Nonetheless, the recognition that the emergence of rationality from nonrational roots has analogues in many processes of self-transcendence in the biological and linguistic realms shows that this problem of more-from-less is not unique to rationality, and need not be fatal to its practice. In fact, the large number of living organisms provides palpable evidence about the results and success of self-transcendence. To the extent that the emergence of life from nonliving systems with very different properties is literally applicable to the self-transcending emergence of reason, we have in teeming life a sort of evidence on behalf of rationality. In any case, the evolutionary analysis provides an alternative to the minimal yet unnecessarily heavy concessions to irrationality such as Popper's, and to unworkable pancritical ideals such as Bartley's. In place of the problem of rationality's self-certification (which would indeed make rationality irrational), and the attempt to do away with all certification in the explanation of reason (which is not possible due to the indispensable certification of background logic), the argument from evolution makes the emergence of rationality a problem of self-transcendence. This problem is difficult and profound, and is not yet solved (as the example of life testifies), but by the same example is not self-defeating or invalidating either. This, I think, is progress.

Evolution is thus a supporter of rationality. However, the emergence of rationality in the biological world introduces the possibility of purposive, planned activity where previously there was none—only the indirect, unplanned flow and ebb of blind proposals and eliminations that constitutes natural selection. Since a central consideration of this book is the way in which rationally directed technology may be supplanting and transcending the naturally selective process which gave rise to it, we turn in the next section to a discussion of the role of rationality in evolution—or the extent to which rationality, hypothetically independent of technology, is a supporter of evolution.

ON THE NECESSARY NONRATIONALITY OF FIRST CAUSES

Of the many philosophic arguments that have been raised against the Darwinian theory of evolution since its invention, the one that has always struck me as the most commanding is the argument from purpose or design. This view, which in one form or another actually precedes the theory of natural selection by several thousand years, asks what we would think if during a walk in a forest we happened upon some marvelously constructed device, such as a timepiece, about whose origins we knew nothing. Despite the natural surroundings of our discovery, we surely would not conclude that this intricate, delicately harmonized mechanism of wheels and gears was an unplanned product of blind mutations and environmental retention—that is, a product of natural selection. Instead, we would see in this timepiece the hand of some foresight, intention, or intelligence. We would see an agency which, however much it used the process of trial and error, took advantage of accidental connections, and worked within the limitations of environmental demands (for example, the requirement of keeping accurate time), still had to have been conscious and directive of its work in a way that natural selection is not.[19]

In the case of the timepiece, of course we would be correct in identifying this system as the product of some deliberate intelligence. But Darwinian evolutionary theory insists that the myriad of interconnected mechanisms of the natural world—an interrelated system far finer than the finest clock—are self-originating and self-perpetuating, and not at all the expression of some preconceived plan or design.[20] Indeed, the unpredictable, open-ended character of the Darwinian world is one of its hallmarks, making it an early alternative to the closed, Laplacean world of classical science, and a precursor of the restless, shifting universe prescribed by twentieth-century physics.[21]

The notion of a divine or otherwise intelligent Designer is more than excess religious baggage for a universe conceived in a Darwinian fashion. If the range and precision of adaptations that constitute the living world are due to the unfolding of some grand blueprint, then the process of natural selection through which these adaptations seem to take shape is but a useful cover story, and we and every other organism within this world no more than puppets dancing on the Architect's strings (albeit puppets with the special advantage of understanding our own puppetry). A preconceived world, even one conceived by a modest First Mover who refrained from any subsequent interventions, is a predetermined world in which freedom, imagination, originality, creativity, and thought—indeed, the fundamental human enterprise of attempting to manage our surroundings and our destiny—become "epi-phenomena" or Platonic shadows to the predetermined plan, with no real chance to influence events beyond the original design. (A First Mover of even greater self-effacement—or wisdom—who deliberately engendered human rationality to understand, alter, and improve upon the originating plan would ease this predicament only slightly, for our rational self-determination might well be a local pocket of free will in a larger predetermined scheme.) Thus, the goal of

Darwinian evolutionists to account for the panoply of natural existence via principles wholly within that existence (without recourse to an underlying purpose or plan) has a result which is fortunate for Darwinian theory, and for all rationally constructed theories, of strengthening the meaningful practice of rationality. The theory of Watchmaker as maker of the living world may also be rationally constructed, but to the extent that free will is a corequisite of rationality, the Watchmaker theory has the effect of undermining the rational process with which it is constructed, and thus is ultimately a form of irrationality.[22]

There are other good reasons (independent of the problem of rationality) why we should view the living world as unplanned. For example, if the world can be explained without reference to a Designer, then on the principle of Occam's razor we have an explanation superior to one that calls upon a supernatural agent. In fact, few evolutionary biologists pursue the natural origins of life for the purpose of safeguarding or supporting the operation of rationality, which most scientists take as an axiomatic given. Likewise, few philosophers concerned with the problems of rationality have consulted the realm of biology for insights, and fewer still have considered the intersection of cognition, evolution, and technology. Nonetheless, the absence of a rational First Cause makes the subsequent practice of rationality possible: we might say that the lack of purpose and contemplative direction in the prehuman universe permits the processes of rational deliberation and choice in the human world.

The converse of the relationship between rationality and nondesigned origins would also hold: if, for reasons other than evolutionary biology, we accept the existence and effective operation of rational decision and choice in human life, then we can reason back from the reality of this rationality to the nondesigned origins of the universe. The spectre of tautology would arise only if rationality and Darwinian theory had no supporters other than themselves, but since the defense of reason has generally been conducted on nonbiological grounds, and the investigation of natural origins conducted in an empirical rather than an epistemological framework, each has ample support independent of the other. The linking of rationality and natural selection thus provides noncircular assistance to both parties, complementing the substantial bases of distinct support that already exist for each. Undesigned origins of life are conducive to the practice of a rationality evidenced by common sense and explicable in part by other biological metaphors such as self-transcendence. (Nondesigned life transcends itself into partially designed life in the case of humans, but the process of self-transcendence occurs in many areas of existence in which the question of design is not central, such as the self-transcendence of simple to more complex nonliving material. Self–transcendence is thus a process larger than, and not tautologically equivalent to, the emergence of design from nondesign, or rationality from prerationality.) Similarly, the practice of rationality presumes a free will or lack of design at the origins of existence consonant with a theory of natural selection that is supported by numerous fossil records and observations in molecular biology. (Again, the reliance of theories of rationality and natural selection alike on rational processes means that the link between rationality and natural selection is not entirely free of the fundamental problems of rationality's accounting for its own operation.)

If the universe is indeed undesigned, we might say that the blind evolution of the cosmos at last finds its purpose in human rationality. Much of the indirect trial and error of natural selection remains in the operation of deliberate rationality—as Karl Popper has made clear in his epistemology of conjectures and refutations, and as we would expect in the self-transcendence of any process—and thus the blindness of the universe is transformed into an imperfect vision in humans. (The imperfection becomes especially prominent when the vision attempts to examine itself, as we are doing here.) The way in which our rationality enables us to back into closer approximations of truth via (fallible) elimination of error is very much in accord with the groping of organisms and their chastisement by the environment in the biological world. Yet we are aware of our trial and error processes and may seek to improve and direct them, whereas the universe, except when acting through us, presumably does not seek to do anything. The self-transcendence of blindness into vision, even if highly impaired, thus makes all the difference, and sets the stage for a physical transformation of the universe via technology.

In sum, we have seen that evolutionary theory supports the notion of a human, active rational faculty in at least three ways: (1) by demonstrating the biological necessity of rationality as a mediator among the many levels of vicarious experience that humans are heir to; (2) by providing a process—self-transcendence—that by analogy explains how an initially prerational condition which is neither rational nor irrational can of its own accord develop into a state of rationality; and (3) by hypothesizing an initially purposeless, designless universe that permits the operation of human free will and meaningful rationality. The first result provides a functional explanation of rationality's existence; the second recasts and relieves the logical problem of rationality's inability to account for its own operation on species-wide and more controversial individual levels; and the third gives a cosmic context to the emergence and impact of rationality.

The anchor in all of these considerations—the terminus of the self-transcending, prerational condition and the blind universe—has been the rationality of common sense and classical, self-evident logic. This, indeed, is the very process we have used to arrive at the conclusions of our deliberations. But does self-transcendence end with the emergence of a rational faculty that can understand and perhaps guide the self-transcendent process? Is rationality, the evolution of evolution into a partially foresighted process, itself immune to further evolution? The postrationality advocated by irrationalists such as Nietzsche and Feyerabend is one type of claim that the evolution of rationality beyond itself has already occurred. We conclude our analysis of rationality and evolution with a brief look at the possible extent and limits of rationality's own further self-transcendence and evolution.

INESCAPABLE LOGIC

The question of whether rationality can transcend itself in a manner akin to the transcendence of natural selection into rationality—the change from unplanned to partially planned proposals that the environment acts upon—can be considered both metaphysically and empirically. The deepest metaphysical question amounts to: can we even conceive of a postrationality that differs from rationality as much as rationality differs from natural selection? The empirical question asks if we have any evidence for the emergence of such postrationalities. The empirical question is easier to answer than the metaphysical, and the answer seems to be negative.

Evidence for postrationality often comes from "ecologies" of rationality that emphasize the need and operation of different modes of discourse (which are sometimes mutually exclusive) in the investigation and navigation of different realms and aspects of existence.[23] If, for example, someone puts a gun to your head and insists that you say that two and two equals five, then the logic of this local situation would seem to dictate that you espouse this irrational conclusion. A comparable situation arises in the study of subatomic particles: here the classic, commonsense rationality which serves us well in the life-size world seems incapable of coming to terms with the paradoxical behavior of subatomic objects. The solution is in the abandonment of classic logic for a new, quantum logic, which allows us, among other things, to accept the existence of "a" and "not a" at the same time (for example, light is both a particle and a wave).[24]

Such local exceptions to commonsense rationality are viewed by many rationalists as well as irrationalists as evidence against the total hegemony of any one type of rationality. Are these and similar cases examples of the self-transcendence of rationality into something quite different? Although the self-transcendence of one process into another by no means annihilates the first process, the result of this evolution is that the first process is subsumed and dominated by the second. Thus, living systems dominate their material origins on Earth in a way that has thus far been irreversible: individual organisms may die but the biosphere grows.[25] Intelligence operates vis-à-vis its living origins in much the same way.

However, the local exceptions to rationality operate differently. Not only does classic rationality guide us to an initial adoption of a postrational process in given situations, but our reason continues to monitor and assess the postrational logic throughout its operation, and may at any time attempt to amend or recall it entirely. For example, we may discover that the gun is loaded with blanks, or that the reports of subatomic behavior which occasioned the new quantum logic are inaccurate due to primitive equipment—in which cases we would reassert that two and two equals four, and that "a" and "not a" cannot be descriptive of exactly the same phenomenon. Thus, rationality plays too active and dominating a role in the postrationalities for these to be considered self-transcendences of rationality in the way that reason and planning have transcended natural selection. As R. I. G. Hughes says after an analysis of some of the "and" and "or" twists of new quantum logic and their necessity:

> One is left, therefore, with a family of logics that includes classical logic, quantum logic, and perhaps other logics as well. Among them one logic still has priority. Although the nonclassical logics have specialized applications, the logic employed for abstract reasoning, including reasoning about logic, will probably continue to be classical logic.[26]

The last sentence suggests not only that rationality has not transcended itself into new postrationalities, but also that it may never do so, and this returns us to the difficulties of even conceiving of such a change. On the one hand, rationality is itself the product of a self-transcending evolutionary process, and thus its own continued evolution seems a reasonable expectation. On the other hand, the very nature of rationality is counterevolutionary in that it provides the means (along with technological implementation) for control and perhaps direction of evolution, and thus reinforcement of its own position. Furthermore, since the only means we have of thinking about a postrational state is employment of our current rational faculties, any projections we make about future postrational conditions are inextricably biased by our rationality: we are analogous both phylogenically to our prerational hominid ancestors and ontogenically to our 1-year-old selves and their capacity to envision rationality when we try to discern the outlines of some future postrational level.

Thus, in addition to the three types of evolutionary support for rationality discussed in this chapter, we have good reason to think that rationality itself has not evolved any further, and have little basis for knowing if, when, and how such further evolution may occur. On the strength of this support, then, we will assume the existence and effective operation of rationality in our considerations of technology and knowledge: we may find circumstances in which technology compromises or even perverts the operation of rationality, but we will not challenge the reality of rationality per se. The problems of rationality independent of technology remain profound and challenging, but I take the evolutionary arguments and evidence cited in this chapter as sufficient to conclude that the problems of rationality are not fatal, and that our common sense acceptance of reason is well warranted.

Before turning to a consideration of technology proper, I would like to look at one additional issue in nontechnological terms: the grounds for optimism and growth of knowledge in an evolutionary (now rational) environment. In terms of rationality, we may say that a discussion of optimism asks whether the common sense we have just documented is good sense—whether it can result in a continuing growth of knowledge. The possibilities of such growth on evolutionary principles independent of technology are crucial for a technological optimism, for they allow technological contributions to the knowledge process to work with rather than against the prevailing evolutionary framework.

NOTES

1. After nearly a lifetime of defining rationality as a willingness to consider all pertinent arguments and evidence, Russell wrote of the tendency to abandon this attitude in some quarters, and the perils of such abandonment: "many people [have] . . . the view that there is no such thing as an ideal rationality to which opinion and conduct might with advantage conform. It would seem to follow that, if you and I hold different opinions, it is useless to appeal to argument . . . there is nothing for us to do but fight it out, by the methods of rhetoric, advertisement, or warfare, according to the degree of our financial and military strength. I believe such an outlook to be very dangerous, and in the long run, fatal to civilization." *The Will to Doubt,* New York: Philosophical Library, 1958, p. 9.

Karl Popper's *The Open Society and Its Enemies,* Princeton, N.J.: Princeton University Press, 1962 (1945), connects the rise of fascism and totalitarianism in the 20th century not only to professed irrationalists such as Nietzsche, but to attempts to enthrone rationality by Plato and Hegel—i.e., to render rationality itself immune from vigorous criticism—to the point of creating an irrational climate.

2. Popper, *The Open Society and its Enemies,* p. 231: "whoever adopts the rationalist attitude does so because he has adopted, consciously or unconsciously, some proposal, or decision, or belief, or behavior; an adoption which may be called 'irrational.' Whether this adoption is tentative or leads to a settled habit, we may describe it as an irrational faith in reason."

3. Presented principally in Bartley's *The Retreat to Commitment,* LaSalle, Ill.: Open

Court, 1984 (1962), and in a series of articles between 1962 and 1984, the most comprehensive of which is "Rationality, Criticism, and Logic," *Philosophia,* February 1982, pp. 121–221.

4. Consider, for example, the following hypothetical conversation between rationalist and irrationalist:

Rationalist: I make one nonrational assumption—to wit, that rational argument counts—and from then on I strive to be rational in all my conduct.

Irrationalist: The fact that your enabling assumption is nonrational shows that you and I are one and the same: irrational.

Rationalist: The very point you just made derives its clout by appealing to our sense of reason or rationality. To rationally deny rationality is inconsistent.

Irrationalist: "Consistency" may be desirable to a rationalist, but it is of no concern to an irrationalist like me. Inconsistency is thus not inconsistent with irrationalism.

Rationalist: The very point you just made derives its clout by appealing to our sense of reason or rationality.

Irrationalist: An irrationalist, by virtue of his or her irrationalism, may reasonably use rationality to defeat rationality.

Rationalist: Whatever merit such an argument may possess, it derives from its appeal to our sense of reason or rationality, and thus if we abandon rationality as you suggest, we abandon with it any strength in your argument.

Irrationalist: Why may I not use rationality in some cases (such as to argue for the abandonment of rationality), and irrationality in others?

Rationalist: Use of rationality to suggest local occasions for irrationality is rational; use of rationality to argue for the complete abandonment of rationality is self-defeating, since it invalidates the very form of argument used to reach this conclusion.

Irrationalist: But defeat of rationality is my goal.

Rationalist: Are you saying that you're consistent?

5. Campbell, "Evolutionary Epistemology," in Schilpp, *The Philosophy of Karl Popper.*
6. Popper suggests that our most fundamental individual contact with external reality comes in the ingestion of foodstuffs (*Objective Knowledge,* p. 37).
7. Samuel Taylor Coleridge defines the appreciation of poetry as a "willing suspension of disbelief," *Biographia Literaria,* ed. J. Shawcross, London: Oxford University Press, 1907 (1817), vol. 2, p. 6—capturing (in the word "willing") the mediating power of rationality in this partial immersion in illusion. See "Flesh and the Growth of Knowledge" in chapter 8 and "The Burden of Rationality" in chapter 9 of the present book for more on the capacity of rationality to adjudicate technology. See also my "Media Evolution and Rationality as Checks on Media Determinism" in S. Thomas, ed., *Studies in Mass Communication and Technology,* Norwood, N.J.: Ablex, 1984, pp. 231–237.
8. See note 2 above for Popper's statement on "irrational faith in reason" in *The Open Society.* Russell, writing of the ultimate roots of science, concludes in *The Scientific Outlook,* New York: Norton, 1931, p. 264, that "when science is considered contemplatively, not

practically, we find that what we believe we believe owing to animal faith, and it is only our disbeliefs that are due to science." Cf. George Santayana, "an introspective critic might be tempted to think [it] self-evident . . . that he himself lives and thinks. That he does so is true; but to establish that truth he must appeal to animal faith." From *Scepticism and Animal Faith*, 1923, reprinted in I. Edman, ed., *The Philosophy of Santayana*, New York: Modern Library, 1942, p. 388. Popper's concession of an irrational faith in reason is less encompassing than Santayana's and Russell's animal faith as prerequisite of belief, but is equally fundamental.

9. See Bartley's works cited in note 3 above.

10. Bartley in his early presentations of pancritical rationality (*Retreat*, 1962; "Rationality," 1982) admits to a logical "core" indispensable to all rational and critical activity. (Indeed, Bartley soundly refutes Quine's argument on the dispensability of every type of logic at given times, i.e., the argument that logic A can be dispensed with by logic B, and vice versa. I agree that Quine is in error, since the very process of dispensing with a logic and replacing it with another presumes a constant background logic in the decisions to dispense and replace.) However, Bartley sees no conflict between the necessity of some logic and the ideal of pancritical rationalism (that nothing is indispensable), claiming that the need for some logic in criticism poses no more of a problem for pancritical rationalism than the recognition that criticism presupposes both a language in which to make arguments and thinking beings to devise criticisms (e.g., "Rationality," p. 186). But the use of logic and rationality on its own behalf, not the requirement of language and life for rationality, is precisely the quandary that pushed Popper into an identification of irrational faith as the basis of reason, and Bartley into the pancritical chase to avoid this. Hence the admission of some degree of logic into a core of indispensable requirements for critical rationality seems destructive of the pancritical enterprise (holding all rational processes open to criticism and possible removal) in a way that the necessity of language, life, and indeed molecules, atoms, and electrons in the critical process surely is not.

In the recent edition of *Retreat* (1984), Bartley responds to several similar objections to pancritical rationalism raised principally and independently of my work by John Post. Bartley's main response is that his critics have taken "criticism" to be something far broader than Bartley had intended; where Bartley was referring to a dispensable justificatory or authority-based criticism, Bartley's critics have focused on the critical process in general, Bartley contends.

I have two responses to Bartley: First, to the extent that he confines his pancritical rationalism to a narrow, esoteric form of criticism, Bartley renders it less relevant to the affairs of the intellect—which encompass all manner of criticism—and indeed less relevant to the problems of rationality. Second, even when confining the object of pancritical rationalism to criticism predicated upon some sort of authority, pancriticism fails because criticism without the minimal authority of being logical (although not necessarily correct) is not taken seriously. As the "Bronx" example in this chapter suggests, criticism devoid of logical structure is scarcely criticism in the philosophic sense.

Peter Munz, *Our Knowledge of the Growth of Knowledge,* London: Routledge & Kegan Paul, 1985, p. 50, supports Bartley's pancritical rationalism as an "attitude" rather than a logical system. But if designation as an attitude removes Bartley's system from possible logical criticism, then this designation on its own defeats the pancritical ideal. If, on the other hand, Bartley and Munz hold an attitude to be as criticizable as a logical system, then why designate pancriticism as an attitude? Note that what I am criticizing here is not the attitude that as much as possible should be open to criticism, but the more extreme claim that everything can in principle be open to criticism and that this answers the irrationalist critique of reason.

11. See, for example, Nietzsche's claim that "an instinct is weakened when it makes itself rational"; or his observation that "the intellect is the error." (Translated and cited by James M. Curtis, *Culture as Polyphony,* Columbia, Mo.: University of Missouri Press, 1978, p. 36.) Paul Feyerabend begins his *Against Method,* London: New Left Books, 1975, with the advisory that "the following essay is written in the conviction that anarchism, while perhaps not the most attractive political philosophy, is certainly excellent medicine for epistemology, and for the philosophy of science" (p. 17).

12. Manfred Eigen and Peter Schuster, *The Hypercycle: A Principle of Natural Self-Organization,* New York: Springer, 1979. See also Ilya Prigogine and Isabelle Stengers, *Order out of Chaos,* New York: Bantam, 1984, pp. 190–191.

13. In the Eigen hypercycle, for example, each protein enhances not only its own reproduction but its partner's, and therein in turn enhances its own. "As Eigen and Schuster put it, mutual enhancement prevails over self-enhancement," Jantsch writes. "The other way around, symbiotic nucleotide-protein pairs would still be able to reproduce themselves, but the hypercycle—and with it, the possibility of evolution toward higher complexity—would break down and evolution would find itself in a cul-de-sac. Evolution is impossible without altruism, even at the molecular level." From "Ethics and Evolution" in *The Responsibility of the Academic Community in the Search for Absolute Values, Volume 1, Proceedings of the 8th International Conference on the Unity of the Sciences, Los Angeles, 1979,* New York: International Cultural Foundation, 1980, p. 376.

14. The leap between the mechanisms of organic evolution and human language is not all that far, and has been explored and mined brilliantly in a series of articles by Lars Löfgren, including "Goals for Human Planning," in R. Ericson, ed., *Proceedings of the Silver Anniversary Meeting of the Society for General Systems Research,* New York: Springer, 1979, pp. 460–467; "Knowledge of Evolution and Evolution of Knowledge," in Erich Jantsch, ed., *The Evolutionary Vision,* Boulder, Colo.: Westview, 1981, pp. 129–152; and "Life as an Autolinguistic Phenomenon," in M. Zeleny, ed., *Autopoiesis,* New York: Oxford University Press, 1981, pp. 236–249. Löfgren's central thesis is that the relationship of genotype and phenotype in biology is analogous to the linguistic processes of description and interpretation: A sentence describes an object or event or an occasion; its interpretation by us causes us to do something—i.e., to think and act in the real world. Similarly, the DNA code is a description that has no consequence in the real world until interpreted into living structures (i.e., the arrangement of molecules to make organs and organisms). Further, although the description leads to and in this sense dominates the interpretation, in both cases (linguistic and biological) the interpretation, once established, has great complementary impact on the description. We may decide after interpreting a sentence to ignore it; the fate of the phenotype in the world will determine the fate of its genotype; in humans, our phenotypes now have attained the power of literally altering our genotypes via gene splicing. See chapter 7, notes 23 and 25, and chapter 8, note 6 and associated text in the present volume for discussion of gene splicing and the level of alteration of natural laws that gene splicing entails.

Löfgren also extends the description-interpretation complementarity to technology in general and computers in particular, writing in "Life as an Autolinguistic Phenomenon," p. 238, that "a technical construction is an interpretation (engineering realization) of a description (engineering plan). Within computer technology, a programming language coordinates descriptions (programs) and interpretations (computations)." The notion of technology as a realization of plans is precisely what I have in mind throughout this volume when I write of technology as embodiments of our ideas (although I also stress the extent to which these plans may be frustrated—the fallible limits not only of our thoughts but our technological implementations). See also chapter 1, note 21 for a statement analogous to Löfgren's about computer languages.

See also Jeremy Campbell's *Grammatical Man,* New York: Simon and Schuster, 1982 for a popular analysis of related themes. Campbell misrepresents Chomsky, suggesting that his model of innate grammar relegates "experience to such small importance" (p. 184), and makes no reference at all to Löfgren. Chapter 11 of the volume, however, provides a clear discussion of DNA as a series of algorithms or grammar-like rule statements operating in a naturally selective environment.

15. The sentence actually discussed by Hofstadter was "If you think this sentence is confusing, then change one pig." It was contributed by Uilliam M. Bricken, Jr., a reader of Hofstadter's *Scientific American* "Metamagical Themas" column, and published there in January, 1982 under a column entitled "Self-Referential Sentences: A Follow-Up" (reprinted along with many other Hoftstadter *Scientific American* columns in his *Metamagical Themas,* New York: Basic Books, 1985, pp. 25–48). Hofstadter quotes this sentence in a delightful analysis of "neurotic" self-reference, not rationality. I changed the form from conditional to infinitive to make the self-transcending aspects, and analogy to the emergence of rationality, a bit more prominent.

16. Nietzsche even had the makings of a Popperian fallibilist at times, writing that "Convictions are more dangerous enemies of the truth than lies." From *Human, All-Too Human* (1878), #483, reprinted in Walter Kaufmann, ed., *The Portable Nietzsche,* New York: Viking, 1954, p. 63.

17. "Can you really imagine a highly intelligent organism that has no idea of object permanence, causality, time, etc.?" Howard Gruber has Piaget ask in "Piaget's Mission," *Social Research,* Spring 1982, p. 257.

The answer is no. But given the rampant irrationality on levels ranging from personal squabbles to nuclear armament, the answer for rationality is not as clear-cut. In Piagetian-Chomskyan terms, we might say that rationality needs more activating and eliciting, and once elicited, requires more encouragement to perform adequately, than do senses of causality and linguistic capacities.

18. Feyerabend, for example, denies that he is advocating the abandonment of logic per se—his objection is to the domination of human affairs by "one" logic (the logic of science), which he would replace by a "whole spectrum" of rationalities including "intuitionistic systems" ("The Gong Show—Popperian Style" in G. Radnitzky and G. Andersson, eds., *Progress and Rationality in Science,* Boston: Reidel, 1978, pp. 387–392). However, without some predominating "meta-logic" to decide which type of rationality or logic to employ in which situations (the background logic referred to in note 10, or Bartley's indispensable logical "core"), the choice of one logic as opposed to another becomes arbitrary or irrational (even if we accept Feyerabend's doubtful proposition that "intuitionistic" systems are rational—doubtful because the fact that an intuition may work or be correct does not mean it was logically derived). See the "Inescapable Logic" section at the end of the present chapter for further assessment of "multiple rationalities."

19. Samuel Butler used the very trial-and-error Darwinian-like development of humanly designed technologies as an argument against the Darwinian view that trial and error in the natural world removed the need for deliberate design as an explanation of nature. See Butler's *Evolution: Old and New,* London: Fifield, 1879/1911, pp. 54 ff.

20. A few Darwinists have attempted to reconcile the dictates of natural selection and divine origins by arguing that the process of natural selection is the deity's way of creating new species. Thus Charles Lyell concludes his general defense of Darwin in *The Geological Evidences for the Antiquity of Man,* Philadelphia: Lippincott, 1871, p. 506 with the observation that "the perpetual adaptation of the organic world to new conditions leaves the argument in favor of design, and therefore designer, as valid as ever; 'for to do any work by an instrument must require, and therefore presuppose, the exertion of more than of less power, than to do it directly.'" Lyell quotes from Asa Gray's *Natural Selection Not Inconsistent*

With Natural Theology, London: Truebner, 1861, p. 55, reprinted in Asa Gray, *Darwiniana,* New York: Appleton, 1876, pp. 87–128. Gray's *Natural Selection* appeared originally as three essays in *Atlantic Monthly,* July, August, and October, 1860.

If this argument amounts to a simple equation or acceptance of the scientific theory of natural selection and its consequences as the workings of "divinity" (with the recognition that, like all scientific theories, natural selection theory is incomplete, imperfect, and under constant repair and possible improvement), then no harm is done to the theory of natural selection, but the notion of divinity is stripped of its usual meaning and consequences. If, on the other hand, the view of natural selection as a divine plan entails any kind of continuing or active divine presence or power, then natural selection is no longer really "natural" selection (for example, the deity could decide at any time to end the process of natural selection). Thus, these attempts to reconcile natural selection and divinity are, from the perspective of natural selection, either pointless or destructive of natural selection. (See also note 3 in chapter 1 for consequences to theology of a hypothetical divine agent that set the world in motion along Darwinian lines and then disappeared.)

With Lyell's further conclusion that the "introduction into the earth at successive geological periods of life—sensation—instinct—the intelligence of higher mammalia bordering on reason—and, lastly, the improvable reason of Man himself, presents us with a picture of the ever-increasing dominion of mind over matter" (p. 506), I am much in agreement, especially insofar as the dominion of mind is human not non–humanly divine. Indeed, the thesis of *Mind at Large* is that humans are changing the universe via the material embodiment of their thoughts in technology. See also my conclusion "that via technology, and notwithstanding its limitations, the future of the universe lies in the human mind," in "What Technology Can Teach Philosophy," in *In Pursuit of Truth,* p. 170. In this sense, evolutionarily emergent humans are a more likely bearer of divinity than is the traditional pre–human, pre–natural agency. (This humanly-emergent view might be termed Emersonian transcendentalism with a technological accent.)

21. The denial of objective reality by strains of contemporary physics, however, goes beyond the indeterminacy of Darwinian theory, and is incompatible with the Darwinian notion of an environment in part external and objective to the organisms it selects. Thus, natural selection posits an objective indeterminism, in contrast to the objective determinism of Laplace and the subjective indeterminism of quantum mechanics as espoused by Heisenberg and Wigner. See Bartley's "Philosophy of Biology versus Philosophy of Physics," *Fundamenta Scientiae,* 1982, pp. 55–78, for more on the conflict between evolution and quantum mechanics.

22. See Bartley in "Rationality," 1982, pp. 185–186, for a discussion of the incompatibility of determinism and rationality.

23. Examples are Henryk Skolimowski's call for an "alternate rationality" necessary for an understanding of "evolution in its higher stages" ("Evolutionary Rationality," in *PSA 1974,* ed. R. S. Cohen et al., Boston: Reidel, 1976, pp. 191–213), and McLuhan's emphasis of a sensory-aware, metaphorical type of reason as superior to traditional "logic" in understanding the complexities of our technological world. See my "McLuhan and Rationality," *Journal of Communication,* Summer 1981, pp. 179–188 and references there for discussion of McLuhan's views on traditional rationality. See also note 18 above.

24. See Guenther Stent, "Limits to the Scientific Understanding of Man," *Science,* March 21, 1975, pp. 1052–1057, for a thesis that our commonsense ways of knowing are inadequate to the task of understanding environments such as subatomic states that were not relevant to our evolutionary development. I refute this view as in fact nonevolutionary in "Evolutionary Epistemology Without Limits," 1982, and in chapter 3 of this volume.

Niels Bohr argues in "Unity of Knowledge," 1954, reprinted in *Atomic Physics and*

Human Knowledge, New York: Wiley, 1958, pp. 67–82, that an understanding of subatomic behavior requires a renunciation of elements of logic and common sense which served us well in comprehension on the macro classical physical level. The implication here is that if, say, light's behavior as both a stream of particles and a wave (physical forms with incompatible properties) strikes us as paradoxical, the source of the problem is our mode of reasoning rather than the poor quality of the data from which we make such observations. Even given the high quality of our 20th-century detection technologies, such a conclusion strikes me as at best premature and at worst as the height of chronocentric conceit. Popper makes a similar point in *Quantum Theory and the Schism in Physics,* London: Hutchinson's, 1982, Preface 1980–1981, arguing against the claims of Bohr and Heisenberg that quantum mechanics constitutes a "final," ultimate explanation of reality. See also note 26 below.

25. Prigogine's emphasis of irreversibility in the chemical emergence of life is applicable here: the vector or direction of chemical reactions (e.g., Eigen cycles) leading to life is toward living systems which in turn produce more complex living systems, rather than revert to their preliving chemical constituents. Preliving compounds are consumed by living systems and serve as a continuing foundation of life; further, even the temporary reemergence of nonliving material upon the death of individuals is absorbed by the living system of the planet in the form of fertilizer, food, and so on. See Prigogine's *Order Out of Chaos* for discussions of irreversibility on physical and chemical levels, and Jantsch's "Unifying Principles of Evolution" in *The Evolutionary Vision,* pp. 83–116, for additional observations on irreversibility in biological evolution.

26. R. I. G. Hughes, "Quantum Logic," *Scientific American,* October 1981, p. 213.

Chapter 3

Fallibilism and Optimism

An assumption of innocence is often made about optimism, an assumption which says that optimism is possible only to the extent that we are ignorant of ourselves, our world, and our knowledge processes which, when truly revealed in their flawed condition, justify a profound and thorough-going pessimism. This "ignorance is bliss" construal of optimism is by and large a legacy of the overconfident Age of Reason, which exalted rationality as a provider of certainty in knowledge and contentment in human affairs. When the epistemological and political turmoils of the nineteenth and twentieth centuries shattered such visions, wounded too was the notion of any sort of wise or knowing optimism.

The fallibilism advocated by such modern philosophers as Peirce and Popper denies the possibility of certainty in all situations, especially the rational, and seems at first glance to support the view of the crippled human knower. Discovery that we can be certain of nothing—not even our own existence—is indeed disappointing. Is this uncertainty necessarily incompatible with the optimistic view that we can continuously improve our knowledge and condition?

If certainty is regarded as a key to the improvement of our knowledge, then the answer is yes, fallibilism is incompatible with optimism. But if we are willing to allow that knowledge can grow without the security of absolute building blocks—that truth acts as a sort of invisible hand that shapes the growth of knowledge (by sounding an alarm on error) without ever clearly revealing itself—then the loss of certainty need in no way compromise our hopes of improving our knowledge and our lot in the universe. In fact, in this chapter I will argue something very much stronger: that lack of certainty is the very characteristic of our knowledge that makes the growth of knowledge possible. Not only is the fallibility of our knowledge not inconsistent with optimism, it is the very basis for the possibility of a better future.

The logic of fallibilism as optimism is tied to evolution, and can be put simply as follows: If the universe and its living and nonliving constituents are forever changing in a way that is at least somewhat unpredictable, then absolute knowledge of these constituents is unnecessary and even counterproductive (assuming such knowledge were possible), since the objects of such knowledge are liable to change in a manner that sooner or later renders such knowledge obsolete. Certainty of knowledge in a constantly shifting world is, we might say, the equivalent of taking perfectly clear and accurate photographs of things that no longer exist.[1] But let us suppose that we were able to take a photograph not only of what is, but of what might be—a photograph of some living or nonliving constituent of the universe, in both its present and all possible future states. Since photographs of this type would necessarily entail a multiplicity of images, such photographs could not provide a perfectly clear or accurate image of an object or process in any one state, such as the present. Rather, in order for such photographs to have any relevance to more than one state of an object, they would need to be vague and ambiguous, roomy and blurry enough to touch upon a variety of possibilities. Such knowledge would, in other words, need to be fallible and conjectural relative to any and all possible states of the universe, including the present.

Even were certainty possible, it would be useful primarily in proportion to the closed or immutable nature of the systems it described; and to the extent that nonliving and living universes are open and indeterminate (as suggested by modern physics and biology), the pursuit of certainty actually obstructs attempts to know existence as it really is. Lack of certainty and ubiquity of conjecture is thus a saving feature of our cognition, an absence of fixity that permits us to keep pace with unpredictability. As Mallarmé put it, "to define is to kill, to suggest is to create."[2] Here we have the makings of an un–innocent epistemological optimism.

Such optimism assumes that we can indeed know something of a widening array of phenomena—that human cognition, independent of the question of certainty, is unlimited in the range or scope of its perusal. Anything less, any absolute boundary beyond which we could not reasonably inquire, would indeed be grounds for pessimism, for it would condemn us to uncertainty without improvement, to fallibility in a fixed, epistemological prison.[3]

In considering to what extent the fallibility of our knowledge permits us to better it and to translate such advances into concrete improvements via technology, we thus must consider to what extent we may hope to extend our knowledge in any case. The evolutionary derivation of our cognition strongly suggests its fallibility (because all evolutionary products are inexact), and this fallibility in turn seems necessary if our cognition is to have any chance of keeping abreast of the evolving universe. We now consider what else the evolutionary origin of our cognitive processes says about the possible range of our knowledge.

EVOLUTIONARY ORIGINS AND LIMITLESS KNOWLEDGE

Since evolution is, if nothing else, a theory of change and growth, one would think that an evolutionary account of cognition would clearly posit an open-ended reach in our ways of knowing. Although I will argue that such openness does exist, the thrust of evolutionary theorizing about cognition has surprisingly suggested just the opposite—that the evolutionary derivation of our mental processes imposes sharp limitations in the range of application of these processes.

Limits in the range of our knowledge have been linked to its evolutionary origin by leaders in evolutionary epistemology such as Konrad Lorenz and Donald Campbell (but not, significantly, by Peirce or Popper).[4] Their arguments run as follows: since our cognitive capacities have been naturally selected based on their performance in the environment in which we as a species historically evolved, these capacities are at a loss in understanding environments with properties radically alien to our own, such as cosmic and quantum mechanical states. In its strongest form, as advocated by molecular biologist Guenther Stent and linguist Noam Chomsky, the "evolutionary limitations" hypothesis predicts absolute barriers beyond which the human intellect will never profitably proceed. Chomsky cites the interior domain of human volition as one for which there was no historical "evolutionary pressure" to explain, and which will thus "always be shrouded in mystery."[5] A weaker form of the argument provided by Lorenz and Campbell calls on evolutionary limitations to explain difficulties in cognition that we have already encountered, such as the apparent paradoxes of quantum mechanics.[6]

Such claims of limitations run into trouble on many points aside from the central argument from evolution. For one thing, any prediction of absolute barriers to human knowledge or of areas which human cognition can never successfully enter is inconsistent with fallibilism, which entails the uncertainty of negative as well as positive knowledge claims. The recognition that we can never know anything with certainty applies, in other words, to what we think or know we cannot know, and thus on purely logical (not necessarily evolutionary) grounds the proscription on certainty actually works to remove or soften any limitations on the scope or growth of knowledge. A fallibilistic universe permits knowledge to grow endlessly, in principle, *because* it grows imperfectly.

Counting more directly against the evolutionary limitations argument are difficulties we have in comprehending phenomena with which we have had long familiarity, such as magnetism, weather, and the nature of human societies. Although our inability to satisfactorily understand such commonalities appears less profound than the problems we face with quantum mechanics, and although the reasons for such ignorance may in fact have no connection to evolution while our problems with quantum mechanics do, the difficulties we encounter with such proximate phenomena undermine the claim that our inability to understand alien phenomena stems from their lack of evolutionary proximity. At the very least, such difficulties with common phenomena suggest that the difficulties we encounter with new phenomena result from something other than their newness—in other words, just like the difficulties with old phenomena, they are consequences of factors having nothing to do with evolutionary familiarity. One nonetheless could maintain that our difficulties with old phenomena are nonevolutionary while our difficulties with new phenomena derive from evolution by contending that evolutionary experience is a necessary but not sufficient requisite of knowledge, whose absence in quantum mechanics accounts for our difficulties in that area, but whose presence in the familiar realm of weather forecasting is not enough to provide satisfactory understanding there. Such a line of argument, however, would have to explain why the problems with quantum mechanics require the postulation of epistemological limitations beyond the obviously formidable ones long in evidence in weather forecasting; indeed, the very newness of our experience with quantum mechanics and other remote phenomena allows that our difficulties in these areas may be less profound than the longstanding difficulties in familiar areas, and may, for all we know, soon be resolved while the familiar problems continue to plague us.[7] The progression of science has to a large extent been just such a process of providing increased understanding of ever more perceptually distant phenomena (such as planetary bodies or microscopic organisms) while leaving crucial aspects of our backyard (like the mind/body problem) sketchily explained if at all. This history runs directly counter to an evolutionary limitations hypothesis.

But the theory that our cognition is restricted by virtue of its evolutionary derivation to areas pertinent to our evolutionary experience founders for a reason far more intrinsic to its own argument: such a theory is wrong in its interpretation of evolution. Biological structures that have evolved in response to one sort of environment commonly perform well in others, living systems perform tasks other than those they were presumably originally selected to perform, and these cases of "preadaptation" or performance transfer constitute one of the mainsprings of evolution, providing a partial answer to the key question of where and how new traits and structures originate. The preadaptive answer is that new capabilities arise from the rearrangement of structures initially evolved for other purposes, or from the juxtaposition of old structures on new sets of environmental problems. The original structure, then, can be said in a retrospective sense to have been preadapted to the new task.

Preadaptation does not deny that most organisms and populations negotiate only small portions of all possible environments, and indeed usually perform poorly (perhaps not at all) in environments sharply at variance from the ones in which they evolved. After all, fish cannot breathe air, we cannot breathe water, and neither of us can survive unaided in a vacuum. On the level of biological organization that we recognize as individual organisms, there is indeed a rather restrictive fit of selectee to selector, with complex combinations of living structures adding up to organisms that usually function best in the environment that shaped this addition. (Humans are in one regard a crowning example of a countertendency in nature, in that we shape alien environments to our specifications via technology.) But on the lower, more general level of living structures and parts that make up organisms—that is, on the level of organs rather than organisms—the picture is quite different: although most mammals function far better on land than in air or water, the skeletal structure of mammalian walking limbs works beautifully as a basis for the bat's flying wing and the porpoise's swimming paddle. Indeed, the observation that parts of organisms can with little or no transformation perform tasks different from the ones they were originally selected to perform, in service of either the original species or of new ones, goes back to Darwin himself, who in his treatise entitled *The Various Contrivances By Which Orchids Are Fertilised By Insects* explains how orchids attract insects (for reproductive assistance) by employing structures earlier evolved for other purposes. More recently, George Gaylord Simpson argued that these and related cases of preadaptation are "practically universal" in evolution, and Stephen Jay Gould has suggested that "most actual events [in evolution] may owe more of their shape to [the] nonadaptive sequelae" of natural selection than to the original selection itself, that is, to the performance of tasks not originally selected for, in environments not initially adapted to.[8]

The lesson of preadaptation for the problem of cognitive limitations is that the evolutionary derivation of our cognitive capacities need not confine the fruitful application of these capacities to areas that are evolutionarily relevant. Indeed, the prevalence of preadaptation and the detachability of structure from originating environment that this entails suggests that the operation of our cognitive abilities is in no sense restricted to the environments in which these abilities evolved. Of course, our cognitive structures may be limited in their scope by factors that are nonevolutionary; or, for evolutionary reasons that we do not yet know, our cognitive traits may be in fact less transferable than their corporeal cousins—exempt, for whatever reason, from the practically universal preadaptive mechanism. But such a possibility flies in the face of the obvious observation that mental capacities are, if anything, enormously more flexible than living physical structures. Our lungs, evolved on land, cannot breathe water; but our minds, also evolved on land, apparently have had little trouble comprehending the aquatic realm.

The situation of mind in evolution thus places no necessary restrictions on the scope or range of our mental operations. Preadaptation is, to be sure, an ex post facto explanation of performance-transfers that already have occurred, that is, an account of how a structure performing well in an actual environment originally performed a different task in a prior environment; thus preadaptation can in no way guarantee or even predict success of any kind for our cognitive processes in as yet undiscovered realms, or in environments of which we are just becoming aware. Preadaptation, in other words, can account for any understanding our Earth-evolved cognition achieves about galaxies and subatomic particles, but it cannot insure or predict such comprehension. This is quite enough, however. For in freeing our cognitive capacities—to succeed or not—from the confines of our evolutionary origins, preadaptation insures that the sacrifice in certainty required by fallibilism is a sacrifice not in vain. The hope that we may infinitely improve and expand our knowledge, made possible on metaphysical principle by the endemic imprecision and uncertainty of our knowledge, is not obviated or compromised by the grounding of our knowledge processes in the flesh-and-blood reality of evolution. The evidence of evolution joins the logic of fallibilism on behalf of the endless growth of knowledge.[9]

RADICAL FALLIBILISM: THE EVOLUTIONARY ADVANTAGES OF ERROR

Thus far we have seen that the uncertainty of our cognition, or the lack of perfection in the fit of our knowledge to any given phenomenon, leaves room for our knowledge to apply (in the same imprecise way) to phenomena not currently at hand. We have also seen that the evolutionary derivation of these cognitive processes need not restrict them to an imperfect understanding only of phenomena at hand (or at hand in the history of our evolution). If epistemological optimism is taken as a potential for a continuing growth of endemically uncertain knowledge in a widening array of areas, rather than as a refinement of knowledge towards perfection or completion in a limited realm, then an evolutionary fallibilism does indeed seem optimistic. But what do we really mean when we say that the uncertainty or fallibility of our knowledge is the very characteristic that allows our knowledge to grow infinitely? Since a fallible understanding of a given phenomenon means that our knowledge in this area is in some important sense incorrect or in error, what we are really saying here is that error itself is the source of growth and thus of our optimism. This generative view of error actually goes beyond the usual interpretations of fallibilism, and requires further explication. (Note that our fallibility as knowers precludes even a partial error-free knowledge, that is, an understanding of a phenomenon which, while wrong for the overall phenomenon, may be perfect for a part of the phenomenon. The incomplete nature of knowledge in a fallibilist context is rather a holographically distributed gestalt that defies reduction to pockets of perfection in a sea of ignorance, and refers to a lack of perfection that more or less pervades every aspect of our knowledge.)[10]

The notion that we learn from our mistakes, and that commission of error should therefore be encouraged not avoided, is a central and oft-remarked on feature of the Popperian program. In this view, error itself is construed as dead wood, excess baggage, or impurities in our body of knowledge, whose identification and elimination leaves our knowledge less incorrect; commission of error therefore is advantageous in that it enables us to identify and eliminate falsities which, if left undetected, would continue to pollute our understanding. However, the more radical fallibilism that I advocate (which I think complements rather than contradicts mainstream fallibilism) views error not as dead wood but rather as raw material from which may arise worthwhile (although never perfect) future knowledge. An incorrect or wrong idea may, in other words, be dead wood relative to our understanding of a given phenomenon or environment, and its identification and removal may thus improve our understanding in this area; but this same error or incorrect idea may, in a sense somewhat analogous to preadaptation, contribute in a positive way to our understanding of a different phenomenon. A wrong theory for one phenomenon or environment may be a right theory (though still imperfect) for another. The Lamarckian model of evolution, for example, is apparently incorrect as an explanation of biological evolution, yet has value as an explanation for aspects of social evolution.

Baldly stated, the thesis of radical fallibilism is as follows: There is no such thing as a false theory. Theories are false relative to given problems, phenomena, and environments, and may or may not be false when brought to bear on other areas. Since we hope that all environments we may come across are expressions of some ultimate, undergirding reality whose properties we may describe in a unified theory—indeed, the goal of most science and philosophy is disclosure of this reality—we should expect that many and perhaps most of our false theories will turn out to be false in every environment we encounter, that is, will be grossly at variance with such underlying reality. There is thus no suggestion of an "anything goes" epistemological relativism here—of all ideas and theories, however outlandish, being in principle equal—since many of our false theories may indeed turn out to be thorough-going clinkers. The point is rather the retrospective, preadaptive observation that today's and tomorrow's more or less correct theories arise from yesterday's and today's false theories.

This direct derivation of benefit from error is actually an important mechanism of biological evolution (as in the case of favorable mutations arising from breakdowns in genetic transfer[11]) and can be better understood when we situate error in the three-part biologically oriented schema for the growth of knowledge described by Donald T. Campbell, and alluded to briefly in the first chapter.[12] This model describes the development of knowledge, like the evolution of living organisms, as consisting of three distinct though interlocking phases: generation or origination of new ideas, for the most part occurring independently of problems these ideas may or may not help solve; criticism and elimination of ideas that fail to help with problems or fail to be in accord with reality, through exposure of these ideas to the real world or by otherwise testing their efficacy;[13] dissemination of the remaining ideas, which are now provisionally viewed as more or less correct knowledge, via education, publication, and so on. The distinctiveness of each phase is discernible in the different intellectual activities appropriate to each of the phases, or the varying degree of utility of a given intellectual procedure in each phase. Conscious, deliberate logic, for example, is useful and necessary in moving us from premises to conclusions (and in critiquing this process and the conclusions), but is usually not the progenitor of original premises. The conception of new ideas rather occurs through sudden (and unconscious) insight, unforeseen brainstorms, metaphor, inspiration, and other essentially alogical processes. However, in the second, eliminative phase, any criticism which was *not* logical in structure would not be taken seriously—thus, if the substance of my criticism of the evolutionary limitations hypothesis was a gut feeling, such criticism would quite rightly be ignored by the intellectual community. (Of course, to the extent that imaginative criticisms are themselves new ideas or ways of looking at things, these may well originate in an alogical manner. The point here is that such ideas are usually taken as criticisms in proportion to their logical presentation.) The final, disseminative stage makes use of both logically planned strategies and propagandistic appeals to the emotions: I remember reading *Brave New World* as a boy because I was attracted to the half-nude woman pictured on the cover.

How does error fit into this differentiated schema? Seen from afar, the three phases tend to blend into one. When we say that a theory is true or false, or that we are mistaken in some piece of knowledge, we are generally speaking of the amalgamated result or consequence of the three phases, rather than the performance of one as opposed to another. Is the mistake of a mistaken theory the equal responsibility of all three phases? Let us assume, for example, that theory "x" is a mistaken theory—for example, that the idea I propounded in the previous section on the preadaptive flexibility of our cognitive structures is wrong and that in reality our cognitive structures are indeed limited to the environments of their evolutionary origin. How have the three phases of knowledge development conspired to bring about such an error?

Since the first phase would be responsible for the initial conception of the "erroneous" idea that preadaptation applies to the performance of cognitive structures, one is tempted to attribute the lion's share of the blame to that phase. Had the idea not been generated to begin with, of course it could not have gone on to plague us as an error. But beyond this fact of no initial idea, no ultimate error, is the first phase indeed the source of my error? The generative stage contributed an idea that turned out to be mistaken; but did the generative stage make the mistake?

Remember that our assumption that the preadaptive idea is mistaken is an assumption about reality, an assumption that in reality cognitive structures cannot be successfully transferred from one environment to another. Error is error, in this case, to the extent that it is contradicted by reality. Now let us consider the contribution of the second phase to the error-producing process. Had this critical, eliminative procedure been successfully applied, we would have discovered that the preadaptive model did not fit the performance of our cognitive structures in reality, and we thus would have removed the preadaptive idea from our body of conjectural knowledge about cognitive structures. Where would this have left us? First, we would be no worse off in our knowledge of cognitive structures than before, and indeed would be a bit better off in the sense that we identified one explanation of cognition that is not the case (this identification of error being the standard Popperian dividend of fallibilism); and second, we would be in possession of an idea about the applicability of preadaptation which, while incorrect as a description of mental phenomena, might be correct for some other aspect of reality. The successful operation of the second stage, then, deprives error of its sting, or removes the erroneous consequences of error.

If the erroneous consequences of an erroneous idea are eliminated, the idea is in a fundamental sense no longer an error. We might thus state the role of the second phase in the error-producing process as follows: the malfunction of the second phase (the failure to uncover the contradiction between the idea generated in the first phase and reality, or to uncover logical inconsistencies) allows an error to be actualized, to wit, makes an error an error. The negative consequences attendant to error—misdirection in our pursuit of knowledge in a given area, practical danger arising from discrepancies between our knowledge of reality and reality as it in fact is—are all due in the final instance to breakdowns in the second, screening stage, to failure, for whatever reasons, of our critical technique rather than of our generative procedure.[14] In the third phase, we do have the opportunity to either neutralize the second-stage breakdown by refusing to disseminate the error, or to exacerbate the problem by spreading the mistaken knowledge. When independent critical faculties are brought to bear in the third stage process, in decisions such as to publish or not publish or to include or not include in a college curriculum, then we are giving ourselves a second chance for discovery and elimination of incorrect ideas. But more often than not the mechanisms of dissemination are governed by nonepistemological factors—will the property sell, will the course attract students, does the proposed book open an old intellectual wound in the publisher's advisor—and offer little of the critical comparison of ideas and reality necessary for error detection. Of course, a mistaken theory may be refused dissemination for the wrong (for example, commercial) reasons, and thus the third stage might inadvertently work to prevent a mistaken theory from polluting our body of knowledge. But the fundamental noncriticality of procedures of the third phase, grounded in economics and psychology rather than the fit of theories to reality, suggests that once a mistaken theory has passed through the second stage censors, it has passed beyond our primary ability to rationally remove. At this postcritical juncture, only the vagaries of buying and selling, and power and reputation, stand in the way of whatever damage might be done by the retention of erroneous theories in our total body of knowledge.[15]

The mistake of mistaken knowledge thus derives mostly from a mistake in the second stage. Returning to our example of my hypothetically incorrect thesis on the applicability of preadaptation to cognitive structures, we can allocate the commission of error in the following way: although my generation of this thesis in the first phase clearly made the error possible, and although the decision to publish this idea surely heightened the likelihood of any harm that might ensue, the true culprit in this story would have been the failure of me and my pre–publication correspondents to adequately criticize my idea in the second phase.[16] Although the first phase created an idea which was out of sync with reality, a proper execution of the second phase would have uncovered this contradiction and averted any ill-consequences that it might have engendered; although the present publication of this erroneous idea insured it some place in our body of knowledge, my scholarly credentials are such that had this idea not been published here, sooner or later I probably would have obtained its publication elsewhere. Thus, by any reckoning, the damage was done in the second stage. (The reader will appreciate, I hope, that though sorely tempted, I've refrained from using as our hypothetical example of error the very analysis of error now under discussion.) In the constitution of error, the first stage proposes and the third stage ratifies, but the second stage actually commits the error.

But how does this pinpointing of error in the three-stage process of the growth of knowledge abet an epistemological optimism? In identifying the erroneous part of error, we also identify the part of error that is not erroneous, and that indeed may be of value. In finding the locus of error in the second phase, we are encouraged to inquire about the possible positive contribution that, say, the first stage of a mistaken theory may make to the growth of knowledge. We no longer need dismiss our mistaken theories, and our fallibility that inevitably engenders such theories, solely as crosses to bear in our quest for knowledge. Now that we know what is wrong about a mistaken theory, we may ask what is right about it.[17]

Any rightness that resides in a mistaken theory is the product of the first or generative stage. Just as in biological evolution, where successful new organisms spring from combinations of genetic material (most of which sustained successful adaptations in the past, some of which did not), so in the growth of knowledge our new (and fallibly) correct notions come from combinations of previously correct ideas and reworkings of previous errors. The history of philosophy and science is indeed a story of Marxes standing Hegels on their heads, and Darwins building on the best of Lamarcks.

Such, then, is the hope of fallibilism. The good news is that we can improve our knowledge through the elimination of error by the successful functioning of the second stage, and gradually parlay these improvements into an increasingly detailed understanding of a wider realm of reality. Better news, however, is that eliminated errors may themselves be sources of future knowledge. The festival of metaphor, brainstorm, and blind and blinding insight yields knowledge that may be wrong for the problem or aspect of reality at hand, but right for the future. We thus need not be downcast about our propensity for error, for in the growth of knowledge, as in nature, little seems to be wasted.

BRINGING IDEAS TO LIFE

We thus have seen that the evolutionary derivation of our cognition has given us an ability to rationally examine and thereby understand an increasing portion of our universe. Rationality and epistemological optimism (grounds for increasing knowledge) are evolutionary endowments of the human condition.

The human species is defined, however, not only by what we have received from evolution, but by what we return to evolution. As James Mark Baldwin and several other biologists have convincingly shown, no biological organism is merely or totally the passive product of evolutionary experience; rather, in the process of navigating, behaving, and living in environments, organisms necessarily affect and change their environments—and thus the evolutionary forces that select these organisms.[18] In other words, they help create their evolutionary experience, which in turn re-creates them. In the human being, this organic capacity for "self-determined" evolution takes on enormous proportions, as we have the ability to imagine and deliberately plan environmental alterations.

Consciousness, rationality, dreaming, and planning, however, are only half of what makes us special as a species; the other half is our capacity to execute our plans to change our environments. We may think about and even understand our environment for a million years, yet the environment will change not one iota due to our thought unless our thought is implemented in some way. We need not accept the political implications of Marx's distinction between knowing and changing to recognize that the epistemological and ontological implications that stem from this distinction are revolutionary and key to our existence as a species.

The mechanism by which we implement and execute our ideas, the conduit of humanly created change, is technology. Unembodied knowledge, after all, is locked into a peculiarly closed system consisting of the neural cortex of our brains and our ability to talk to one another. This place of detention is fluid, elastic, and remarkable, but ultimately closed nonetheless.[19] The embodiment of a human idea in a technology instantly breaks this hold, however, and sends our ideas cascading among the infinite atoms of the material universe. The interface of thought and matter via technology is apparent not only in sophisticated cases such as atom smashing and the creation of new elements, but also in the operation of the most primitive technologies. An idea about building a thatched hut changes the world in no way whatsoever, but its implementation in the arrangement of sticks, reeds, and so on does result in a modestly changed environment. Through technology, then, we return to the physical universe; our ideas have intercourse with the material realm, the world of atoms and molecules, of rocks and trees and breezes, from which we, and thus our technologies, originated.

This volume considers some of the epistemological consequences of our technological existence. We examine this existence always against the backdrop of our evolutionary derivation—our evolutionarily derived cognitive structures, our rationality, and our grounds for optimism. Yet we will be on the lookout for ways in which the technological embodiment and implementation of knowledge may radically alter even the evolutionary process and the backdrop from which it arises. (The growth of knowledge implemented in nuclear weapons, for example, can scarcely be considered positive.)[20] When we look at the technological alteration of the universe, we are looking not only at an alteration of material, but also at an alteration of the process of the universe, an evolution of the rules of the game by which the universe heretofore has played. Earlier we saw that the emergence of rationality may be construed as an evolution of unintentional evolution into deliberative direction; in the phenomenon of technology we will find that this direction is given substance, and the evolution of evolution is actualized.

NOTES

1. Perfect knowledge of the past would of course be immensely valuable in the historical sense of helping us better understand the present and plan the future. The counter-productive nature of such certainty—were it possible—comes from our tendency to mistake knowledge of what was for knowledge of what is and what will be. A defining characteristic of certainty or absolute knowledge is that it is in some sense timeless, yet timeless is precisely what knowledge in evolution cannot be.

2. Quoted by McLuhan and B. Nevitt, *Take Today: The Executive as Dropout,* New York: Harcourt Brace Jovanovich, 1972, p. 10.

3. The only limitation that lack of certainty imposes on the scope of our knowledge is prevention of complete knowledge: if our knowledge in any area grew to the extent of being complete for that area, this knowledge would then be certain for that area (that is, the area would possess nothing more to be known); thus if certainty is unattainable, completeness must be also. Limitless growth in a fallibilist context therefore entails continuing expansion of knowledge without achievement of completeness. A mathematic expression of this relationship would be an asymptotic curve, in which knowledge grows ever closer to completion but ever short of its attainment.

4. Lorenz writes that "just as the grain of the photographic negative permits no unlimited enlargement, so there are limitations in the image of the universe traced out by our sense organs and cognitive apparatus. . . . Where the physical image of the universe formed by man has advanced to the atomic level, there emerge inaccuracies. . . . It is though the

'measure of all things' was simply too coarse and too approximate for these finer spheres" ("Kant's Doctrine," p. 203); and Campbell chides Herbert Spencer for viewing "human cognition as validly encompassing all reality, rather than just those aspects behaviorally relevant in the course of human evolution" ("Evolutionary Epistemology," p. 437). Note that what is being suggested here by Lorenz and Campbell is not the endemically approximate or less than valid nature of our knowledge in general—indeed, although Campbell holds such a fallibilist view, Lorenz occasionally suggests that our knowledge may be "absolutely valid" when our research does not demand "greater precision" (pp. 191–192)—but rather the disability of our cognitive capacities when applied to areas beyond their evolutionary origin.

No such disability is suggested by Popper, and Peirce explicitly argues that the positing of unknowables in principle is inconsistent with fallibilism and the growth of knowledge ("The Scientific Attitude," p. 55). Peirce cites Auguste Comte's discussion of the chemical composition of stars being permanently beyond our knowledge as an example of the folly of making absolute claims about what cannot be known: "The ink was scarcely dry on [Comte's] printed page," Peirce writes, "before the spectroscope was discovered and that which he had deemed absolutely unknowable was well on the way of getting ascertained." Peirce here anticipates one of the central arguments of the present volume: the role of technology in extending the realm of the fallibly knowable. Neither Lorenz nor Campbell seems to have considered the impact of evolutionary technology on the evolutionary knowledge process.

See my "Evolutionary Epistemology Without Limits," and "Cosmos Helps Those Who Help Themselves" (paper presented at Conference on Space: Agenda and Opportunity, MIT, April 5, 1985, to be published in C. Mitcham, ed., *Research in Philosophy and Technology*, vol. 9, Greenwich, Conn.: JAI Press, 1988) for further discussion of the pitfalls of negative dogmatism. See also Colin Cherry's citation of Marconi's observation that "Long experience has . . . taught me not always to believe in the limitations indicated by purely theoretical considerations. These—as we know—are based on insufficient knowledge of all relevant factors." Marconi developed radio wireless communication in the face of warnings by Heinrich Hertz, discoverer of electromagnetic waves, that these were incapable of supporting widespread telecommunications. See William Edmondson, ed., *The Age of Access: The Posthumous Papers of Colin Cherry*, Dover, N.H.: Croom Helm, 1985, pp. 23 ff.

5. See note 24 in chapter 2 for Stent citation. Chomsky approvingly cites Stent and speaks of the eternal unknowability of human volition in *Reflections on Language*, pp. 24–25. See note 5 in chapter 1 of the present volume for Bertrand Russell's counter to Chomsky's argument about the lack of evolutionary pressure for understanding of volition; Russell made this point in 1914.

6. Examples of apparent paradoxes in quantum mechanics are the view that light is both a stream of particles and a wave (forms of existence with mutually exclusive properties), and test results which seem to show that measurement of the properties of one particle affects the performance of another particle at distance from the first simultaneously. See chapter 2, note 24 for more on light, and chapter 9, note 12 for further discussion of the action-at-a-distance controversy.

7. The more remote or new the phenomenon under study, the more uncertain our knowledge of this area, including presumed barriers to our knowledge of this area. Similarly, the more familiar and repeatedly examined the phenomenon, the better corroborated the obstacles we may encounter in our knowledge there.

8. Darwin, *The Various Contrivances By Which Orchids are Fertilised By Insects*, Chicago: University of Chicago Press, 1984 (1862/1877), pp. 282–284: "When this or that part [of an orchid] has been spoken of as adapted for some special purpose, it must not be supposed that it was originally always formed for this sole purpose. The regular course of events seems to be, that a part which originally served for one purpose, becomes adapted by

slow changes for widely different purposes. . . . On the same principle, if a man were to make a machine for some special purpose, but were to use old wheels, springs, and pulleys, only slightly altered, the whole machine, with all its parts, might be said to be specially contrived [that is, preadapted] for its present purpose. Thus throughout nature almost every part of each living being has probably served, in slightly modified condition, for diverse purposes, and has acted in the living machinery of many ancient and distinct specific forms."

Other prominent discussions of preadaptation: A. Dorhn, "Der Ursprung der Wirtbelthiere und das Princip des Functionswechsels," 1875, excerpted and translated in E. S. Russell, *Form and Function,* Chicago: University of Chicago Press, 1982 (1916), pp. 274–278; D. D. Davis, "Comparative Anatomy and the Evolution of Vertebrates," in G. L. Jepsen et al., eds., *Genetics, Paleontology, and Evolution,* Princeton, N.J.: Princeton University Press, 1949, pp. 64–89; G. L. Stebbins, *Variation and Evolution in Plants,* New York: Columbia University Press, 1950; Simpson, *The Major Features of Evolution,* New York: Columbia University Press, 1953, p. 159 (source of quote); W. J. Bock, "Preadaptation and Multiple Evolutionary Pathways," *Evolution,* June 1959, pp. 194–211; Ernst Mayr, *Animal Species and Evolution,* Cambridge: Harvard University Press, 1963; Gould, "The Problem of Perfection," *Natural History,* January 1977, pp. 32–35, and "Darwinism," p. 384 (source of quote). Gould points out in this last essay that the numerous operations of the human brain in a modern world whose complexion is far different from the world in which our brains evolved must far exceed the evolutionary pressures which shaped the initial development of our brains. See my "Evolutionary Epistemology Without Limits" for further discussion of the history of preadaptation and its implications for human cognition.

Gould recently suggests that the term "preadaptation" be replaced by "exaption," owing to the former's incorrect connotation of prior fit to a future environment (when in fact the fit is ex post facto—that is, the characteristic only becomes useful in the new environment after the new environment arrives). See Gould and Elizabeth Vrba, "Exaption: A Missing Term in the Science of Form," *Paleobiology,* 1982, pp. 4–15; and Roger Lewin's discussion of this essay in "Adaptation Can Be a Problem for Evolutionists," *Science,* June 11, 1982, pp. 1212–1213.

9. See my "Evolutionary Epistemology Without Limits" for further analyses of the nature and logical limits of limits, especially the see-saw relationship between limits in certainty/precision and limits in scope. See also chapter 7, note 20 in the present volume for refutations of several presumed limitations to knowledge of our interior selves.

10. Perfection in a local zone would mean that our knowledge in this zone would not be fallible; hence localized perfection is incompatible with fallibilism.

The distribution of fallibility through all parts of a knowledge system is "holographic" in the sense that components of a hologram exhibit all properties of the hologram regardless of how finely it may be divided. See Karl Pribram, *The Languages of the Brain,* Englewood Cliffs, N.J.: Prentice–Hall, 1971 for discussion of the human brain as a community of holographically functioning systems.

11. The overwhelming majority of breakdowns in genetic transfer result in unsurvivable products; yet noise in transfer is at the same time a major source of successful variations in a universe Darwinianly conceived. (Preadaptation—change of environment to fit already formed characteristics—would be another.)

12. Campbell, "Unjustified Variation."

13. Logical consistency would be an example of another type of check upon a theory. See Bartley, "Rationality," 1982, for a full discussion of various types of criticisms to which our theories can be usefully subjected.

14. Incorrect rejection of a corroborated theory—one that matches reality—is the other possible consequence of malfunction in the critical, testing phase. See Imre Lakatos,

"Criticism and the Methodology of Scientific Research Programmes," *Proceedings of the Aristotelian Society,* 1968, pp. 149–186; "Falsification and the Methodology of Scientific Research Programmes," in I. Lakatos and A. Musgrave, eds., *Criticism and the Growth of Knowledge,* Cambridge: Cambridge University Press, 1970, pp. 91–196; and the subsequent literature on his ideas (e.g., G. Radnitzky and G. Andersson, eds., *Progress and Rationality in Science,* Boston: Reidel, 1978) for a discussion of the complexities entailed in deciding whether to reject a theory or reject the test results when test results seem to contradict a theory.

15. Since the three stages often operate recursively in large groups of knowledge producers—that is, person or group two may criticize and test ideas that person or group one has disseminated—an incorrect idea can be so identified even after it has passed an initial series of tests and is disseminated. However, books are rarely recalled after publication because their ideas have been found wanting; and a mistake in the critical process which prevents publication of a worthwhile idea is more difficult to correct, for lack of dissemination may keep everyone other than the original creators, critics, and negating disseminators (who rejected the idea) ignorant of this contribution. The ease of amendation and even primary publication in electronic dissemination of texts may help this problem. See discussions of electronic text processing and publication in chapters 5 and 8 of this book.

16. Since effective critical processes usually entail the participation of more than one person—especially when that one person is the creator—phase two of the knowledge process also requires a degree of dissemination. We might distinguish between dissemination for the purpose of criticism and dissemination for the purpose of preserving ideas (a hallmark of the third stage), though as suggested in the last note, dissemination in the third stage often serves as prelude to another round of the second stage. The value of self-criticism becomes apparent as the only unambiguous second-stage process in this three-part model. See the "Meta-Cognitive Technologies" section of chapter 5 and discussions in chapter 8 for further consideration of criticism as an essentially group process.

17. Thus Popper's fallibilism ultimately provides the occasion for implementation of Wittgenstein's advice to ask not what is wrong about an activity, but what is right about it (Wittgenstein, quoted by Alexander Alland, Jr., *The Artistic Animal,* New York: Anchor, 1977, p. 92: "Do not say he is playing the game badly, but rather find out what game he is playing well").

18. Baldwin's principle of "organic selection" (later termed the "Baldwin effect") holds that organisms evolve not only in response to the external environment but in response to their own behavior: in other words, the activities of organisms select additional organic structures or characteristics supportive of the already successful activities. The behavior of organisms thus constitutes a level of selective pressure that operates prior to and in conjunction with the selection of the external environment. Baldwin uses the example of learned versus inherited aspects of intelligence: the former creates an environment which encourages or positively selects for development of the latter (*The Story of the Mind,* New York: Appleton, 1898/1907, pp. 33–36). In such a manner, learned or acquired characteristics influence the course of evolution without transgressing the indirectness required by the theory of natural selection.

Popper describes such an evolution of organs "under the influence of their use" as an example of how Lamarckian evolution can be simulated in natural selection; see *Objective Knowledge,* p. 268 for Popper's discussion of Baldwin. See also J. M. Broughton and D. J. Freeman-Moir, eds., *The Cognitive Developmental Psychology of James Mark Baldwin,* Norwood, N.J.: Ablex, 1982, for detailed analyses of organic selection and other aspects of Baldwin's work.

Fairfield Osborn and C. Lloyd Morgan are also credited with discussion of a similar principle around the same time as Baldwin. According to Alistair Hardy, *The Living Stream,* London: Collins, 1965, p. 164, Baldwin and Osborn first broached the theory in a discussion following a lecture by Morgan in 1896. Jacques Vonèche points out in "Evolution, Development, and the Growth of Knowledge," in *The Cognitive Developmental Psychology of James Mark Baldwin,* p. 52, that Baldwin's first published mention of organic selection appears in 1896, however, whereas Morgan's is dated 1897.

In general, Baldwin's principle focuses on organic behavior patterns and not on enduring organic alterations of the external environment as a central factor in natural selection. The latter is a central concern of the present volume's inquiry into technology, and is highlighted nicely by R. C. Lewontin's "Organism and Environment," in H. Plotkin, ed., *Learning, Development, and Culture,* New York: Wiley, 1982, pp. 151–170. "Organisms assemble their environment out of the bits and pieces of the world," and later alter them, Lewontin writes; "indeed, an environment is nature organized by an organism" (p. 160). Lewontin also draws an apt analogy between evolutionary theories which fail to take into account the active role of the organism in shaping its selecting environment and philosophies (ranging from Plato to British empiricism) which fail to see the way our understanding of the world changes the world. See also Jeremy Campbell's discussion of Julian Huxley and Bernhard Rensch's principle of "anagenesis," or the increasing control of organisms over their environment, *Grammatical Man,* p. 134.

The revolution in philosophy which begins to take the active human knower into account is the subject of chapter 4 of the present book and the basis of a philosophy of technology.

19. See chapter 6 for discussions of speech as a communication technology (that is, the action of vocal chords on the material environment of air). In this sense of speech and even nonverbal expression and gesture being a material rearrangement of external reality, the only thoroughgoing example of unembodied knowledge would be knowledge which is uncommunicated. Since humans do communicate and engage in other technological activities, however, the closed system of purely unembodied knowledge is hypothetical. (Such a noncommunicative, nontechnological system would be open to the environment as a passive receiver of knowledge—open for input, but not for output. This inability to produce—to "export entropy," as Jantsch puts it [in "Unifying Principles," p. 93]—would likely result in an accumulation of entropy leading to the system's total closure or demise, thus rendering the system unviable or hypothetical for this reason as well.)

See also Stéphane Mallarmé's observation that "thinking is to write without accessories, or whispering" ("Crisis in Poetry," 1886 in *Selected Poetry and Prose,* ed. and trans. M. A. Caws, New York: New Directions Books, 1982, p. 75). Mallarmé here profitably reverses the usual trajectory of speech extending thought, and writing extending speech—for which, see Freud's characterization of writing as "in its origin the voice of an absent person" (see note 2 in chapter 8 of the present volume). Mallarmé's implication that the processes of thought and speech be comprehended in terms of the technological act of writing is consistent with the "technological" philosophies of Marx and Dewey discussed in chapter 4 of the present volume.

20. At least one observer has publicly argued that massive armament with nuclear weapons may be a necessary step in demonstrating the futility of war and helping us live together. See Douglas Martin, "McLuhan Center Says A-Bomb May Be Good," *New York Times,* February 12, 1984, p. 20. (Note that this was the view of the acting Director of the McLuhan Center in 1984, Derrick de Kerckhove, and not of Marshall McLuhan, who died in 1980.)

Chapter 4

Technology: The Embodiment of Human Ideas and the Unnoticed Philosophic Revolution

Among the many exquisite predicaments to which the human species finds itself heir is the discrepancy between thought and material. As we contemplate the world around us, and of which we are a part, we are acutely aware that this very process of thinking about the world is in some fundamental sense radically different from the object of this thought, the world itself.

The human drive for uniformity has attempted to reduce or resolve this difference in any way possible. For some, whom we might call "materialists," mind is nothing more than brain matter, a convenient illusion that stands in the same logical relation to thought as the notion of God does to the motion of planets. For others, the intensity with which we are aware of our own mentality in all our dealings with the world warrants an "idealistic" explanation of existence, a view that all that seems to be material is really a creation of the mind. And then there are those who, strive as they might, are unable to satisfactorily resolve this difference, and therefore employ it as a fundamental orientation in their subsequent investigation of reality. They are called dualists.

In the sense that I too see a profound difference between the world and thinking about the world, I am a dualist. Dualism, however, need not commit us to a view that the difference between matter and mind is utterly or absolutely irreducible. Rather, I see dualism as a recognition that the difference between matter and mind, whatever its origin, will be of continuing epistemological significance. Thus, in the likely case that the phenomenon that we call mind will someday be explained satisfactorily (albeit fallibly and, as is the case with all explanations, incompletely) in terms of the atoms and energy exchanges of our brains, this would not obviate the dualist insistence on the all-important distinction between the ways that atoms and energy make the world, and the ways that atoms and energy make our thoughts about the world. (The same would apply to the less likely possibility that the material world would someday be explained as a special sort of mind-stuff.) In other words, whether the mind is someday explained as some incredibly special type of material, or the world as some incredibly special type of thought, the difference between the ordinary and special cases of the "one" substance would continue to warrant our significant notice.

The historical development of the divergence of mind from material can be accounted for by evolution. Indeed, the preceding chapters in this volume were concerned with, first, sketching a plausible account of mind as an evolutionary product of the material world (or, better, the special material that we call mind as a product of the more ordinary forms of material that we associate with the rest of the world), and, next, with discussing the implications of an evolutionary account of mind for such perennial epistemological problems as rationality and optimism. However, as pointed out at the close of the last chapter, the human mind is not only a product of evolution, but, in the impact that we exert upon the material world, the human mind is manifestly also a producer of evolution. The dualism that we employ here in our investigation of technology is thus a doubly interactive or open dualism, in which material gives rise to mental in the first, naturally selective instance, and mental recreates material in the second, technological interchange.

UNIDEATED MATERIAL AND IMMATERIAL IDEAS

Let us consider three types of objects or aspects of reality: a tree, an idea about how we might use this tree in the service of communication, and a piece of paper. The first two entities are easily classified in terms of the material/thought dichotomy discussed above: the tree is a material object, and the idea a product of mind or thought. Where in this schema would we locate the piece of paper?

Further inquiry into the nature of the tree and the idea provides an answer. The tree is not merely material, but, in the context of a universe in which ideas are possible, the tree is an example of what might be called unideated material—that is, material whose existence is wholly independent of any human idea or thought. (I am assuming here that material reality exists independent of human perception, and that the tree we are speaking of was not deliberately planted. I also assume that the tree is not the product of divine thought.) The classification of our idea about the use of the tree can be similarly refined: in the context of a universe in which material existence is a possibility, an idea is not merely an idea but rather an immaterial or unembodied idea—that is, an idea lacking embodiment in any material form. (The idea, of course, is contained in some manner in the material of our brains, but this containment is not embodiment, since the brain is in no sense an expression or product of our idea about how to use a tree.) The fine tuning of material and idea into unideated material and immaterial idea pinpoints the nature of our piece of paper: it is neither pure material nor pure idea but a combination of both, a materialization of our idea about how to use the tree for communication (in this case, by cutting the tree into very thin slices that we can write upon). Paper, then, is ideated material or a materialized idea—the material of the tree impregnated by our idea about its use, material rearranged according to the specifications of our thought; or, the embodiment or realization in material form of our idea about use of the tree. This is the hybrid, dialectical nature of all technology, whether paper, toothpick, or skyscraper.

The epistemological implications of this toothpick hybridization are enormous. The difference between an unideated piece of material (the tree) and the same material touched by the human mind (the paper) is such that even the most recalcitrant materialist must be impressed with the reality and potency of the force responsible for the difference: the human mind, dependent upon yet surpassing the very special type of matter located in the brain. Similarly, the difference between an idea unembodied and the idea realized in tangible form testifies to the role of material in a way that even the most dogmatic idealist would be at peril to ignore. The mingling of mind and matter in technology thus counts equally against simple materialism and simple idealism, dramatically demonstrating the dualist contention.[1]

Moreover, the mingling of mind and matter that constitutes technology demonstrates a very special type of dualism, one in which both partners interact in major and continuing ways, as opposed to the segregated dualism of mind-and-matter parallelism or the unbalanced interactive dualisms in which one of the partners (usually the material) utterly dominates the other.[2]

Yet despite these and many other possible lessons that technology holds for philosophy and cognitive sciences, and despite the biological antiquity of the technological phenomenon (which is at least as old as the human species, and, if we count beaver dams and anthills, a good deal older), the philosophical implications of technology have been generally unexplored until the last century. (There have been a few notable exceptions, such as

Aristotle's discussion of techne.) Thus, in all of Kant's heroic strugglings to reconcile the conflict of mind and matter we find nary a mention of that aspect of reality in which the reconciliation had already taken place: technology.

The reasons for this philosophic avoidance of technology are no doubt worthy of investigation themselves, although they may boil down to the simple fact that before the Industrial Revolution the rate of technological change was so slow that technological existence, though pervasive, was in a sense invisible, taken for granted, and thus not noticed by philosophers. Less than a century after Kant's technological innocence, and shortly after the onset of the Age of Invention and its application to communication and industry, we encounter Marx's highly significant and groundbreaking (though, as I shall argue, flawed) treatment of the technological/epistemological phenomenon. We will look first at what technology can do for the dilemma of Kant.

THE MISSING LINK IN KANT'S INTERACTIONISM

A question often considered by people of a philosophic bent is this: if a person with a sound mind were locked away in a room at birth, how much of the world, its objects, and its relations would that person be able to deduce? Since we now know that in order to have a sound mind, one needs some contact with the environment—if only in the minimal Chomskyan sense of eliciting innate knowledge, or in the negative Popperian sense of showing the limits of (and thereby shaping) our innate expectations—we might rephrase the question to read: after we had enough contact with the world to develop a sound or normal mind (whether in the Chomskyan, Popperian, or Piagetian sense), how much of the remainder of the universe and its attributes could we deduce with no further worldly contact? In other words, to what extent is external reality prefigured by our intellect?

Kant's work is remarkable not only in the way it attempts to answer this question, but also because the work itself is evidence of an affirmative answer, an example of knowledge of reality prefigured by logic. Three-quarters of a century before Darwin's theory of evolution, and more than a century before any publicized knowledge of genes, Kant devised a model of innate knowledge which is highly compatible with the modern notion of cognitive structures as genetically transmitted evolutionary adaptions to the environment. And well before the fireworks of the Industrial Revolution, Kant laid out a system for the interaction of mind and matter which, while lacking in a crucial respect precisely because of its ignorance of technology, nonetheless leaves us on the verge of a technological epistemology.

The problems that Kant inherited from the history of philosophy were daunting. One explanation of human knowledge, developed by Descartes and his followers, held that we know what we know about the external world because we already have knowledge of this world, in a quite explicit sense, inside our heads. Not only did this approach depend on the implausibility of a myriad of clear and distinct innate ideas accurately relating to (or even matching imperfectly) each of the myriad of our potential experiences in the outside world, in addition the Cartesian concept of knowledge smacked of the tautology that we know what we know because we already know it.[3] Thus, against this view of the total prefigurement of the world by our intellect, the British empiricist school of Bacon and Locke proffered a model of knowledge with no intellectual prefigurement, that is, no innate basis of knowledge. In this model we come to understand the world through the impression and conjunction of repeated sensory and perceptual experiences on our passive yet absorbent minds. It fell to Hume to notice that, while the empiricist critique of Cartesian a priorism was warranted, the empiricist alternative was not. No amount of repeated experiences in themselves could ever logically support a conclusion that we had obtained something beyond these mere repeated experiences, for example, a general law that would hold for experiences we had not yet encountered. One *could* have held that our innate mentalities supply the missing leap from experience to theory or law, the leap that is not to be found in the experiences themselves; but the empiricist rejection of all innate knowledge foreclosed this option. Thus, in denying innate sources of knowledge, the empiricists had crawled out on a limb—a limb which Hume reluctantly sawed off. Into this eighteenth-century prelude to nihilism stepped Immanuel Kant, spurred on by the additional problem of explaining how, in the wake of Hume's convincing demonstration that scientific knowledge was impossible, modern science such as Newton's had achieved such palpable success.

Kant's solution is a masterwork of subtlety and delicate balance.[4] Agreeing with the empiricists in their rejection of the Cartesian model of numerous discrete mental pictures innately matching external reality, Kant nonetheless insisted that in order for us to understand anything at all, we must have some capacity to understand prior to the understanding. The empiricists actually had already admitted as much in their view of the mind as an absorbent organ, but the notion of an absorbent mental capacity was insufficient to account for the leap from specific observations to general theory, and thus save the empiricist program from the devastation of Hume's attack. It was Kant's genius to see that a capacity to understand had to be, on logical grounds, far more than a passive recipient of knowledge—that a capacity to understand must in the process of understanding inevitably impose its characteristics upon that which is being understood. These characteristics of our cognitive capacities are not really knowledge in themselves—rather, they are potentialities of knowledge, expressed only when an external object goes through the cognizing process—and thus they bear little resemblance to Descartes' notion of innate ideas. However, in logically demonstrating that knowledge must be more than the mere sum of our sensations and experiences, Kant speaks to Hume's point that no number of repeated experiences in themselves ever logically add up to a general theory or scientific knowledge: the missing ingredient that Hume noticed, says Kant, is precisely the additional ingredient that our cognitive capacities impose upon the objects of their understanding. Thus, to understand any aspect of external reality is inevitably (and regardless of the properties of this aspect itself) to place this aspect in space, time, and relationship with other aspects of external reality: in other words, to supply the missing link between instances and theory.

The nature of this missing link (the way our cognitive capacities leave their imprint on our knowledge of the external world) can be better grasped in a few concrete examples. Consider our vision, which is itself a type of knowledge process. We look with unaided eyes at the outside world, seeing shapes and lines and colors, and naively think that our vision passively picks up or reflects the world as it really is. Yet we know from our modern technologies that the world is also alive with all manner of ultraviolet, infrared, and other types of radiation that our naked eyes cannot detect and therefore fail to include in their image of the world. Therefore, what we see with our eyes (and, for that matter, with our extrasensory technologies, which also have their biases) is not the world as it really is, but rather the world as it is processed through our visual apparatus—molded, edited, and selected by the dictates and constraints of our visual apparatus, such as our sensitivity to certain wavelengths and not others. Much the same occurs when we hear a voice on the telephone. Although what we hear originates as a natural human voice, by the time it reaches our ears the voice has been altered by its encoding and decoding to and from electricity and the many miles of cable it transversed to get to us: it has been bent, distorted, and compressed by the contours of its journey. (The pejorative connotations are not necessary to the general principle—the contours of telecommunication could conceivably make a weak original clearer. In other words, alteration can either diminish or enhance the original.) Thus the voice that we hear is actually a contribution of the original plus the shaping it receives from the apparatus of telecommunicated hearing. Applying these examples to Kant's notion of knowledge, we may define knowledge as follows: human knowledge is the alteration of our experience or knowledge of the world that inevitably occurs when we experience or know the world. In the realm of science, this alteration occurs both in the experiencing of discrete events and in the leap from discrete instance to abstract, general theory.

In so resolving the Humean crisis, Kant conjured an exquisitely interconnected interplay of mind and matter. The human intellect, our mind, does indeed prefigure our knowledge of the material world by providing a necessary and encompassing master mold for knowledge. However, without the material of the outside world to work upon, our cognitive mold provides no knowledge at all. In the absence of material to experience, our cognitive structures would be subjects without objects, vainly trying to consume themselves, or, to elevate the metaphor a bit, trying to make sense of their own reflections in an endless hall of mirrors. Yet for all the indispensability of the material world to the knowledge process, it too can have no meaning to humans without a cognizing agent to actively digest it, transforming it in the digestion. In Kant's schema, a world without mind (or cognizing structures existing in some unique type of matter/energy interface in the brain) is quite literally a senseless place, an insane countervacuum of excrutiating density and dizzying motion, all very much impossible for humans. So in solving the epistemological problem about the source of knowledge, Kant contributed an answer to the ontological question of the distribution of mind and matter in the universe.

Yet despite the ingeniousness and plasticity of Kant's solution—and for all its influence on subsequent philosophizing—the solution itself has never really taken hold. No sooner had Kant laid down his pen than there arose a school of idealists such as Fichte who interpreted Kant to mean that the mind creates not only the relations among material objects but the material objects themselves. Materialists, for their part, were all too eager to seize upon Kant's insistence on the a priori necessity of induction as justification for a disembodied, supralogical principle of induction that validated the empiricist program, ignoring Kant's view of induction as part of a comprehensive mental package that pervades and shapes our experience of material in ways that the empiricists disallowed. Thus Kant's philosophy has functioned as a sort of grand philosopher's Rorschach test, ironically demonstrating its own claims about the nature of knowledge by providing whatever philosophers have wanted to find in it, serving as the raw material upon which thinkers have unconsciously imposed and then discovered their own impressions.

Certain details of Kant's schema have been especially liable to misinterpretation. Lorenz, for example, as we saw in Chapter 1, argued that Kant's notion of the absolute logical priority of cognitive structures to external experience is not really compatible with the biological view of genetically transmitted innate modes of knowing, since anything genetically transmitted must be the product of some organic experience with the environment, and thus exists a posteriori rather than a priori to this external experience. Nevertheless, an evolutionist can respond in Kant's favor that in each individual (as opposed to species) inherited cognitive structures are indeed in place prior to all external experience (this point is usually acknowledged by Kant's evolutionary critics), and that even on the species level, the encounter or experience with the environment that results in the cognitive structure is a process in which the environment acts upon or selects from something that already exists in the gene pool of the species. Thus this "preenvironmental" something (arising in the first or generative stage of evolution, independent of environmental encounters) is indeed a priori to environmental experience. (As far as I know, this point has not been considered by Kant's critics.) Attacking the material side of Kant's equation, Lorenz and other evolutionists also contend that Kant's notion of material reality as utterly sealed from humans independent of our ability to experience it (in other words, Kant's notion of the incomprehensible "thing-in-itself," as opposed to the aspects of things that we are capable of perceiving) does not fit the evolutionary notion of external reality as existing independently of human experience yet having great and continuous impact upon human evolution.[5] However, if an environment (whether immediately perceivable or not) has any evolutionary impact on us whatsoever, then in a very profound sense we have experienced and known it, whether or not we are consciously aware of this experience; thus this material reality falls well within Kant's model of mental/material interaction. On the other hand, if there is indeed, as Kant suggests, an ultimate material reality utterly closed to humans, then by definition it could not have any impact at all on our evolution, and thus would not so much contradict evolutionary theory as be entirely irrelevant to it. Or, as Ernst Haeckel says, "Why trouble about this?"[6] Thus, despite these seeming inconsistencies, Kant's theory does indeed nestle quite comfortably within an evolutionary framework.

Then we have the problem of the confidence Kant placed in our cognitive structures, which he deemed capable of yielding absolute truth, as in Newton's theory of the universe. But Newton's theory turned out to be partially wrong (or at least not anything like absolute truth), and as philosophers such as Peirce and Popper have made abundantly clear, absolute knowledge of any sort is quite simply beyond the human power to possess.[7] On this point of the ability of our cognition to generate absolute truth and certainty, Kant is grievously in error. But has this over-optimism been the reason that Kant's mind-and-matter interactionism has not been taken fully and seriously? I think not. To begin with, Peirce and Popper are, sad to say, very much in the minority in their insistence on the inevitably fallible nature of human knowledge; indeed, the knowledge claims of many of the empiricisms and idealisms that have dominated nineteenth and twentieth century philosophy make Kant's claims seem tame by comparison.[8] More importantly, Kant's interactionist epistemology translates well into a fallibilist framework. Thus Popper has done an excellent job in demonstrating that capacities like the cognitive structures described by Kant are necessary in the production of knowledge but not necessarily accurate: they are absolutely needed for the production of knowledge, but the knowledge they produce is conjectural rather than certain. Hence our prefigurement of knowledge is not merely a process of stamping our mold on the knowledge we produce, but, more significantly, a testing, editing, and stretching of the mold as we encounter genetically unforeseen experience. A fallibilist interpretation of Kant is thus not only possible, but fruitful.

What, then, has been the stumbling block in the reception of Kant's philosophy? I would suggest that the problem with Kant's philosophy has been that, for all its integration of mind and matter, it offers no evidence in its own behalf other than Kant's very penetrating (and, I think, largely correct) reasoning. After reading Kant we see the world through new eyes, and realize that what we see is to some degree a product of our own eyes, a mixture of us and the outside world, and in a sense a reflection of ourselves. Yet this vision is tenuous, difficult to sustain in a commonsense everyday world of seemingly discrete material objects and human mentalities, and prone to degenerate, even in philosophers' minds, into simpler forms of materialism and idealism. The trouble with Kant's system is that on its own it is ultimately a closed system, entirely contained within the realm of unembodied reason and ironically unable to indicate a tangible manifestation of the very mind and matter interaction it professes.

This is where technology can be of assistance. Unlike trees and ideas about how to use them, the technology of paper, log cabins, and all material inventions is in essence a mixture of mind and matter. Technology, unlike natural material objects and ideas, has no existence at all other than as a product of mental/material interaction. In other words, technology simply would not be were it not for the interaction of mind and matter. Moreover, technology, as we have seen, is the material of the external world rearranged to the specification of thoughts produced by the human mind, and thus substantiates the peculiarly Kantian interactionist view of the human mind leaving its mark on all our experience of the external world (as opposed to other possible interactionist views which might give the material world a more dominating role in the partnership).[9] But most importantly, technology springs the Kantian notion of mental/material interaction from the closed, self-contained system of intellect alone. In the nontechnological, original Kantian model, human mentality or intellect arranges our experiences or mental percepts of the external world (not the material of the external world itself) into an understanding of the external world (that is, ideas and theories about the external world in our intellect); thus, although material reality obviously figures significantly in this system, all the action and consequences take place in the intellectual realm. In technological interactionism, however, human mentality arranges components of the external world itself (material, not experiences of material) into an altered external world (that is, a tangible, physical manifestation of the ideational process); here the action commences in the mental and concludes in the material realm. Technology is therefore truly and fully interactional in a way that Kant's naked schema is not: the toothpick mixes intellect and matter not only in our understanding, but in an observable change of external reality as well.

Technology, then, vividly fulfills Kant's interactionist schema. A piece of paper, a building, or a rusty nail by its mere existence does more to demonstrate and substantiate the interplay of mind and matter than all the weight of Kant's exquisitely reasoned arguments. Had more of Kant's successors focused on the furniture in their rooms—had they seen the epistemological dynamite in the very armchairs in which they were seated—the course of modern philosophy might have been very different.

But in substantiating Kant's unsubstantiated model, technology necessarily goes one step beyond it. The source of this step lies in the location of the outcome of the mental/material interaction. When the outcome is confined to a change in our understanding, as it is in Kant's original model, then the outside world continues as is, and the result, for all the active mentality that engendered it, is essentially a passive one. But when the outcome is expressed in the external world, as it is with technology, then the world is necessarily changed and re-rendered, and the result, however minor or inconsequential, is thus a fundamentally active and (in the colloquial sense of the word) a real one. The point here is not so much that we change the world through application of our technological tools and machines—though this is surely the most important ultimate outcome of technology—but the more primary observation that technology unapplied, in its mere existence, changes the nature of external reality. A single toothpick, an iron nail, or a blank piece of paper adds something ontologically new to the world in a way that all the trees, iron ore, and ideas in the universe on their own cannot.

External change, then, may be considered the prime characteristic that distinguishes technology from unembodied knowledge. In the sense that change or rearrangement is necessarily a consequence of or a type of application, we might say that there is no such thing as unapplied technology—that technology is by its very nature an application of ideas to external reality, which can then be further applied in secondary, tertiary, and other ways ad infinitum (for example, we can use the nail in the construction of a house in which we write late at night on a piece of paper . . .). At the same time, all technologies originate in acts of understanding, in the wholly intellectual, purely Kantian interplay of mind and matter that is knowledge.[10] Thus, another way of describing the special type of mind and matter interaction that constitutes technology is to say that technology straddles and combines the worlds of knowledge and application, or is a phenomenon of knowledge and its application rolled into one.

The shift in our discussion from understanding to change gives us our cue to take leave of Kant and the vivification that technology provides for his interactionist theory of knowledge, and turn to a thinker for whom the difference between understanding and alteration is everything. For Karl Marx, change was not only the signal characteristic of material reality come into contact with human beings, but the lifeblood of social existence. In what follows, we confine our consideration to Marx the techno-epistemologist.

MARX ON MIND IN MATERIAL

The opposition of mind and matter, as we have seen, has been and continues to be one of the central concerns of philosophers and those who attempt to make deeper sense of our lives. Equally prominent, and in various ways paralleling, overlapping, and transcending the mind/matter dichotomy, has been the opposition or distinction between stasis and change, being and becoming, and existence and alteration.[11] This opposition parallels the first in that mentalist philosophies such as Plato's have usually been champions of a static world view, whereas empirical philosophies such as Bacon's have often served as prescriptions of change. Moreover, the stasis-change polarity continues in interactionist philosophies, depending on how and where the mind/matter division is resolved. Thus, although Kant's interactionist formula mixes mind and matter and in this sense transcends the mind/matter conflict, the placement of this resolution in the realm of the intellect results nonetheless in a static world picture, whereas technology, in embedding the mind/matter mixture in the material world, necessarily confronts us with the prevalence of change.

In the second part of the nineteenth century, change was very much on everyone's agenda. Darwin's theory of evolution (and its philosophic analogue in Herbert Spencer's work[12]) was central in calling attention to the mutability of existence, demonstrating that the very nature of things is that they change—quite independent of any external directive such as the Deity's—and attempting to catalogue the principles and vehicles of these endlessly occurrent transformations. Yet almost as important, and serving as a complement to Darwin's theory of natural or spontaneous change, was Karl Marx's study of humanly-created change. This relationship between these two systems has been often overlooked. But as Darwin's work on spontaneous alteration in the world concludes with the human being at the apex of natural evolution, Marx's focus on humanly induced and directed change in the world seems a logical continuation of Darwin.[13]

To speak of the ways in which humans change the world is actually to speak of two types of change: the change that results, in the first instance, from human action upon the raw, material world; and the change that occurs, in the second instance, when humans act upon or behave in response to the results of the first type of change. This second type, the change of previously created change, amounts to humanly created social structure, and entails all manner of alteration and channelling of the effects of the first type. As a social theorist and historian, Marx was primarily concerned with this second-order change. Nevertheless, Marx fully appreciated the fundamental epistemological import of the first type of change—the human utilization and alteration of the natural environment—and this technological concern makes Marx's work complement not only Darwin's examination of wholly natural change, but also Kant's consideration of wholly cognitive human interaction with the environment.[14]

As is well known, Marx saw his contribution to philosophy as providing a materialist remedy for the over-idealization of Hegel and Kant (that is, their emphasis on intellect rather than material reality). Marx's view of the human alteration of the natural world is thus a palpably materialist one in which human hands (or the material parts of the human body) reshape the natural environment in order to satisfy material human needs such as food, clothing, and shelter. For Marx, then, the technological enterprise or the contact point between humans and the natural world is the act of labor, with the thoughts and ideas that precede this act (and, as we have already seen, in a sense make this act possible, in that an alteration of the environment is usually an implementation of a prior idea about such change) taking a backseat.

Marx's emphasis of the technological satisfaction of material needs similarly minimizes the ideational component of technology. A house certainly can be reasonably construed as elements of the material world rearranged to fulfill the tangible human needs of shelter and warmth, and less tangible psychological needs such as privacy. However, such a needs orientation pays insufficient attention to the fact that a house is also a material embodiment of numerous ideas of form and mathematics describing how to build a house and its many parts—the contribution of the architect as opposed to the consumer.[15]

Marx's concern with the worker and the consumer rather than the thinker, however, should not be interpreted as a simple or vulgar materialism—a view of mechanistic hands rotely reshaping the environment in response to the stomach's churning demands, exclusive of the finer and more sensitive aspects of human existence. To the contrary, we find in Marx a view of the act of labor as a fully human, almost spiritual endeavor, which, when conducted under the right circumstances (free of the capitalist's yoke), becomes an act of love. Thus, despite the short shrift given mental conception, Marx's view of technology is a very rich one, with humans leaving their touch on the external world in ways that far transcend the mere physical manipulation of material objects. In making the act of labor a labor of love, Marx restores much of the quintessentially human aspects of technology which his downplaying of mind loses.

Still, a nonideational view of technology has consequences that differ sharply from those that follow from a view of technology as first and foremost an embodiment of human ideas. Since labor and consumption are accomplished just as readily by groups as by individuals (indeed, in the case of many types of work, much more effectively by groups), the labor-consumptive view of technology quite naturally supports collective social orders. On the other hand, since thinking or origination of ideas is fundamentally though not exclusively an individual activity,[16] the ideational view of technology encourages exaltation of the individual in social systems, for example, capitalism.

In as much as selection and dissemination in evolution (the second and third evolutionary stages, as described in the previous chapter) operate via groups rather than individuals, the weight of evolution may appear to fall with the collective or nonideational approach to technology. However, when we consider how profoundly the first stage of evolution—the generation of new possibilities—is an individual process, with new arrangements of genes occurring within individuals and first expressing their characteristics to the world through individuals, we see how indispensable the individual is to an evolutionary framework (at least a naturally selective, Darwinian one).[17] Thus, while an evolutionary treatment of technology can certainly accommodate—and indeed must accommodate—the complex dynamics of groups and social forces, an approach that diminishes the individual to the extent that Marx's does is, in the last analysis, non- or even counterevolutionary.

Moreover, the labor-consumptive (materialist) interactionist view of Marx runs the risk of becoming as closed a system as the self-sealed ideational interactionism of Kant. Deprived of the infinitude of possibilities of human mentality (or at least deprived of the full flow of such possibilities), even the sensitive human reshaping of the environment that transcends mere materialism becomes a closed system. After all, the act of love (despite its obviously very potent generative possibilities) is not as widely generative as the act of unbounded imagination and mental creativity (although mentality ultimately derives from the act of physical love, in the sense that mentality emerges from evolution, which is propelled by acts of generative procreation). A technological or human-material interaction operating wholly within a material world, however multi-faceted and subtle this materialism, is ultimately stunted unless it takes into major account the unique phenomenon of the human mind.

Marx's materialism, however sophisticated, does not make such an ideational accounting, and thus his epistemology of the stomach cannot serve on its own as an appropriate model of technology construed as an evolutionary embodiment of human ideas.[18] Like Kant's wholly intellectual interactionism, Marx's wholly material interactionism does not do justice to the fullness and depth of the mind and material interaction that is technology. Yet in moving the action into the material world—in recognizing that the ultimate expression of human existence lies not in understanding but in changing the world—Marx takes a huge and all-important step beyond Kant, and brings philosophy into the technological age. We next consider some of the first fruits of this philosophy in the twentieth century.

A TECHNOLOGICAL ADJUSTMENT OF POPPER'S "THREE WORLDS"

Thus far we have considered the relationship of technology and philosophy from the two simultaneous, complementary perspectives of first, what philosophic systems can teach us about the nature and function of technology, and second, what the reality of technology can teach us about our philosophic systems. Thus, Kant's system of mind imposing its characteristics on our understanding of matter lays the groundwork for a view of technology as mental expressions in matter, while, in so materializing our mental impositions on matter, technology provides a much needed tangible instantiation of Kant's delicate interactionist schema. Similarly, Marx's view of the human experience as quintessentially a confrontation with the environment resulting in environmental alteration highlights the ontologically and epistemologically revolutionary aspects of technology, while technology conceived as an embodiment of ideas provides an opening to the infinite world of mental generation that Marx's system of change occludes.

The unavoidability of technology in the twentieth century has resulted in numerous interesting and useful examinations of the meaning of technology in human life, but few as epistemologically relevant as the systems of Marx and Kant. One exception is the work of John Dewey, who held that we best reveal natural laws by doing things with nature rather than reflecting upon it. Tools and instruments in this experiential view are our primary and only authentic cognitive vehicles, providing knowledge that is more useful and true (the two are more or less equivalent in Dewey's system) than unembodied ideas we may have about natural events in which we have no tangible involvement.[19] Martin Heidegger also emphasized the technological capacity to reveal fundamental workings of the universe. In a sort of Nietzschean amalgam of Kant and Marx, he explored technology as a material expression of the human will upon the world.[20] Much less known is Friedrich Dessauer, who considered technology an embodiment of ideas in a Platonic sense—that is, an embodiment not of ideas of human origin but of ideas that human inventors obtain from an eternal, preexisting world of ideal solutions.[21]

Each of these systems performs the much needed service of making technology a central philosophic issue. Each correctly critiques Kant for failing to locate the interaction of mind and material in technology, and each sees technology as the most profound cognitive experience available to humans. However, each in its own way is incompatible with a naturally selective evolutionary epistemology. Dessauer's approach of preexisting, eternal solutions is not only nonevolutionary but also nonorganic in that the ideas embodied in his technologies are not of living, human origin (humans function as conduits rather than creators in this system). Heidegger's heritage from Nietzsche results in a depiction of technological expression as a single-minded, monolithic, overarching force, which is inconsistent with the tentative, multidimensional quality of technological revelation we would expect from an evolutionary, fallible phenomenon. Dewey's perspective is the best of the three, but his analysis runs into difficulties on the other side of the relative–absolute spectrum, where his equation of truth and utility tends to obscure the role of the reality that exists independent of humans, and in fact gave rise to both the human mind and the tools it produces.[22]

The twentieth century philosophic system that provides the most assistance in furthering our understanding of technology as an evolutionary way of knowing and interacting with the world is ironically one which, like Kant's, is not centrally concerned with assessing the technological phenomenon. This system is the "three worlds" schema of Karl Popper, designed to clarify the relationship between human mentality and the material world in a cognitive rather than an explicitly technological sense. Attempting to account for the objectivity and autonomy of human knowledge once it has been created, Popper posits in addition to the material world of atoms, trees, and brains ("World 1") two domains of human intellect—"World 2," a subjective realm, wholly internal to the human mind, consisting of the very processes or acts of thinking, feeling, and imagining; and "World 3," the products or results of these processes: ideas, symphonies, and strategies which, once created, have a life of their own.[23] In Popper's system, products of the human intellect are as ontologically distinct from the human intellect (cognitive processes) as the human intellect is from the material world. All three arise and operate in an organic, evolutionary, fallibilistic framework.

The relationship between World 1 (the material world) and World 3 (products of the intellect) pinpoints the importance of technology's interactive character. In the absence of technology, the denizens of World 3 (unembodied ideas and so forth) have no direct connection to the material world: although our ideas result from the cognitive digestion (World 2) of our experiences of the material world (experiences which, as Kant showed, are themselves products of our perceptual apparatus), these ideas once formulated have no direct, continuing relationship to the material world they describe. Should we wish to review, criticize, or test the accuracy of these ideas as descriptions of the material world, we can do so only by reintroduction of the cognizing agent (World 2). Thus, in a nontechnological setting, intercourse between Worlds 1 and 3 can occur only through the mediation of World 2, the acts of human mentality.

Technology changes all of that, since technology is in its very essence a union of the material of World 1 with the ideas of World 3. An iron nail is at the same time part of the material world and a product of human mentality. By its very existence it constitutes a continuing test of the theories it embodies about the malleability of iron, an ongoing confrontation between human ideas about the material world and the material world itself. Even the humblest of technologies thus bridges the cavernous gap between nontechnological Worlds 3 and 1.

A special advantage of casting this interaction in terms of Popper's three worlds comes from the autonomy of World 3, which highlights the distinctiveness of technology from its human creators as well as from the natural material world. Popper's distinction between Worlds 2 and 3 helps us see that technology is not merely a mixture of material and mental, but a fusion of material with products of the human intellect—ideas—which are already objective and autonomous of their individual human creators before they gain tangible embodiment in technology. This distance of technology from its inventors is crucial to technology's role as a vehicle of human species-wide self–transcendence: we transcend ourselves only through creations that go substantially beyond us, not that function as dependent or subjective appendages. In the case of technology, we have products that may exist and work in our total absence.

However, the very fact that Popper's three-world model can accommodate technology only through a straddling of two worlds—which has the important result of emphasizing the interactive nature of technology—suggests that Popper's model as it stands is inadequate to the task of clearly accounting for the scope and subtlety of the technological phenomenon. Does not a reality as transcendent as technology deserve a world of its own?

The problem with Popper's model is that its World 3 is drawn in terms too primarily ideational: the fundamental criterion for World 3 citizenship is being a humanly produced *idea,* with the material expression of ideas awarded a second-class or derivative status. Indeed, even Popper's World 2 seems cast in an overly mentalistic framework, paying insufficient attention to the interflux of matter and energy in the brain that supports and in part constitutes the activities of mind. Accordingly, I suggest the following techno-materialist reformulation of Popper's three-world system: "T-World 1" of natural material, consisting of nonliving material and all living material save human beings (or human brains); "T-World 2" of human beings, in particular that part of humans (the brain) which gives rise to, supports, or even constitutes an activity which, as far we as know, is unique in nature and the universe—the activity of thought or the human mind;[24] "T-World 3" of humanly touched or artificial material, ranging from the fleeting disturbances in the air made by our vocal chords when we speak, to new elements with half-lives of millions of years created in our nuclear chambers.[25]

The location of unembodied knowledge or ideas in this technological rendering of Popper's three worlds becomes a problem, just as the placement of technology was a problem in Popper's original, more ideational model. The question is whether unembodied ideas, as products of the human intellect, are special cases of artifactual T-World 3, or whether as entities contained wholly within humans, they rightfully belong in T-World 2. I would make the decision as follows: when ideas are not communicated through language or even through nonverbal gesture, they remain utterly private and internal to the person who has them, and thus fall within the domain of T-World 2. On the other hand, the instant an idea is communicated, it is freed from its internal human origin, and the very act of communication makes an imprint on the material world (even in the ephemeral, minimal sense of vocal chords making imprints in the air): this imbues all communicated ideas with T-World 3 status. Communication, then, is a mover of T-World 2 ideas to T-World 3 emdodiments, however insubstantial. This suggests an intricate relationship between communication and technology, to be explored in Chapter 6.[26] Although the distinction between an unembodied, uncommunicated idea and a thought minimally communicated and represented in a raised eyebrow or a whisper may be thin indeed, the distinction seems cleaner and more satisfactory than the technological conjunction of Worlds 3 and 1 required in Popper's original system.

Thus, through retooling Popper's three worlds, we arrive at a sharply drawn and responsive accounting of technology as a concrete expression of the human mind, forged in the application of human mentality (T-World 2) to the natural material world (T-World 1). As the sole constituent of T-World 3, technology enjoys a unique ontological status commensurate with its unique role in the universe: with the exception of humans themselves, nothing is as special in the universe or as different from all other things as technology. (Among other reasons, this is because technology is the product of planning and direction—even though many specific technologies and applications arise from accidents or unintended consequences[27]—whereas all other things in the universe, including even living and human forms, are as far as we know the product of blind, undirected evolution.[28]) Technology owes the uniqueness of its existence to the uniqueness of the human species that produces it, and the T-World 3 designation calls attention to this double originality by placing technology in a category distinct from human mentality (T-World 2), which in turn is distinct from the natural material world (T-World 1) from which mentality arose. (The evolutionary priority of the material world to mind, and of mind to products of the mind, is of course already present in Popper's original three-world schema.) Thus, in the profoundest possible sense, technology or artificial material is something new under the sun: skyscrapers, tissue papers, spears, and rubber bands are the beginnings of a new world or new universe, as the material of the natural, original universe reinvents and extends itself through the fantastic creative channel of the human mind.

PRODUCTIVE KNOWLEDGE

We have seen that eighteenth, nineteenth, and twentieth century major philosophic attempts to explicate the relationship of humans to the material world provide a basis for our view of technology as a material embodiment of human ideas, and that technology, construed in this fashion, can in its turn help develop and improve these philosophic formulations. The three major systems considered here—those of Kant, Marx, and Popper—are by no means the only philosophies with something to contribute to a technological epistemology. In succeeding chapters other approaches will be brought to bear on the technological constitution of our ways of knowing, including Bacon's view of human technological dominion over nature, Dewey's functionalism, and the "authenticity" of being in existential philosophies as it relates to artificial intelligence.

The recognition of technology as a material embodiment of human ideas and a tangible interaction of mind and matter opens up numerous pathways for further consideration of technology. We might, for example, endeavor to rate technologies according to how much mind and how much matter has gone into the mental/material mix. For example, a stone picked up as it is and used as a weapon has much less mind in the mix than a stone chipped to human specification. Similarly, a spoken word has much less material in the mix than the same word written on paper. We might profitably inquire into the qualitative ways that mental expression differs from technology to technology, noting, for instance, that while all technologies are material expressions of human ideas, some technologies are embodiments of ideas about how to convey other ideas quite apart from the ideas primarily embodied in the technology. Thus a book, regardless of its content, is an embodiment of a human idea or strategy about how to communicate ideas—in this case through print rather than speech or film—and at the same time a carrier of ideas in its content, for example, the plot of a novel (the plot presumably being quite distinct from the embodied decision to publish in a book rather than produce in a film). The special performance of these "double-cognitive" communication technologies is the subject of Chapter 6.

Keeping such issues in mind, we will proceed in the next chapter to consider a question more fundamental to the relationship of technology and knowledge, and central to this entire volume: to what degree and extent does technology, an embodiment of our ideas or a material expression of human knowledge, constitute, transform, and further our knowledge? Exploration of the technological enablement of knowledge brings us to a view of knowledge that goes beyond the traditional philosophic choice of objective versus subjective knowledge, both of which assume the operation of naked, unaided faculties as the basis of cognitive activity. Recognition that human knowledge presupposes not only intellect, senses, and external reality but also the tools that engage, mediate, and vivify these processes—and may materially alter the objects of their understanding—points to a knowledge that is neither subject nor object but product.[29] Implicit in our consideration of productive knowledge will be not only the weak claim that all knowledge is in some sense conditioned by technology, but the stronger assertion that human knowledge would not exist as such were it not for technology. This suggests that notions of pure, technologically innocent knowledge are but wishful abstractions (made possible, as we shall see later, by various types of technology), and the factors of evolution, rationality, and fallibility that shape our knowledge operate on an increasingly prevalent technological basis.

NOTES

1. See my "What Technology Can Teach Philosophy" in P. Levinson, ed., *In Pursuit of Truth: Essays on the Philosophy of Karl Popper,* Atlantic Highlands, N.J.: Humanities, 1982, for a fuller discussion of the technological refutation of simple materialism and idealism. Note also that evolutionary theory leads to the same conclusion (see "Evolution, Reality, and Truth" in chapter 1 of this book).

2. See Mario Bunge, *The Mind-Body Problem,* New York: Pergamon, 1980, for a ten-part taxonomy and discussion of monistic and dualistic treatments of mind and matter. The evolutionary-technological dualism that I suggest here lies somewhere between the "emergentist materialism" advocated by Bunge and the interactionism of, for example, Popper and John C. Eccles (*The Self and Its Brain,* New York: Springer, 1977) on Bunge's scale: like the emergentist materialist, I view mind as a state or process that arises in certain activities of the human brain (which in turn has emerged via self-transcendent evolution from nonthinking living material); however, unlike Bunge, and like Popper, I stress the profound impact that mind in turn has upon the world (see also note 9 below). My differences with Popper's type of interactionism are, first, whereas Popper sees ideas themselves as the primary vehicle of the mind's influence upon the world, I see technological embodiment of ideas as the main or even sole mode of mental impact, and second, whereas Popper sees a major realm of ideas devoid of material, I cannot find ideas anywhere in the world lacking the material substrate of the human brain and/or material embodiment or representation in technology. See "A Technological Adjustment of Popper's 'Three Worlds' " in this chapter and "Turning the Universe Inside Out" in chapter 9 of this volume for more on my differences with Popper's interactionism and Bunge's emergentist materialism. (The theories of Plato and Gregory Bateson are examples of what Bunge calls "animism," or what might be called "emergentist mentalism": matter derives meaning and even existence from form and idea. See, for example, Bateson's *Mind and Nature,* New York: Bantam, 1979, p. 4: "In the beginning was the idea." This works as an explanation of technology only when technology is detached from its human, evolutionary origins—ironic since Bateson is an evolutionist—because the ideational origin fails to address the evolutionary emergence of mind, the producer of ideas, from nonthinking material.)

3. The overconfident Cartesian program nonetheless provided an important corrective to

the Christian-Platonic pessimism (which holds that our knowledge is not only fallible but in general unimprovable) that dominated European epistemology for more than a thousand years. And the Cartesian approach offers a philosophic context for Chomsky's work in innate rules of grammar (for example, see Chomsky's *Language and Mind,* 2nd edition, New York: Harcourt Brace Jovanovich, 1972, ch. 1), much as Kant provides a philosophic environment for Piaget.

4. The best explication of Kant I've ever seen is in Julius Seelye's translation of Albert Schwegler's *History of Philosophy,* New York: Appleton, 1888, pp. 269–307. The edition of Kant I'm using in discussions of his epistemology is J. M. D. Meiklejohn's 1855 translation of the second (1787) edition of *Critique of Pure Reason,* London: Dent, 1934.

5. Lorenz, "Kant's Notion of the A Priori in the Light of Contemporary Biology," and *Behind the Mirror.* See also Gerhard Vollmer's "Kant and Evolutionary Epistemology," Proceedings of the 7th International Wittgenstein Symposium, Kirchberg am Wechsel, Austria, August 1982, for a comprehensive analysis of conflicts and agreements between Kantian and evolutionary epistemologies. Vollmer accepts Lorenz's criticisms of Kant, and points out that Kant rejected a protoevolutionary account of reason available in his time (a "preformation system of pure reason") in favor of an absolute or eternal a priori. However, if we accept (a) a distinction between organism and environment and (b) the naturally selective initiative of the organism in organism-environmental relations, then "preformation" and always prior may indeed be the same—and Kant's rejection of an "evolutionary" approach may be evolutionary nonetheless.

6. Ernst Haeckel, *The Riddle of the Universe,* trans. J. McCabe, New York: Harper, 1900, pp. 380–381.

7. The presence of noise or discrepancies in all representations or reproductions vis-à-vis originals suggests that the human inability to know with perfection is not a peculiarly human trait (as suggested for example by Christian humility-fallibilism) or even an intrinsically organic characteristic, but rather a feature of all transfers of information (including the ones we call knowledge) in the universe. See also chapter 5, note 15 below.

8. Hegel's idealism, Comte's positivism, and some strains of logical positivism in the twentieth century are examples of post-Kantian systems with knowledge claims more extravagant than Kant's.

9. An eliminative or reductive materialist (who believes mind is either nonexistent or atomistically reducible to material)—and certainly an emergentist materialist—could explain the technological reordering of the world as just another consequence of the brain. (The terms "eliminative," "reductionist," and "emergentist" are used in Bunge's taxonomy discussed in note 2 above.) But as the consequences of this special thinking material become greater and more apparent—moving, with the possibilities of space travel, even into the reaches beyond Earth—so too does the very specialness of this material, until it coincides with what I take to be the epistemological purpose of the concept of mind: to call attention to a difference between thinking material and other material that is so great even the term "thinking material" might not do the difference full justice. In this sense, I think the technological rearrangement of the world provides vivid evidence of the mind (uniquely thinking and productive material) and its impact.

10. As suggested in chapter 3, note 19, the shift from unembodied knowledge to technology seems closely tied to the act of communication. The unspoken, unexpressed idea becomes technological the instant it is spoken (through a slight rearrangement of the material of air) or indeed often the instant it is expressed (even a nonverbal gesture has neurophysiological impact, however minor, on those who perceive it). Speech is thus more unambiguously technological than gesture, since speech whether perceived or not entails a fleeting impact on the air. We might therefore put the knowledge-technology continuum as

follows: first, uncommunicated and unembodied thought (nontechnological); next, thought expressed in nonverbal gesture (sometimes technological, sometimes not); and finally, thought expressed in speech (always technological). Material rearrangements more lasting than speech—such as writing, thatched huts, and so on—of course are also always technological.

11. Franklin L. Baumer presents a good overview of this opposition in thinking since the Middle Ages in *Modern European Thought,* New York: Macmillan, 1977.

12. As Spencer points out in the Preface to the Fourth Edition of his influential *First Principles,* New York: Appleton, 1864/1880, pp. v–vi, chapters appearing in the first edition were in fact written before the publication of Darwin's *Origin of Species* in 1859. Spencer, however, goes on to acknowledge a debt to Darwin's "luminous" work in the refinement of Spencer's arguments in later editions of Spencer's work.

13. The story that Marx wanted to dedicate the English edition of *Das Kapital* to Darwin and that Darwin refused is apparently apocryphal, however. Bartley reports in *The New York Review of Books,* October 27, 1977, p. 45, the news in *Annals of Science,* July 1976, that Darwin's letter declining dedication was sent to Edward Aveling and not to Karl Marx, and concerned Aveling's *The Student's Darwin,* not Marx's *Capital.* Typical of the belief that Darwin declined Marx's honor is Isaiah Berlin's account in *Karl Marx,* New York: Oxford, 1939/1963, pp. 247–248: "He [Marx] offered to dedicate his book to Darwin, for whom he had a greater intellectual admiration than for any other of his contemporaries, regarding him as having, by his theory of evolution and natural selection, done for the morphology of the social sciences, what he himself was striving to do for human history. Darwin hastily declined the honour in a polite, cautiously worded letter, saying he was unhappily ignorant of economic science. . . ."
Nevertheless, the very fact that many, including Marxists, subscribed to such an account demonstrates in itself the appropriateness of a Darwin-to-Marx succession in one branch of the history of ideas. See Jacques Barzun, *Darwin, Marx, Wagner,* Boston: Little, Brown, 1941, especially part II, chapter 4, for detailed analyses of the many connections between Marxists and Darwinians, including Engels' eulogy of Marx as the "Darwin of sociology."

14. See Daniel Bell's "Technology, Nature, and Society" in *The Frontiers of Knowledge: The Frank Nelson Doubleday Lectures at the National Museum of History and Technology,* Garden City, N.Y.: Doubleday, 1975, pp. 27–78, for a thorough and sympathetic analysis of Marx's materialist improvement of Kant's intellectual interactionism. My "What Technology Can Teach Philosophy," note 16, and what follows below in the present volume offer some criticisms of Marx's improvement, specifically its shortcomings in helping us understand the relationship of technology and abstract knowledge.

15. Alfred Sohn-Rethel seeks in his *Intellectual and Manual Labor,* Atlantic Highlands, N.J.: Humanities, 1978 to reconcile Kant's "idealism" and Marx's materialism by positing an abstraction that arises from commodity exchanges and which indeed dominates human affairs in a manner analogous to the domination of nature by categories of thought in Kant. However, Sohn-Rethel's situation of this "abstraction" in economics, and his explicit rejection of Kant's asocial origins of cognition ("the conceptual basis of cognition is logically and historically conditioned by the basic formation of the social synthesis of the epoch," p. 7; see also pp. 37 ff.), deprive his analysis of cognition of the biological, species-wide links which make Kant's approach so amenable to natural selection theory (evolutionary epistemology). See chapter 8 in the present volume for a discussion of knowledge as an exclusively or overwhelmingly social event (the view of Sohn-Rethel and sociologists of knowledge) versus knowledge as an intrinsically or initially individual, species-wide property with important social determinants (the view of this volume). See also "Meta–Cognitive Technologies" in chapter 5, and note 16 in chapter 3.

16. See "Meta–Cognitive Technologies" in chapter 5, and note 16 in Chapter 3.

17. The species may be the unit of selection, as Gould (for example, in "Darwinism" p. 384) and others suggest, but the individual is nonetheless the unit of generation and first expression of traits. As Stebbins and Ayala put it in their "Evolution of Darwinism": "The fate of species depends on the ability of the individuals making up the species to cope with the environment, and such an ability can only result from the natural selection of genes" (p. 81).

18. Frederick Engels offers a comparable nutritional approach to technology, evolution, and cognition in his *Dialectics of Nature,* 1882, where he suggests that human cognitive processes capable of advanced tools and technology arose as a result of the human shift to a meat diet, which in turn was made possible by primitive hunting tools. "Adaption to a flesh diet . . . has considerably contributed to giving bodily strength and independence to man in the making. The most essential effect, however, of a flesh diet was on the brain. . . . A meat diet led to two new advances of decisive importance: to the mastery of fire and the taming of animals." An unfortunate consequence of our mental-technological development for Engels, however, is that "all merit for the swift advance of civilization was ascribed to the mind. . . . Men became accustomed to explain their actions from their thoughts, instead of from their needs." So, like Marx, Engels substituted needs and hands for thoughts and minds as the animating forces of technological and cognitive evolution. The above excerpts were reprinted in Larry Hickman and Azizah al–Hibri, eds. *Technology and Human Affairs,* St. Louis, Mo.: Mosby, 1981, pp. 215–220.

A more plausible nutritional approach to knowledge and evolution is taken by Guenter Waechtershaeuser, who in his "Light and Life: On the Nutritional Origins of Perception and Reason" (paper presented at 150th Annual Meeting of the American Association for the Advancement of Science, May 27, 1984, New York City) challenges Campbell's view ("Evolutionary Epistemology," 1974) that vision is a derivative of touching or locomotion on the grounds that the earliest organic interactions with light were photosynthetical or food-producing. Thus the eye may be a descendant of structures designed not to scout the environment but to eat it (or at least draw energy from it).

See also Popper's conjecture that ingestion of foodstuffs is at the basis of our feeling of reality (*Objective Knowledge,* p. 37).

19. See, for example, Dewey's *Experience and Nature,* Chicago: Open Court, 1925, p. 161, where he argues that "What is sometimes termed 'applied' science may then be more truly science than what is conventionally called pure science." Dewey also adopts a Marxist epistemological approach regarding thought and change, to wit, "It is not thought as idealism defines thought which exercises the reconstructive function. Only action, interaction, can change or remake objects," p. 158. Larry Hickman's "Making and Doing in a Democracy: Dewey's Experience of Technology" (paper presented at American Philosophical Association Meeting, December 1984, New York City) provides an analysis and bibliography of Dewey's work about technology. See also Webster F. Hood, "Dewey and Technology" in P. Durbin, ed., *Research in Philosophy and Technology,* vol. 5, Greenwich, Conn.: JAI Press, 1982, pp. 189–207.

20. Heidegger's views on technology are principally presented in his *The End of Philosophy,* trans. Joan Stambaugh, New York: Harper & Row, 1973, and *The Question Concerning Technology,* trans. William Lovitt, New York: Harper & Row, 1977. Both are reprints of works written several decades earlier. See Michael Zimmerman, "Technological Culture and the End of Philosophy" in *Research in Philosophy and Technology,* vol. 2, 1979, pp. 137–145, for discussion of these works. See also my comments on Heidegger in "What Technology Can Teach Philosophy."

21. Dessauer, *Philosophie der Technik; Das Problem der Realisierung,* Bonn: Cohen,

1927, and *Streit um die Technik,* Frankfurt: Knecht, 1956. The second part of *Philosophie* appears in translation as "Technology in Its Proper Sphere" in C. Mitcham and R. Mackey, eds., *Philosophy and Technology,* New York: Free Press, 1972, pp. 317–334, 375–377. Dessauer recognizes that "Kant did not investigate [the] encounter of mind with ideas which results in empirical realization" (p. 376) and views this empirical realization in technology as "the greatest earthly experience of mortals" (p. 331).

22. In short, Dessauer and Heidegger are too metaphysical for an evolutionary epistemology, and Dewey is not metaphysical enough. Dewey's strategy was to demystify the metaphysical notion of external, detached reality by rendering it in terms of instrumentation or human usage. Evolutionary epistemology demystifies reality by giving it an organic history and impact—this approach does what Dewey's does without weakening the notion of transhuman autonomous reality which was the strong and valuable suit of classical metaphysics. On Dewey's derogation of external autonomous reality, see *Experience and Nature,* p. 88, where he writes that "objects are distorted when their affiliation with the epic, temple and drama are denied, and there is claimed for them a rational and cosmic status independent of piety, drama, and story." Evolutionary epistemology asserts that objects have both a connection to drama (human perception and reconstitution) and an independent cosmic status.

The difference between Dewey's and my view of technology can be further put as follows: In my view, the nail is epistemologically and ontologically significant whether used or not, because its existence constitutes a new type of reality in the universe, the mixture of a human idea with iron ore material, and because it has ready potential for use. In Dewey's view the nail attains significance to the extent that it is actually hammered and then does a job (such as holding two pieces of wood together).

Our views about unembodied thought having little impact on the world outside of its embodiment in technology are virtually the same, however, as are our views of the centrality of instrumentation in the growth of science and the fundamentality of communication in human life. See chapters 5 and 6 in the present volume for treatments of these last two issues, and also see note 10 above. For Dewey on communication, see *Experience and Nature,* ch. 5, which begins with the sentence: "Of all affairs, communication is the most wonderful."

23. Popper's "three worlds" philosophy is presented primarily in *Objective Knowledge* and *The Self and Its Brain.*

24. See "Artificial Intelligence and Real Life" in chapter 7 for my arguments that even the most advanced current "expert systems" artificial intelligences do not think in the autonomous, self-generative way that the simplest humans do, although someday we may create technologies that think in such a way.

25. I recently came across a similar taxonomy in Marcus Hartog's "Mechanism and Life," 1908, (reprinted in his *Problems of Life and Reproduction,* New York: Putnam's, 1913, pp. 216–242), which suggests a division of matter into "things-at-large" (material conditioned only by antecedent inorganic causes), "organisms" ("which grow and store energy and matter for their own needs"—these include humans and human minds as well as other living forms, and are self-propelled in behavior) and "machines" (products of organisms created to serve organic needs). Hartog, p. 252, credits Samuel Butler with anticipating such a taxonomy in his Notebooks, now available as *The Notebooks of Samuel Butler,* ed. H. F. Jones, New York: AMS, 1968. The only difference between Hartog's schema and my own is that Hartog sees a greater distance between natural nonlife and life than between life and intelligent life (and thus he places the last two in the same category), whereas I see the greater distance between life and intelligent life (and thus I place natural nonlife and life in "T-World 1"). Actually, both distances are crucial, which suggests the need for a four-part model that apportions existence into natural nonliving material, natural living material

(biological life operating via natural selection), thinking material (our brain/minds) and technological material (which includes artificial living material and may include artificial intelligence material). I develop such a four-part model in "Technology as the Cutting Edge of Cosmic Evolution" in *Research in Philosophy and Technology,* vol. 8, 1985, pp. 161–176, and in "Artificial Intelligence and Real Life" in chapter 7 of the present volume. On the treatment of technology per se (as distinct from natural material—nonliving, living, and intelligent), however, the three-part models suffice.

26. See also notes 10 and 22 above.

27. On the prevalence of unintended invention, see my "What Technology," note 8, and my "Human Replay: A Theory of the Evolution of Media," Ph.D. dissertation, New York University, 1979, pp. 79–80, 196–197, and the references to Arthur Koestler's *The Act of Creation,* New York: Macmillan, 1964, pp. 192, 214, 216 ff. discussed there. Examples of unintended invention explored in "Human Replay" are the telephone (initially developed by Bell to help the hearing-impaired), the phonograph (intended by Edison as a recording device for telephone conversations, not primarily as an instrument of musical reproduction), and motion pictures (again intended by Edison as an adjunct to his phonograph, which by this time had become primarily musical). The difference, however, between these unintended devices and the products of natural selection is that the human devices were the results of conscious planning and design that performed in ways other than originally planned, whereas in natural selection there presumably is no planning at all. See "On the Nonrationality of First Causes" in chapter 2 of the present volume.

28. Deliberate breeding of organisms by humans—what Darwin called "artificial selection" (*The Origin of Species,* Hammondsworth, UK: Penguin, 1968 [1859], p. 153; see also chapter 7, note 25 in the present volume)—thus constitutes a technology and a precursor of genetic engineering. By the same reasoning, deliberate or planned procreation of humans also introduces a technological element in our lineage. In this sense, humans become both creators and products of technology, and again preview a possible time when we may create ourselves with greater specification via gene splicing. (Note that this conception of human as product is far more real—less metaphorical—than the frequent discussion of how humans become products of their ideas and technologies by virtue of living in environments which are created by these ideas and tools. See "The Burden of Rational Technology" in chapter 9 for a consideration of to what extent technological environments may deprive us of rational initiative, and a criticism of views such as Jacques Ellul's—for example, *The Technological Society,* trans. J. Wilkinson, New York: Vintage, 1964 [1954]—that technologies change humans into hapless technological appendages.)

29. The treatment of human knowledge as intrinsically "productive" rather than subjective or objective is consonant with Dewey's argument on behalf of instrumental knowledge (knowledge embodied in actions and tools that fulfill ends) as opposed to the subjective knowledge of idealism and the objective detached knowledge of Baconian science (*Experience and Nature,* pp. 139 ff.), and in accord with Marx's (and Engels') emphasis of labor. The "productive" knower is indeed implicit in all philosophies such as Kant's that focus on cognition as an active process that transforms qualities of the objects it seeks to understand.

Chapter 5

Technology as an Agent of Cognitive Evolution

For millions of years, innocent eyes peered up at the interplay of lights in the night sky that we now know as stars, galaxies, and constellations. When at last human intelligence peered through these eyes, they not only saw but wondered what these nightly light shows really were—but the lights were so distant, so untouchable, so removed from living experience and thus the fruitful operation of our intellect, that even the brightest minds were wrong or baffled. How on Earth could we know what we were seeing?

In ancient times, people played with colored lenses, and, occasionally directing the lenses up at the stars, were able to see the lights with a bit more clarity as the lenses filtered out distracting colors. Using sundials and nonoptical instruments available in Ptolemy's time, Copernicus developed a theory (briefly proposed by Aristarchus nearly two millenia earlier) in which the sun, and not the Earth, was the center of the universe. That humans could devise such theories on the basis of such poor quality experience is a testament to the power of the human intellect, and, as Kant so aptly explained, the preeminent role of our mental capacities in the growth of knowledge. The generation of ideas in the first phase of the knowledge process, operating almost blindly or with little reference to the outside environment, often hits upon descriptions that subsequent observations show to be surprisingly in accord with the initially obscure environment.

But the decisive breakthrough in our understanding of the heavens waited on the invention of the telescope. What happens when we peer at the sky through this marvelous concoction of lenses and mirrors? In some way those minuscule, ephemeral, faraway specks of light get pulled into the human orb where they can be perceived, digested, and (fallibly) understood by our evolutionarily derived cognitive capacities. In one sense, the tiny pinpoints of light are magnified to a size where they become humanly comprehensible. In another sense, the awesome distance between human beings and celestial objects is reduced so as to bring these objects within range of our intellect's constructive engagement. In yet a third way of looking at things, telescopes transform humans into the dimensions of the cosmos, where our eyes and minds may roam the stars, perchance to understand them. In every case, the telescope is an artificial cognitive appendage, an embodiment of an idea about extending our cognition which, when attached or hooked in to our evolutionarily selected cognitive structures, extends their productive operation to areas far and wide of our evolutionary origin.

Telescopes, microscopes, computers, and all such embodiments of ideas about extending our cognition are thus epitomes of technology as an agent of cognitive evolution—and their existence and performance provides yet another clear refutation of the view that our cognitive capacities are restricted in their success to environments pertinent to our evolutionary past (see Chapter 3). But before fully examining these spectacular technologies specifically designed to further our cognition, we first consider how all technologies, as material expressions of human knowledge, work to extend our knowledge and thus figure in its and our evolution.

THE UNINTENDED TECHNOLOGICAL LIBRARY OF SUCCESSFUL KNOWLEDGE

In the sense that bird nests and beaver dams may be considered prehuman technologies or organically directed rearrangements of the material world, the technological impulse has great biological antiquity, and vastly predates the appearance of human knowledge. Evolution is a two-way process, consisting not only of environments selecting organic characteristics for survival, but also of organisms actively influencing and even creating environments, and thereby contributing to the factors responsible for their own survival. The technological pattern (or the organic reshaping of the environment that to some extent self–determines every organism's evolution) therefore reaches back to the very origins of life. Indeed, Popper points out that even genes produce enzymes that help create the environment in which genes function.[1]

Thus technology is tied into evolution in a way that precedes and transcends the relation of technology to evolving human knowledge; accordingly, we might say that technology is an agent of evolution before it is an agent of cognitive evolution. On the other hand, to the extent that life itself is a knowledge process—a proposing, selection, and accretion of genetic information—the intimate connection of technology or environmental reshaping to life is "cognitive" from the very beginning. And yet there is a world of difference between the most intricate and elaborately constructed spider web, termite nest, or beaver dam, and the crudest humanly used implement. The first sort of organic manipulation of the material world, however elaborate, is wholly unplanned and genetically fixed or instinctual; the second, however coarse or simple, is heir to the human intellect and all its capacities for rational direction, self-criticism, and limitless imagination.

The filling with human ideas of the organic funnel already inserted into the material world—that is, the human ideation of the organic manipulation of the environment that constitutes our technology—is thus an innovation in the history of evolution. Indeed, as suggested in an earlier chapter, the joining of human mentality with the technological impulse may be considered the evolution of evolution itself: with the literal unleashing of human mentality into the material world via the vehicle of technology, a new chapter is written in the story of the universe, in which the unplanned, autonomous universe turns itself inside out and becomes a product of human planning. We leave the cosmic implications of the cognization of the organic manipulation of the environment, however, for the final chapter, and turn now instead to the more immediate question of what impact the fusion of human knowledge and organic manipulation has on human knowledge.

The expression of our knowledge in material form has at least two interrelated consequences for the constitution and growth of knowledge: first, knowledge is permanentized, that is, it becomes much more autonomous of its human creators and of time than is immaterial knowledge or ideas internal to the human intellect (T-World 2); second, knowledge is implemented, that is, it is put to some tangible test, which among other effects allows us to better judge its fit to external reality. Since every technology is at once both a permanentization (concretization) and implementation (application) of the knowledge it expresses, these two factors operate in tandem in every technological act. For the purposes of our analysis we can consider these factors one at a time.

Permanence has always been a central concern of mortal humanity. In the realm of human knowledge, some degree of permanence is necessary if any growth is to occur at all: ideas generated in the first phase must be preserved long enough to be criticized in the second phase, and survivors of the second phase are disseminated or preserved in the third phase. In the next chapter, we will examine a genre of technology, communications media, whose specific function is to preserve or record knowledge as in the content of books, films, and computer memory banks. The point to be made now, however, is that in a sense more fundamental than the special recording function of communications media, all technology acts to preserve or record knowledge: whether the washing machine, the automobile, or a blank piece of paper, technology externalizes, materializes, and therein preserves ideas (in this case, ideas about how to clean, travel, and communicate) whose origins are unembodied and thus highly perishable—trembling on the verge of disappearance, to paraphrase Dewey.[2] Even at the very threshold of the technological phenomenon, an idea shouted is externalized (materialized in air) and thus capable of lasting forever, whereas a thought unspoken or otherwise unexpressed is doomed to die with its thinker.

Most technological expressions are, of course, a good deal more durable than spoken or gestured ideas. In the extreme case of, say, the Egyptian pyramids, they transcend not only their individual human creators but also the societies and civilizations in which they were born. Thus when we step out for a moment and look at all the technologies that currently exist on our planet, we see a far-flung living catalog of human knowledge past and present, a global encyclopedia or carnival of human ideas throughout the ages. The sum total of our technologies is in many ways the ultimate Library of Congress.

But technologies are more than mere passive material markers of human ideas. Each technology is also a tool, a way of doing something, whose very existence demonstrates that the idea embodied in it *works,* that is, in some sense is appropriate or right for the environment to which it relates. Had the pyramids not been constructed on principles which recognized the ravages of time and entailed strategies to avoid them, we would now not have pyramids and the ideas they embody in our catalog of concretized knowledge; similarly, the legions of flying machines at the turn of this century that never got off the ground or crashed shortly after takeoff were obviously material expressions of ideas about air transport that were incorrect, and thus failed the test of technological application. It follows, then, that technologies are significantly more than records of ideas: technologies are records of ideas that have survived confrontations with external reality and are thus, to some degree, accurate descriptions or explanations of reality. The technological library is hence more selective than a simple repository of materialized human thought, admitting only those thoughts whose embodiments have performed with some degree of success. When we encounter technologically embodied knowledge, we know at the very least that this knowledge cannot be wholly inaccurate.

The equation of technology with minimally accurate or fit knowledge again raises the question of what knowledge exists that is not technological, or what is its nature? Is an unembodied idea that proves false in a technological confrontation with the environment (for example, the many early ideas about flying that flopped) knowledge in any sense? The value of error in the knowledge process (see Chapter 3) suggests that failed technologies can at the very least provide worthwhile negative knowledge—identification of what does not work, so that these inappropriate ideas will not be embodied again in pursuit of the problem at hand—and may even provide embodiments that prove to be of serendipitous use for other tasks. Lessons garnered from crashed flying machines and rockets that exploded on the launch pad were in large part responsible for our eventual successes in air flight and space travel.[3] Thus, we may distinguish among three types of knowledge relative to general technological embodiment: unembodied knowledge; embodied knowledge that fails or does not work; and embodied knowledge that in at least some minimally accurate way does its job. The third type is by far the most prevalent and accessible, because it exists in everything technological around us. The second is enormously useful in an instructive sense, but it is rightfully avoided with the utmost priority in practical application, for no one wants to be a passenger in a crashed vehicle. The first type of knowledge—the unembodied kind—may be considered not knowledge at all, or merely potential knowledge, if we choose to make material embodiment a defining criterion of knowledge. In either case, whether potential or unembodied, this type of knowledge may be fashioned and developed to some extent wholly within the intellect, as when we examine ideas for logical inconsistencies, or perform "thought" experiments.

Thus we ought to stop short of equating technology with all knowledge, and settle instead for a view of technological performance as successful knowledge, specifically knowledge which has survived a material confrontation with the external world. This settlement, however, recognizes the predominant role of technology in the constitution and growth of human knowledge. For when we consider the evolutionary antiquity of the technological impulse, we find that long before the erection of formal science and even superstition to develop and safeguard knowledge; long before the literal libraries of ancient Alexandria and even earlier deliberate storehouses of ideas; long before, that is, humans became personally and socially aware of knowledge as the central flame of human existence to be stoked and treasured, we quietly ratified, preserved, and extended this flame a myriad of times in innumerable technologies. The emergence of science and knowledge-aware activities of course has greatly increased, not diminished, the technological contribution. Thus technology continues to be, as it always has been, the primary constituent of human knowledge and the only constituent of its transpersonal growth.

The birth of the recognition of knowledge as a staff of human life, whenever and however that momentous discovery (or series of discoveries) took place, had the effect of enormously increasing our knowledge for many reasons. Among the most important is the impact of this event on the technological phenomenon. On one obvious level, the more knowledge that results from thinking about knowledge, the more knowledge there is to embody and implement in technology, and thus the more knowledge there is to be technologically tested and preserved. Beyond this exponential expansion of the technological basis of knowledge—the knowledge, that is, that results from any and all technology by virtue of technology's embodiment/implementation function—the coupling of knowledge-awareness with the technological act had a consequence that may be even more revolutionary. For once knowledge becomes consciously pursued, we may construct technologies that not only embody and implement knowledge by reason of their being technologies, whatever their tasks, but that are deliberately designed to help in the pursuit of knowledge. In other words, technology in addition to its primary function may serve as a deliberate material vehicle for the full force of human rationality and imagination directed toward the pursuit of knowledge. We will now begin to consider the consequences and possibilities of this most potent techno-epistemological union.

THE BIOLOGICAL ANTIQUITY OF TECHNOLOGY AND THE PURSUIT OF KNOWLEDGE

The technological age, as we have seen, is in a sense synonymous with the age of life on this planet, since the manipulation of the environment or rearrangement of physical material that characterizes all technology is also a function of even the simplest living systems. The appearance of human beings, as we also have seen, is a breakpoint not only in the evolution of life but in the technological age as well, for the injection of human imagination and rational direction into the technological pattern increases the quantity and quality of material rearrangements in a manner that results in something profoundly new. Indeed, the technological implementation of human ideas has remade the Earth. We turn now to a second breakpoint in the technological age, a breakpoint that occurs when technology is directed not toward reshaping the material world but rather toward expanding the knowledge and ideas whose implementation in other technologies reshapes the world. If technology is the implementation of knowledge, technology applied to the pursuit of knowledge is a meta- or second-order technology: technology acting toward its own creation. If first-order or general human technology transforms the external, physical world, this second-order technology transforms the internal, mental world from which our technology originates, that is, the internal world of our intellect and ourselves. Cognitive-intended technology thus remakes the cognizers: we ourselves.

In a strictly chronological sense, technologies used in pursuit of knowledge are probably almost as old as human technology in general. Indeed, if we agree with cognitive anthropologists such as Alexander Marshack that a series of V-shaped scratches on a 300,000 year-old ox-rib are remnants of a primitive computational or recording system, then we have evidence of cognitive technologies nearly as ancient as material reformed for more practical purposes such as rocks flaked to make cutting edges.[4] In fact, the very growth of general purpose technology throughout the ages has surely been due in some part to the expansion and organization of knowledge made possible by coexisting cognitive technologies. Such has certainly been the case in the past few hundred years.

While cognitive and general technologies may have coevolved, however, cognitive-intended technologies have clearly been very much the junior partner in the relationship until modern times. One reason for the historically low profile and small number of cognitive technologies may be that our senses and mentalities, unaided and naked, perform much more effectively as producers of knowledge than do our bare hands, feet, and teeth as manipulators of the material world. Consequently, practical technologies such as hoes, ships, and weapons advanced much more rapidly than cognitive technologies because implements of work were more sorely needed by our naked parents than implements of reason. Moreover, even when the pursuit of knowledge became formalized, the efficiency of our unaided senses and mentality was such that many early scholars considered technologically gathered and produced knowledge inferior to knowledge generated wholly by our natural faculties. Thus, Socrates decried writing as a misbegotten image of the spoken original in the *Phaedrus* (a claim which, fortunately or unfortunately for Socrates, his pupil Plato troubled to write down),[5] and a reading of Aristotle suggests that in conflicts between evidence gathered by natural senses and artificial devices, the testimony of our natural organs should be deemed the more reliable.[6] While the poor quality of measuring instruments in Aristotle's day may have justified such an attitude for ancient scientists, its dogmatic adherents retarded the growth of knowledge for nearly two millenia after; indeed, the presumed superiority of unmediated observation was used by Galileo's opponents as a weighty argument against his telescopically developed theories.[7]

Of course, Galileo won the argument—if not in the court of Rome, at least in the court of history—and the triumph of Galileo represents the triumph of the telescope, and the coming of age of technologies deliberately designed for cognitive purposes. This triumph of technological over natural modes of knowledge procurement has been due in large part to the fact that modern technologies gather and process information not so much by supplanting or replacing natural faculties, but by replicating and extending natural modes to areas such as distant galaxies where they were formerly unable to operate. Thus the technological victory is not at all a triumph against nature but rather a triumph of nature (or of nature plus) and results primarily because of this inclusion of nature. The pattern has been especially vivid in the development of communications media, in which technological representations of reality have become increasingly similar to the reality we perceive and construct with our naked senses. Thus photography, for example, since its origin as a still, silent, colorless, two-dimensional medium, has become successively more like human vision with the attainment of motion, synchronized sound, color, and, most recently, the recreation of the third dimension in holography. This humanization or naturalization of technology, especially communications media, will receive more attention in the next chapter.[8]

For now, however, we will consider the ways in which two types of technology have amplified our natural cognitive processes. These technologies may be viewed as cognitive-intended in that, although many of the consequences of their operation may have been unintended and unpredictable (and indeed remain so), they were specifically designed from the very beginning, unlike technology in general, to assist in the pursuit of knowledge. One class of such cognitive-intended technology, typified by telescopes and microscopes, has helped increase our knowledge by dramatically expanding the realm of raw sensory experience on which the faculties of our intellect feed. The other class, tracing its lineage back through libraries and writing systems to the 300,000 year-old incised oxrib—and epitomized today by the computer—has spurred our knowledge both by speeding and by enlarging our internal mental faculties themselves. Recalling Kant's conclusion that knowledge is the product of external experience acted on by innate cognitive structures, we can see that cognitive-intended technology has touched both necessary bases. We look first at technology as an exponent of external experience.

TELESCOPES, MICROSCOPES, AND HUMAN EQUATION TO THE COSMOS

"The principles of the pure understanding," Kant says, "are not applicable to any object beyond the sphere of experience."[9] It follows, then, that although we may increase our knowledge by improving the action of our cognitive capacities (memory for example) upon what we already experience, a more direct expedient is to expand our sphere of experience or the data that our cognitive capacities are given to act upon. Knowledge-intended technologies have long attended to both tasks, but the more immediately striking breakthroughs have usually resulted from the technological augmentation of sheer experience.

When we speak of experience, we are referring to the contact that we (or our mentalities) have with the external world through our senses. In the case of the growth of scientific knowledge, the pertinent sensory experience has been primarily visual, with acoustic and tactile experiences a distant second and third, and olfactory and gustatory sensations virtually negligible. Indeed, with the exception of the "sniffing" and "tasting" conducted by remote probes such as those on the surface of Mars in 1976, and the huge receivers "listening" for extraterrestrial messages emanating from the far corners of the universe ("listening" is a metaphor in any case, as the receiving devices are attuned for electromagnetic and not sound waves), the technological contribution to the sensory basis of knowledge by and large has been to the sense of vision. So central is seeing to human understanding, that even when the quarry of technological detection is something unseeable such as radiation, the technology usually transforms its discovery into a reading on a dial or some similarly visible manifestation.[10] Thus, technology deliberately applied to increasing external experience in service of knowledge is most often a technological extension of sight to realms whose size, distance, or physical constitution renders them invisible or poorly visible to the naked eye.

The most primitive way of extending our vision to areas not at hand but accessible in principle to the naked senses of our species is to converse with someone who has had physical access to the area beyond our vision. A desert dweller, for example, could do much to extend the vision of jungle inhabitants, and vice versa. In much the same vein, the literal physical travel of human beings to far-off places has an obvious cognitive relevance in that human eyes are necessarily transported along with human bodies. Such corporeal, transportive extensions of vision have made major contributions to the growth of knowledge, as in the case of Marco Polo's trips to China and Darwin's voyage to remote parts of the world in the Beagle. Indeed, where the impediment to vision is purely one of distance, technologies of transportation may, in an unintended way, prove to be the ultimate knowledge-gathering devices.[11] Spaceships traveling at nearly the speed of light may well obsolesce the telescope.[12]

But since nothing other than light and associate forms of energy now travels at the speed of light, and since the unviewability of many real realms has nothing to do with distance (as in the cases of viruses and infrared light), technologies of transportation can provide only a partial remedy for unviewability. What is needed are technologies that move information and images, not people; that transform rather than merely transport; and that extend vision not beyond what one person can see, but beyond what any person can see. The telescope and the microscope are exemplars of such technology.

The telescope is the less radical in its powers of transformation, in that the telescope extends vision across distance which in principle is bridgeable by physical transportation. Moreover, although since the time of Galileo telescopes have discovered new planets and, more recently, previously unseen galaxies, the fundamental function of the telescope is making clearer that which is already visible—namely, the lights in the heavens. We make discoveries with the telescope because, like the protagonist in Plato's *Meno,* we already know to some degree what we are looking for.[13] (The knowledge preceding telescopic extension is, of course, sensual or empirical—the knowledge of the skies obtained by naked vision—rather than wholly internal or innate as in the *Meno.*) The microscope, on the other hand, opens up worlds to our eyes which, as in the case of microorganisms, were utterly invisible to us beforehand, and to which no real or projected physical transport could ever bring us closer. Extension through the microscope thus operates primarily as a discontinuous, qualitative leap into the unknown (no one knew about microorganisms before Leeuwenhoek), whereas the stars revealed to us by telescopes are the result of more continuous, quantitative extensions of vision. Devices that detect invisible forms of radiation and energy such as infrared light and electromagnetic waves are, in this sense, like microscopes. Still other technologies, like chemical reagents and particle chambers, extend our perception to unseen areas by actively stripping away or rearranging the environment so as to reveal the processes and objects that may be operating below the surface. These types of active technological interventions often operate in conjunction with microscopes.[14]

To what extent are the objects of telescopic and microscopic vision artifacts of the technologies, and to what extent are they reports of objects as they really or naturally are? To the degree that technology operates by extending and not supplanting human functions such as vision, the products of telescopic/microscopic vision need be no less real or natural than the world we construct with our naked sight. However, telecommunicated extensions of vision entail levels of encoding and decoding (or transformation into energy forms suitable for transmission) not operating in unaided vision, and these additional processes are inevitably subject to their own translation errors. Even a single, simple addition to the natural perception pipeline such as a lens of a telescope inevitably introduces a new noise component or propensity for distortion.[15] Moreover, the problem of artifactual knowledge is heightened when we observe via deliberate disturbance of the environment, as in subatomic particle chambers.[16] Thus we are entitled to ask: in what ways does technologically extended perception distort or render unnatural the realms it brings into human focus?

Campbell's hierarchy of vicariousness model, discussed at length in Chapter 2, helps clarify our question. Vision itself, as Campbell points out, is an extension of touch, a sort of physical contact from afar, which bequeaths us both the safety of distance and the likelihood of error—a misreading of the environment—that plain touch cuts through.[17] Protoplasmic probes neither lie nor hallucinate, though they often die from the truth they discover. Vision, on the other hand, is biologically advantageous despite its propensity for falsehood, as the high survival rate of sighted organisms including ourselves so vividly demonstrates. Remembering Simpson's observation that our ancestors' view of the branches they were jumping toward must have been at least somewhat veridical, or else we would not be here today, we have reason to conclude that the recreation of the world in our vision corresponds in significant part to the world as it really is, or at least as would be revealed to us by our fingers. Our survival as a species, in other words, indicates that the natural extension operant in vision is in some fundamental way nondistortive of what it reflects. The question, then, is whether the artificial or rationally directed extension of technology, which stands in the same vicarious, extensional relation to vision as vision does to touching, is equally nondistortive in its projections.

Aristotle, as we have seen, thought not. Moreover, the fallibility of our intellect must pervade our technologies as it does our theories, and thus we should know without even looking that telescopes and microscopes are prone to error. But prerational evolution and its products are imperfect and fallible as well, and thus our question about the accuracy of experience-extending technology boils down to this: is there some reason for us to think that the increment in fallibility entailed in the telescopic/microscopic extension of vision is significantly greater than that entailed in the visual extension of touch? In other words, how does the fallibility of rationality and technology compare with the fallibility of unplanned evolution?

If the success of our rational, technological species to date is any indication (of course if we blow ourselves and our world to bits with our technologies this epistemological success will be a total moral and ontological failure), then the answer is clear: human rationality and technology are manifestly less fallible or error-prone than the undirected processes of evolution in dealing with the objects of the real world. The technology of space travel provides a significant example of this epistemological success and its practical consequences. Many denizens of Earth have lived here longer than humans and have populated the world in greater numbers; this speaks well for the understanding of their environment encoded in their genes. But only one organism of Earth has understood this planet and its situation in the cosmos so well as to be able to leave it in space ships, with the result that this organism alone will likely have the ability to leave its home before it burns in the sun: rational, technological understanding has given us the unique opportunity of tying our fate to the infinitely larger domain of the universe. Thus rationality and technology are beginning to fulfill their promise of transcending their blind evolutionary parentage, of replacing eyes that see the here and now with eyes that also see (imperfectly) beyond and into the future. Applying this conclusion to the question of experience-extending technologies, we may conclude that the products of telescopes and microscopes are, if anything, more truthful representations of reality than the pictures supplied by our naked eyes; or, in extending our eyes to areas formerly unviewable, these devices allow our eyes to operate in technologically opened domains with a degree of truthfulness at least the equivalent of that obtained in realms naturally accessible to human vision. Either way, technologically extended vision adds up to a net gain in imperfectly truthful experience. Although the addition of an artificial lens to our retinas brings a new potential for error, the quality of these technologies apparently is such that mediated vision results in a more successful plying of reality nonetheless. Much the same can be said for observation via deliberate molecular or subatomic disturbance of environments: to the degree that the knowledge obtained works in human interactions with the world, this knowledge may be presumed to be at least a partially accurate description of the world as it is.

The improvement of experience via technologically extended perception has thus been evidenced not only in the beginnings of space travel, but also in countless independent tests here on Earth in areas of bacteriology, geology, and all the sciences that employ lenses, literal and figurative, in their work.[18] Of course, we must never become so arrogant about our technologies that we forget their fallibility and capacity for distortion—a capacity quite inescapable, in that any action or technology we apply to rectify the imperfection will be imperfect as well, that is, will introduce a distortion component of its own.[19] At the same time, we can be aggressive about identification and amelioration of sources of technological error, and in this way derive a maximum benefit from our technologies and their errors.

The harvest of the benefit of experience-extending technologies has been large. We get planets, stars, and galaxies at incomprehensible distances, teeming worlds of tiny protons, atoms, and proteins right under our very noses, rendered in proportions viewable to the human eye and thus digestible by human cognition. The banquet is partaken by all aspects of human mentality, not only our intellect. E. H. Gombrich reminds us that single snowflakes are mere dabs of white, nondescript and unprepossessing, until a magnifying glass brings to life their delicate latticework and beauty.[20] Magnification transforms the snowflake to a size that can engage the human aesthetic faculty. Experience-extending technologies thus allow us to produce knowledge and beauty, to make our antientropic mark in areas formerly immune to the human transaction. The "disproportion of man" to the cosmos that so awed Pascal[21]—our minuscularity in comparison to the immensity of the universe, our ungainliness relative to the tiny subatomic constituents of existence—is both revealed and repaired by telescopes and microscopes that bring the infinite and infinitesimal into the human orb, or perhaps transform humans into universal proportions.

Thus, just as technologies of agriculture greatly increase our supply of comestible commodities, so the deliberate engines of sensory experience increase the food for thought on the cognitive table. We have no guarantee that every item on this stunning smorgasbord will be cognitively digestible or even edible, nor is there any reason to assume that our cognitive faculties would be able to adequately process the sheer quantity of new data that our sensory technologies provide. The preadaptive principle discussed in Chapter 3 suggests that although our cognition evolved as a processor of the quantities of experience provided by our unaided senses, our unamplified mentalities need not be precluded from sooner or later comprehending vastly greater magnitudes of experience; but preadaptation entails no encouragement to suppose that this will in fact happen.[22] Fortunately, we have long had technologies which do for our internal cognition what telescopes and microscopes do for our senses—which allow us to handle huge increases in data by expanding our capacity for memory and speeding our powers of calculation. Without such direct enhancers of mentality itself, our sensory technologies might well leave us, cognitively speaking, like the Ancient Mariner, drowning in a sea of water that we are unable to drink. We turn now to a consideration of ox-ribs, libraries, and computers.

COMPUTERS AND TRACTABLE IMMENSITY

Numbers are an amazing invention. The real physical world has no numbers, only objects and events which may be counted and ordered by humans. When we abstract these countings and orderings from their real objects and begin manipulating these relationships according to the laws of logic, we enter the world of numbers.[23] When we write these numbers down, we are taking the first step in the deliberate technological extension of our computational and reasoning processes.

The first step may have been taken as long as 300,000 years ago, as we have seen, in a piece of ox-rib incised with repeating zigzag patterns.[24] Even if we decrease this time estimate tenfold, we are still left with the likelihood that written numbers or computations preceded written words and texts by many years. This suggests that, in the hard material extension of intellectual processes, numerical relationships came before linguistic ones—the prototypical computer preceding the first intended libraries, even if we take those libraries to be paintings on the walls of caves in Lascaux, France.

A prime effect of number systems is to increase the quantity of sensory experience or reports of events in the real world that our cognitive faculties can process; this in turn increases the speed with which we may collate these events, and the possibility of uncovering unexpected relationships and patterns among them. A striking example of unexpected benefits resulting from numerical abstractions is found in the advantages of the Arabic as opposed to the Roman numeral system, which I regularly use as a case in point in my classes in Introduction to Mass Media. I often begin these classes by promising an A to any student able to do a simple multiplication wholly through Roman numerals, with no conversion to the Arabic system either mentally or in writing at any time in the exercise. I make this offer with no fear of contributing to grade inflation, however, for I know full well that this apparently easy exercise cannot be done: even the simplest multiplication is impossible to execute in the Roman numeral system, or indeed in any number system without a placeholder, zero, such as that found in the Arabic system. (Of course one could achieve the results of multiplication wholly within the Roman system by converting the multiplication problem into an addition problem and then proceeding to do the lengthy addition. But this would then be an exercise of addition, not multiplication.) Thus, whatever the reasons for the invention of the placeholder zero, its presence in the Arabic number system has permitted the development of a series of increasingly complex mathematical procedures, beginning with multiplication and division and culminating with the calculi invented by Newton (and Leibniz independently)—the requisite for Newton's and subsequent theories of mechanics and the cosmos. Historians long have wondered why the Roman Empire, with its penchant for engineering and gadgetry, nonetheless remained an unmechanized, unindustrialized society. Perhaps its cognitively limiting number system was part of the reason.

Electronic computers do for quantities, speed, and relationships of numbers what numbers do for arrangement of experiences of the real world; indeed the computer in the twentieth century has extended the properties of numbers in much the same way as the printing press since the fifteenth century has extended the properties of letters. An indication of the impact that an increase in the mere speed of handling numbers might have had on the growth of knowledge is suggested by the disclosure that Newton spent the vast majority of his time, not in the creation of his original theories, but in writing out the lengthy equations and computations needed to clarify and refine his ideas.[25] What further revolutionizing ideas might Newton have had the time to contribute, had he been freed by even a pocket calculator from the drudgery of computations performed by hand? (On the other hand, Newton might have squandered his technologically found time on unproductive frivolities—though his somber bachelor existence suggests not.)

The contribution of computers to the growth of knowledge in the second part of the twentieth century has been far from hypothetical, and indeed far exceeds the simple saving of time through a speedup of processing. Lightning speed in the handling of data skyrockets the quantity of data that may be handled, and this exposes the human mind to bigger chunks and smaller details of the universe. The huge quantity of information processed by computers encourages and even makes possible grand, unifying theories of existence, long the goal of science, by sorting and classifying vast and variegated expanses and myriads of minute facets of existence in the time we ordinarily would take to attend to a single event or aspect of reality. Computers, then, open new worlds to human cognition in much the same way as telescopes and microscopes—alleviating our myopia where we need the big picture, keeping track of details we might otherwise overlook—though the impediment overcome with computers is neither distance nor size per se, but rather one of quantity or numerical complexity. As Bell Laboratory executive W. O. Baker and his colleagues have put it, "computers have significantly expanded the domain of tractable complexity . . . [making] possible extensive explorations, validations, and refinements that in some cases would have been literally impossible."[26] Relating this performance to the problem of experiential glut, we can see that computers are not only analogous to telescopes and microscopes, but also redemptive of the promise of increased knowledge provided by these experience-extending technologies: with a computer in the kitchen, orchestrating and organizing the flow of dishes, no sensory food on the cognitive table need go to waste.

Subjecting the process-extending computer to the same scrutiny as its experience-extending partners the telescope and microscope, we need to ask in what ways the computer may be distortive or dysfunctional in the pursuit of knowledge. Prototypical criticism of what computers do precedes even Aristotle, and is found in the teachings of the mentor of his mentor, Socrates, who warns that writing will result in an atrophy of memory and the ruination of dialogue.[27] Much the same spirit is present today in the criticism of computers and even pocket calculators, which some fear will result in a society of mathematical illiterates, or of people who no longer know how to add and subtract but only how to push buttons.[28]

This type of drawback is at once more trivial and more serious than the possible distortion of reality attendant to telescopes and microscopes: more trivial in that the drawback would not distort or call into question the knowledge obtained through computers, more serious in that our cognitive faculties themselves would be damaged or compromised by computers. Technology is thus in the odd position of being suspect here because it is doing its job too well—the speed-up of data processing wrought by computers increases our knowledge so effectively as to raise the specter of obsolescence of certain aspects of our precomputer cognition. A Faustian air hence hangs around the computer, with an immense and immediate gain in knowledge thought by some to be purchased at the cost of our cognitive souls.

This objection to computers touches on a central problem of progress, in which advances in life, knowledge, and power are often accomplished through the sacrifice of systems which performed well enough, and indeed even made possible the creation of the new systems which replace them. Socrates was, in one sense, entirely correct in his critique of writing: one certainly cannot have a dialogue with a written page, and the availability of extrapersonal memories on written pages has no doubt resulted in individual human memories not being sharpened as fully as they could have been. Yet as someone interested in the growth of knowledge, surely Socrates would be obliged to acknowledge the increase in our understanding of the world and (albeit to a lesser extent) ourselves which has ridden on the deliberate permanent recording of knowledge—an accomplishment which, by any reasonable standard, has been progressive rather than degenerative, and eminently worth any loss of dialogue or memory.[29] Moreover, the swell of scientific knowledge made possible by writing has even engendered a rebirth of live, immediate dialogue on a scale far more vast and democratic than in Socrates' time, first with telephone and now with interactive computer text and video hook-ups. In the case of electronic mail and computer text "conferencing" especially, we see the progeny of writing restoring the very interactive, responsive quality that commitment of words to a written or printed piece of paper had lost. Readers can indeed ask questions of texts available via computer networks, and expect to receive answers on their personal computers from authors on the networks either immediately or within hours or days.[30]

Writing, then, despite what it sacrificed, apparently has been an unqualified success for civilization, and in fact has made much of our civilization possible. What about the computer? In assessing its possible dangers, we have to look carefully at what it sacrifices, and in particular to what extent the losses are likely to be irretrievable. Taking the most extreme possible scenario (which the force of cultural habit likely will prevent from ever happening), let us assume that in 200 years the ubiquity of computers results in no one knowing how to add or subtract. What would be the ultimate cognitive calamity that could befall this world in which computations could be done only with machines? By this time computers would be so omnipresent (they practically are already) that nothing short of a complete nuclear or other holocaust (which would destroy everyone in any case) could possibly disable every computer in the world. Nor, for the same reason, could any dictator manipulate all computers in the world to follow his, her, or its bidding—even in 1986, pocket calculators and personal and mini-computers are in too many nooks and crannies in the world for any one central authority to program.[31] But even if the seemingly impossible were to happen—if all computers in a world utterly dependent upon computers for mathematical computations were somehow destroyed—where would that leave us? We still would have old books from which we could relearn the rudiments of mental arithmetic. And if by some ill fate the books too were destroyed, the more intelligent among us, working backwards from the electronic computations they recently had witnessed in computers, would no doubt be able to rediscover the art of mental computation in short order. Our ancestors, after all, once discovered and developed the techniques of mathematics all on their own, with no examples of computers, no recent memories of $2 \times 4 = 8$ in electronic display, to guide them.[32]

Thus the contraindications of computers, like the distortions of their sensory partners the telescope and the microscope, and the drawbacks of their printed and written predecessors, seem small indeed. The amount of knowledge gained through these technologies, on the other hand, has been enormous.

In discussing the enhancement of our internal mental faculties by number systems and computers, we have made only parenthetical reference to the extension of our mental faculty of memory through the technologies of written language, books, and libraries. Spoken language in effect is the first of such technologies,[33] and the contribution of these memory-extenders to the growth of knowledge and indeed to our very existence as a species has been overwhelming and crucial. However, their contribution has been directed primarily toward the criticism, dissemination, and storage of knowledge, rather than to its creation per se—creation in the interaction of sensory experience and cognitive faculties amplified so dramatically by telescopes, microscopes, and computers. Criticism and dissemination (which includes storage and the opportunity for further criticism) are part of the process of knowledge creation in the larger sense, but the technologies that extend these activities, especially the disseminative function, operate somewhat differently from the technologies that extend the initial generation of knowledge by improving our sensory experience and our mental digestion.[34] The disseminative technologies work primarily as transferrers or conduits of information or knowledge previously created, whereas the generative technologies—telescopes, microscopes, and computers—work primarily to create this knowledge in the first place. Consideration of books and libraries thus shifts our focus from the generative phase of cognition to the closely related, overlapping area of communication that includes telephones, televisions, and, once again, computers. This is the world of communications media, the subject of our next chapter.

META-COGNITIVE TECHNOLOGIES

To briefly summarize, we have thus far looked at three ways in which technology contributes to the growth of knowledge:

1. All technologies are embodiments of knowledge, and thus in a fundamental sense constitute knowledge. Moreover, technological embodiments contribute to the growth of knowledge by being both durable manifestations and workable implementations of knowledge, that is, unintended libraries of knowledge that has survived some encounter with the real environment.
2. Some technologies, such as telescopes and microscopes, are specifically designed to increase our knowledge by expanding the sensory experience upon which our cognitive faculties feed. These technologies may be assisted by techniques that perturb the studied environment for the purpose of revealing underlying processes or objects.
3. Other technologies, such as computers, are specifically designed to increase our knowledge by augmenting the operation of our cognitive faculties themselves.

Technologies of the latter two types may be considered double- or meta-cognitive devices in that they share the fundamental embodiment/duration/implementation function of all technologies, while performing an additional task deliberately in service of the pursuit of knowledge. A telescope is a durable implementation of an idea about how to increase our knowledge, which, when applied to its task of extending our sensory experience of the stars, helps produce knowledge (such as theories of the cosmos) entirely distinct from the knowledge embodied in its construction (for example, theories of optics or lens making).

The cognitive-intended or meta-cognitive technologies described in the second and third categories above share a further similarity in that both are directed toward the primary or generative stage of the knowledge process. As we have seen, ideas are produced, in the first instance, in the confrontation between our sensory experience (itself prefigured by our natural sensory structures, such as eyes, and extended through telescopes and microscopes) and our cognitive faculties (such as recall and logical ordering of experience, both assisted by computers).

Knowledge proceeds to develop through the selection of ideas via criticism and testing, and the dissemination and preservation of those ideas which have passed the tests. These subsequent stages—criticism and dissemination—are, unlike the generative stage, intrinsically social in execution. Thus, an individual may look at the stars with or without technological assistance (such assistance is usually, but not necessarily always, the product of a group endeavor: one may look at the stars through a piece of quartz crystal found in a stream), and produce an idea of what the stars are like, wholly on his or her own. This idea may, for all we know, be highly accurate. Certainly the isolated circumstances of creation would have no bearing on whether the idea was correct—highly on-target theories have originated in the past from guesses made outside of the mainstream scientific community (such as Wallace's conjectures about natural selection).[35] Moreover, even when (as is often the case) observation and thinking take place in a literal or effective group context (in a literal group, other human beings are physically participating in the process; in an effective or virtual group, we are alone but observe and think in conjunction with ideas supplied by others), the actual conception of an idea remains a wholly personal event, occurring entirely inside an individual's skull.[36] The circumstances change drastically when we enter the criticism or testing stage: although an individual may attempt to criticize his or her own ideas wholly on his or her own, such internal criticism would understandably be regarded as insufficient and even suspect by other people who might entertain this individual's ideas. Thus, though internal criticism may be quite successful in principle and practice—an individual may correctly identify a flaw in his or her own thinking—the logic that error detection works better with more detectives on the job, as well as the psychological consideration that an individual may be too close to or enamored of his or her own idea to see its errors, combine to situate the critical stage of the knowledge process in the social arena. The possibility of a private knowledge crumbles entirely in the third or disseminative stage of the process, which cannot take place at all unless an individual has another human being to whom he or she can disseminate.

Thus criticism and, to an even greater degree, dissemination are essentially group or social activities. What sorts of technologies would be able to assist us in these latter stages of the knowledge process? In the primary or discovery stage, technological assistance entailed either a one-way movement of information from object to perceiver (as in the telescope), or a spiral situation in which technology receives information from an individual and returns that information to the individual after sorting, classifying, or otherwise clarifying the information (as in the computer). Thus, the telescope conveys information from the stars to the viewer but not vice versa, and the computer (which can also be used as a social technology) can be operated entirely and profitably by a single person who feeds data into the preprogrammed system and collects results. But since social activities such as criticism and dissemination of knowledge involve more than one person, they require of technological assistance the capacity to transfer or exchange information between and among individuals. Such technologies or media of communication have been around a lot longer than microscopes and telescopes; speech itself is presumably at least as old or even older than abstract number systems.

Communications media bear interesting similarities and dissimilarities to their meta-cognitive cousins, the telescope and (sometimes) the computer. Unlike these primary cognitive technologies, communications media are of minimal function without at least two individuals, one on each end of the pipeline. A book bereft of readers arguably is not a book, and a telephone call without a receiver is certainly not a telephone call.[37] (A device like the tape recorder could be used by a single individual to record a message for his or her own use later, but the technology in this case would be functioning as a cognitive extender of memory rather than a vehicle for social criticism or dissemination, that is, it would not be functioning as a communicator.) Moreover, unlike meta-cognitive technologies, whose sole or primary purpose is the pursuit of knowledge, communications media are as often as not used for noncognitive, or at least nonintellectual, ends. Thus, with the minor exception of the Peeping Tom, telescopes are constructed for the prime purpose of increasing practical and scientific knowledge of the Earth and the heavens, whereas books and surely television can as easily be used to entertain as to inform. Cognitive-intended and communications technologies do, however, share at least one highly important characteristic: both embody knowledge or information in their structure quite distinct from the information they convey in their content or in performance of their tasks. Thus, as already noted, a book is a paper and print embodiment of a strategy of communication as well as a carrier of the information on its pages, just as a telescope is an embodiment of theories of optics and lens making as well as a reporter of information about the stars.

Communications media, then, work on numerous levels as a double-edged sword in the service of cognition: they assist (and, in the case of speech, allow) the second two stages in the knowledge process (criticism and dissemination); their very function depends on the participation of more than one person (a social essence which renders them ideal media of social criticism and dissemination); they serve realms of the intellect and nonintellect alike; and like meta-cognitive devices, communications media convey separate knowledge in their structure and content. Moreover, to make matters even more interesting, one of the key meta-cognitive technologies, the computer, is a central communications medium as well.

We turn now to a detailed consideration of the function and evolution of communications media, the vehicles of knowledge criticism and dissemination.

NOTES

1. Popper, *The Self and Its Brain*, pp. 451–452. Popper suggests here that human technology may be considered "a higher stage of something that goes back to the very beginning of life" inasmuch as enzymes produced by genes constitute a "self-created artificial environment" and thus are "vaguely analogous to tools produced by the human brain."
We might take the origins of technology back yet a further step and identify the second, "transcatalytic" polymer in the Eigen cycle as a tool produced by the first polymer that enhances the survival of both polymers. See chapter 2, note 12. See also chapter 6, note 15 for the role of language in the distinction between human technology and technology as a phenomenon of life in general.

2. *Experience and Nature*, p. 148. Dewey is speaking here of the evanescence of unapplied objects; I find the observation more applicable to unembodied objects. See chapter 4, note 22 in the present volume.

3. Thus the tragic explosion of the U.S. space shuttle Challenger in early 1986 will likely yield lessons that will help propel us to the stars.

4. Boyce Rensberger, "The Oldest Works of Art," *The New York Times Magazine*, May 21, 1978, pp. 26–29 ff. discusses the 300,000 year-old ox–rib and Marshack's interpretation of its V-shaped scratches. See Alexander Marshack, *The Roots of Civilization*, New York: McGraw-Hill, 1972 for consideration of the development of mathematics in more recent prehistoric times (20,000 to 30,000 years ago). See also Jeremy Campbell, *Grammatical Man*, pp. 152 ff.

5. *Phaedrus*, 275–276. Socrates also attacks writing here as destructive of human memory and incompatible with genuine dialogue: when questioned, the written word "always

gives one unvarying answer". Whether these views were Socrates' or Plato's (or the degree to which each may have held such views) will likely never be known. Socratic attribution is suggested by the fact that Plato wrote and Socrates did not (indeed, were these critiques of writing Plato's, we would be obliged if we took them seriously to hold all of Plato's writings suspect, including the *Phaedrus*), and by Plato's decision to place these views in Socrates' mouth rather than his own, which he might have done in another dialogue. At the same time, an attack on the derivative quality of writing is consistent with Plato's low regard of art as a doubly removed degradation of the primary realm of ideal forms (the physical or external reality captured in art being one step removed from the ideal). See "Computers and Tractable Immensity" in the present chapter for a discussion of writing and memory, and "Electronic Facilitations of Print" in chapter 6 for a discussion of writing and interactive dialogue.

6. Aristotle held that, although natural and artificial processes were capable of error (*Physica*, II:8), one natural process—perception of primary qualities such as color through the senses—"is never in error or admits the least amount of falsehood" (*De Anima*, III:3). Such sensory realism (what we perceive via our unadorned senses is true) ironically provided a foundation both for the new empirical science of the Renaissance (including subsequent empiricist philosophies such as Francis Bacon's and John Locke's), and for religious-philosophic attacks on its technological, "non–natural" instrumentation (see note 7).

7. Stillman Drake, ed., *Discoveries and Opinions of Galileo*, New York: Doubleday, 1957, p. 73, reports that at least one of Galileo's opponents (Giulio Libri) simply refused to look through a telescope. Feyerabend relates that others accepted the telescope as a superior mode of observation on Earth, but doubted its reliability in the heavens (*Against Method*, p. 123).

Telescopic observations on Earth were accepted because these were confirmable by one's travelling to the distant object and then using naked vision—but telescopic observations of the sky were crucial precisely because they could not be duplicated by unaided sight (this has changed in the case of the moon in the twentieth century due to space travel). The telescope at first was thus accepted where it was a convenience rather than a necessity (on Earth), and rejected where it was a necessity (in the heavens). See note 18 below for more on this technological version of the Meno paradox, or the tendency of independent empirical corroboration of technological observation to decrease as the necessity for the technological observation becomes greater.

The point of Feyerabend's brief is that those who opposed Galileo were not operating out of irrational faith, but from consistency with a commonsense, logically defensible Aristotelean position (unaided senses are least prone to error). Indeed, Feyerabend concludes that Galileo's eventual triumph—despite its more accurate portrayal of external reality—was a victory not of logic and evidence over religious obstinacy, but rather of "clever techniques of persuasion" and emotional manipulation (p. 142). As we saw in chapter 3 of the present volume, propaganda and nonlogical appeals indeed play a role in the dissemination of knowledge; however, the refusal of many of Galileo's opponents to even entertain a possible contradiction of Aristotle shows that propaganda (of which appeal to authority is a primary vehicle) was evident on both sides. An explanation of the triumph of Galileo more likely than Feyerabend's thesis is that Aristoteleans and Galileans alike furthered their views through mixtures of logic and propaganda, but that Galileo's won because his was by and large less propagandistic and more consonant with the nature of external reality.

See also Gunnar Andersson, "Naive and Critical Falsificationism," in P. Levinson, ed. *In Pursuit of Truth*, pp. 50–63, for additional criticism of Feyerabend, and discussion of Galileo's corroboration of telescopic planetary observations by naked-eye scrutiny of the stars and planets under special dawnlight conditions.

8. The pattern of media replication of natural communication beyond biological boundaries of sight and sound is the subject of my "Human Replay: A Theory of the Evolution of Media," Ph.D. diss., New York University, 1979.

9. *Critique of Pure Reason*, pp. 209–210. Kant considers this the central lesson of his critique, to wit: "The Critique of the pure understanding, accordingly, does not permit us to create for ourselves a new field of objects beyond those which are presented to us as phenomena. . . ." (p. 206, "Critique" with capital "C" in the original), and uses this conclusion to refute arguments for knowledge of the soul, the Deity, and the totality of the cosmos, none of which is available as an object of sensory experience.

10. Marshall McLuhan attributes the dominance of the visual in knowledge to the ascension of the alphabet and then the printing press in the West, both of which abstract reality into linear, visual forms. See, for example, *The Gutenberg Galaxy*, New York: Mentor, 1962, pp. 72 ff. Whether the undeniable relationship between alphabetic communication and visual knowledge is due, however, to the visual bias of alphabetic media, which selected visual knowledge as its most appropriate content (McLuhan's view), or to a premedia tendency toward the visual in human cognition, which selected alphabetic media as its most appropriate vehicle (my view), is unclear. Against McLuhan, we have the nonalphabetic Chinese inventions of such visual devices as the printing press and ceremonial rocketry (though underutilized by Western standards), and Piaget's demonstration of visual cognition (discrimination of short and tall glasses filled with water, and so on) in preliterate children. With McLuhan is Patrick Heelan's *Space-Perception and the Philosophy of Science*, Berkeley, Calif.: University of California Press, 1983, which argues that our commonsense Euclidean conceptions of space and time are consequences of our rectangularly "carpentered" world (Heelan, however, does not inquire into why we build our technologies with right angles in the first place, and whether this is indicative of an innate cognitive Euclidean visual tendency).

See also chapter 4, note 18 for Waechtershaeuser versus Campbell on nutritional (photosynthetic) versus locomotive origins of visual perception and knowledge.

11. See "Flesh and the Growth of Knowledge" in chapter 8 for a discussion of the role of transportation in building communities that generate, criticize, and disseminate knowledge.

12. See chapter 9, note 12 for arguments, contra special and general theories of relativity, for the possibility of faster-than-light travel.

13. *Meno*, 80 ff. But see also note 18 below, and note 8 in my "Information Technologies as Vehicles of Evolution" in C. Mitcham and A. Huning, eds., *Philosophy and Technology II*, Boston: Reidel, 1986, pp. 44–45, for discussion of a variant of the Meno paradox that arises in telescopic and indeed in all technological cognitive exploration, to wit: we apply technologies to precisely those areas of which we have the least nontechnological corroborating knowledge (for if we had this nontechnological, naked sensory knowledge, we would not need the technologies).

14. Aharon Kantorovich ("Quarks: An Active Look at Knowledge," *Fundamenta Scientiae*, III, 1982, pp. 297–319) discusses this more actively produced type of technological knowledge, such as that obtained in particle chambers where technological pertubation of reality causes a shake-up that reveals hitherto unseen elements and relationships. Such extension of experience by active disturbance rather than passive observation of the environment is consistent with Dewey's view of knowledge as obtained via application, which permits "more extensive interaction of natural events with one another, an elimination of distance and obstacles, [and] provision for interactions that reveal potentialities previously hidden" (*Experience and Nature*, p. 162).

On the subatomic level, such discoveries may be considered "microscopic" in terms of the present schema, to the extent that the objects and relationships uncovered were wholly unknown prior to the technological intervention. Macrochemical experimentation in laboratories would by the same criterion be telescopic.

15. The ubiquity of noise is recognized and formalized in the Shannon-Weaver "mathematical" theory of communication (Claude Shannon and Warren Weaver, *The Mathematical Theory of Communication,* Urbana, Ill.: University of Illinois Press, 1949), and is the equivalent of the fallibility of representation and understanding recognized by philosophers. See, for example, the discussion of Peirce and Popper in chapter 3. See also chapter 4, note 7. Noise may also be construed as the engineer's equivalent of that which renders Kant's "thing-in-itself" ultimately unknowable; see his *Critique,* pp. 180 ff.

The inevitability of noise may also be cosmologically equated with entropy (as by Norbert Wiener, *The Human Use of Human Beings,* New York: Avon, 1950/1967)—though one need not equate the unavoidability of noise with an ultimately increasing and overpowering entropy (as has Wiener). See "Turning the Universe Inside Out" in chapter 9 of the present volume.

16. In the realm of quantum mechanics, the counterintuitive behavior of technologically disturbed environments has been accounted for by a variety of theories ranging from views that the results are entirely artifacts of technological manipulation and are not descriptive of reality as it is, to views that the realities observed are indeed artifacts of manipulation as is all reality in general (that is, reality is intrinsically artifactual).

See Fritz Rohrlich, "Facing Quantum Mechanical Reality," *Science,* September 23, 1983, pp. 1251–1255, for summaries of some of these views and exposition of the generally accepted alternative view that quantum mechanical behavior is not artifactual but real, that is, indicative of technologically independent relationships, albeit a reality that is profoundly different from the one we perceive through naked senses. Kantorovich's view of reality revealed via technological perturbance would be in agreement with this, as presumably would Dewey's epistemology, which holds that natural laws only become fully apparent when nature is technologically stimulated. See also chapter 3, note 6 for brief descriptions of some of the apparent paradoxes of quantum mechanics.

But see also chapter 9, note 12 for yet an additional view of quantum mechanical results—one which suggests that although the behavior is real, the prevailing interpretations of that reality are incorrect—and for a further suggestion about how a new interpretation might open the door for faster-than-light travel.

17. Note that this dual relationship of vision's increased safety and likelihood of epistemological error in comparison to touch or physical contact obtains even if vision originated as a photosynthetic rather than a locomotive system (see chapter 4 note 18 for Waechtershaeuser's suggestion in this area). Vision thus now functions as a substitute for physical movement regardless of whether the substitute was preadaptive (the origins of vision were not locomotive) or literally descended from processes of locomotion.

18. A new species of the Meno paradox (or at least perversity) afflicts us, however, in that the areas of greatest independent corroboration (here on Earth) are usually those least in need of technological exploration, and areas most in need of technological exploration (furthest from Earth) are least likely to admit to independent, nontechnological corroboration. The chemical composition of stars, for example, may be deducible via spectroscopic analysis of light, but we have no way other than these analyses, or techniques similarly dependent on very long-range observation, of corroborating our findings. The Meno paradox points out that in order to find knowledge, we must have some idea of what we are searching for, yet to the extent that we already know what we are searching for, we have no need to search. Thus knowledge is either impossible or unnecessary. Various solutions to this problem suggest that we have some sort of incomplete prior or innate idea of what we might

find, which is then "fleshed-out" in our actual knowledge discoveries. See Larry Briskman's "Articulating our Ignorance: Hopeful Scepticism and the Meno Paradox," *Et Cetera,* Fall 1985, pp. 201–227.

About the best we can do with our technological problem is to endeavor, wherever feasible, to test the technological performance against nontechnological (or different technological) experience. We can point a telescope at a town across the sea, and if our telescopic observations are later confirmed by in-person observation, we may reasonably conclude that the telescope provides nondistortive observation (though not necessarily across distances far vaster than a sea). The discovery of microorganisms via microscope, and the later observation that a certain disease occurs when the microorganism is present and does not occur when the microorganism is absent, provides a similar nontechnological corroboration of microscopic perception.

See also Walther Zimmerli, "Who Is to Blame for Data Pollution," in C. Mitcham and A. Huning, eds., *Philosophy and Technology II,* pp. 291–305, who points out that as technologies grow more sophisticated and powerful, they grow beyond our ability to regulate with individual, ethically accountable responsibility. We might characterize this problem as another Meno variant, since sophisticated powerful technologies are precisely those most in need of individual accountability.

19. This is why accidents such as those at Three Mile Island and Chernobyl are in principle unavoidable, whatever the level of safety features we may install. In view of this inevitability of error, we must ask ourselves whether the ultimate risks of a given technological activity are worth the benefits (since these ultimate risks can never be fully eliminated). In most cases, I think they are. For example, the loss of seven lives in the Challenger space shuttle, though tragic and unnecessary, was a loss of life confined to those people who were willing to take the risk (albeit perhaps not fully aware of its proximity). In the case of a nuclear power plant accident, the risk is far greater both in numbers of people exposed and in duration, and the risk is generally not voluntary. I thus have reservations about fission power technology. (See also notes 10 and 17 in chapter 9.)

20. E. H. Gombrich, *The Sense of Order,* Ithaca, N.Y.: Cornell University Press, 1979, p. 9.

21. Blaise Pascal, *Pensées,* H. F. Stewart, trans., New York: Pantheon, 1950 (1669), pp. 18–31.

22. Indeed, biology suggests that immense quantitative changes might be less amenable to preadaptive response than are long range qualitative shifts: mammalian walking structures would likely fail long before they became capable of moving 1,000 miles per hour, but they have adapted over time to perform successfully in water (porpoise paddles) and air (bat wings).

23. My assumption here, however, is that the world of numerical relationships is not a Platonic world—that is, it does not exist independent of human thought. A theorem may have logical implications not initially seen—indeed, the history of mathematics has to a large degree been the "discovery" and proof of new relationships—but these relationships are potentials of the original theorem that come into being only via the application of human thought. In this sense, their existence is no more real than any logical possibility. (The Kantian "thing-in-itself," unlike a theorem, has existence in a non-human world; reality unperceived by humans can and does affect other living organisms, whereas mathematical theorems have no impact outside of human circles.)

See Raymond L. Wilder, *Mathematics as a Cultural System,* New York: Pergamon, 1981, pp. 27–28, for a discussion of Platonism in mathematics.

24. Rensberger, "The Oldest Works of Art."

25. See I. Bernard Cohen's "Newton's Discovery of Gravity," *Scientific American,*

March 1981, pp. 167–179, for descriptions of the arduous and "repetitive" honing of mathematic models and real world observations that constituted this "discovery."

Indeed, Newton's response to Robert Hooke's claim to have discovered the law of gravity was, "Dr. Hooke could not perform that which he pretended to: let him give demonstrations of it. I know he hath not geometry enough to do it" (Frank E. Manuel, *Portrait of Isaac Newton,* Cambridge, Mass.: Belknap, 1968, pp. 152, 422).

26. W. O. Baker et al., "Computers and Research" in *Electronics: The Continuing Revolution,* P. H. Abelson and A. L. Hammond, eds., Washington, DC: AAAS, 1977, pp. 56–61. See also Heinz Pagels, "Fires in Space," *The New York Times Book Review,* August 21, 1983, pp. 9, 18, who explains that, although the laws governing physical processes inside stars were hypothesized in the 1920s, "astrophysicists were hampered by the sheer complexity of the mathematical equations that describe nuclear and thermodynamic interactions. Then after World War II, high speed computers were built. Using computers, astrophysicists could manage the mathematics involved and make detailed models simulating the interior of stars."

Improvements now occur not only in the hardware of computers but in techniques and scopes of computer modelling almost on a yearly basis. Better computer models in 1986, for example, led to dramatic moderation of the "nuclear winter" hypothesis proposed in 1983 on the basis of models available then. (See James Gleick, "Less Drastic Theory Emerges on Freezing after Nuclear War," *New York Times,* June 12, 1986, pp. 1, 20.) (A less severe "winter" scenario may or may not make us less concerned about a possible nuclear exchange—but if the less severe scenario is closer to the truth, I'd hold its generation, criticism, and dissemination to be nonetheless advisable. See also chapter 1, note 14.)

27. *Phaedrus,* 275–276. See note 5 above.

As McLuhan points out (for example, in *Understanding Media,* New York: Mentor, 1964, p. 9), new media are regularly regarded as aesthetically and often morally inferior to their predecessors. See Elizabeth Eisenstein, *The Printing Press as an Agent of Change,* New York: Cambridge University Press, 1979, pp. 14–15, for discussion of an early treatise on why "monks should not stop copying because of the invention of printing" (reasons included keeping idle hands occupied). I also came across a marvelous example in William McKeever's "The Moving Picture: A Primary School for Criminals" in *Good Housekeeping,* August 1910, pp. 184–186. So sure was McKeever of the pernicious effects of the movies in comparison to the less odious dime novels that he did not even trouble to end his title with a question mark.

I've no doubt that had our distant hominid ancestors been able to express themselves, they would have decried spoken language as destructive of its prelinguistic beginnings—and they would have been right.

28. J. W. Wyatt et al. report in "The Status of Hand-Held Calculator Use In School," *Phi Delta Kappan,* November 1979, pp. 217–218, that 43 percent of the teachers they surveyed thought pocket calculators would diminish the memory and mathematical ability of children. The authors also report that more than 100 studies on the effects of pocket calculators show no immediate adverse consequences (the technology was too recent for longitudinal studies).

Criticisms of computers in general as destructive or demeaning of human life (mainframe computers, personal computers, attitudes associated with computer "culture", and so forth) span more than a decade and include Joseph Weizenbaum's *Computer Power and Human Reason,* San Francisco: Freeman, 1976 and Theodore Roszak's *The Cult of Information,* New York: Pantheon, 1986.

The discussion in the present chapter addresses computers as enhancers of cognitive data processing. Computers as word processors are discussed in "Electronic Facilitations of Print" in chapter 6, artificial intelligence is considered extensively in "Artificial Intelligence and Real Life" in chapter 7, and computer telecommunication networks and personal computers are assessed in "Communities of Representations" in chapter 8 of the present volume.

29. See "The Agony and the Ecstasy of Abstraction" and "The Book in History" in chapter 6 for discussions of the impact of writing and printing on human civilization.

30. "Asynchronous" or nonimmediate communication in which dialogues among a few or many individuals take place over several days or longer (for example, participant A asks a question on Monday, participant B responds on Tuesday or Wednesday—after thinking about the question, and at a time of B's choosing) may produce exchanges of richer intellectual quality than those resulting from immediate face-to-face dialogue. See "Communities of Representation" in chapter 8 and the works cited there for a fuller discussion of computer "conferencing" and its contribution to the knowledge community.

31. The Future Computing Company of Dallas, Texas, lists 28 million personal computers in use worldwide in 1984, and The International Data Corporation of New York City finds 31 million personal computers in worldwide use in 1985. Future Computing Company expects their figure to quadruple by the end of the decade.

Personal computers use 8-, 16-, or 32-bit microchips, range in price from several hundred to six or seven thousand dollars, and are employed primarily for word and graphics processing, data management, and telecommunication. (A "bit" is the smallest unit of computation, and describes a 0 or 1 state, or an above or below level of voltage. Eight bits are usually needed to produce a letter or other visible character—hence word processing becomes easily feasible with machines capable of processing 8 bits at one time.) The IBM (and IBM–compatible) and Apple lines of personal computers are the most popular.

Minicomputers are much more powerful than personal or microcomputers, and are essential for management of large data bases. They range in price from five to fifty thousand dollars, and enjoy only a fraction of the current use of personal computers.

Pocket calculators sold for a few dollars in the 1980s, and were universal in the Western world by then. They are "dedicated" devices that can do only the simple mathematical tasks for which they are programmed.

32. Complex mathematical modelling achievable only via computers would likely not be retrievable in a world suddenly denied computer help. However, the fact that a technology gives us a unique capacity which the withdrawal of the technology would lose is surely no reason not to employ the technology in the first place—indeed, technologies are valuable precisely because they give us abilities we otherwise would not have. See notes 7, 13, and 18 above for discussion of the technological species of the Meno paradox.

On retrievability of basic arithmetic abilities, Wilder (pp. 150–152) conjectures that most if not all ancient Greek mathematic work destroyed in the fire at Alexandria has been recovered by Renaissance and more recent mathematicians.

33. See chapter 3, note 19 and chapter 4, note 10 for brief discussions of speech as a technology. See "The Agony and Ecstasy of Abstraction" in chapter 6 for a further discussion of speech.

34. As suggested in note 28 above, computers are remarkably multiapplicational technologies in that they operate both as generators of fundamental knowledge and as disseminators and storers of knowledge. Thus in the case of computers we have what amounts to the same technology performing different types of knowledge tasks.

35. Ancient Greek conjectures about atoms and so forth may similarly be considered accurate knowledge that arose independently of any well–developed scientific social fabric.

36. In the sense that all original ideas are derivations, refutations, or other reflections of ideas supplied by other people, all thinking is social. By "effective" group context I do not mean this historical connection of ideas, but rather temporal exchanges of information which

may occur during a generative process—say, in letter–writing among colleagues working on similar problems. The social dynamics of such temporal exchanges seem sufficiently different from those of one person working alone with books that they warrant the designation of "effective" groups. See chapter 8 for further discussion of representational and in-person groups.

The asocial character of the generative process may be necessary for its healthy functioning—that is, premature exposure of the creative act to group criticism may impair or derail it. Popper, for example, writes of the value and need of some dogmatism in the early phases of the knowledge process, or the dangers of caving in too quickly in defense of our theories, in *Conjectures and Refutations,* New York: Harper & Row, 1963/1965, p. 49, and Munz mentions the evolutionary value of a "cushion" which allows new ideas to develop free of external harassment (*Our Knowledge of the Growth of Knowledge,* Boston: Routledge & Kegan Paul, 1985, p. 85; and "DNA, Falsificationism, and Dogmatism," *Et Cetera,* Fall 1985, pp. 254–271). We might even extend the concept of such a nursery to exclude all group interaction—supportive as well as critical—in the early phases of knowledge generation.

See also Harlan Cleveland, *The Knowledge Executive,* New York: Dutton, 1985, p. 72, who argues that "early openness is bad for bad ideas, and bad for good ideas, too."

See also "Flesh and the Growth of Knowledge" in chapter 8 in the present volume for a discussion of privacy and the knowledge process.

37. More specifically, a book never read is unfulfilled or incomplete as an act of communication, and if we define "book" as a communication, the never-read book is not a book. Such a book of course would exist nonetheless as a physical object, and thus would have ontological significance and epistemological import as both a material expression of a strategy of communication (publishing as opposed to speaking, for example) and an abstract representation of the ideas discussed on its pages. Further, the never-read book has the continuing potential to become a communication act—that is, to be read. The unreceived phone call has far less ontological and epistemological import, and exists not at all other than in the memory of the caller (and perhaps in a record in the telephone system).

This emphasis on the completion of the communication act in the function of communications media is in accord with Dewey's focus on technological significance arising not so much from the existence of a tool or instrument, as from its use (see chapter 4, note 22). The present view continues to differ from Dewey's, however, in that the present view also emphasizes the importance of a potential for implementation or use, whereas Dewey's does not. (In *Experience and Nature,* p. 156, for example, Dewey writes that "The Norsemen are said to have discovered America. But in what sense? . . . unless the newly found and seen object was used to modify old beliefs, to change the sense of the old map of the earth, there was no discovery in any pregnant intellectual sense, any more than mere stumbling over a chair in the dark is discovery till used as a basis of inference. . . .")

I hold that potentials are philosophically and practically noteworthy, whether an unreported continent or unread books, because in most cases they are implementable in principle, and sooner or later usually are actualized. Indeed, I cannot think of any circumstances other than the physical destruction of a book (that is, its ontological elimination) which would render it forever unreadable. (The physical elimination forever of all sentient beings in the universe would have the same result.)

If we distinguish among three possible fates of books—destroyed, unread, or read—we might put the difference between Dewey's and my view as follows: Dewey draws the central demarcation between read and unread, with a far thinner line between unread and destroyed; I see the most significant gap between unread and destroyed, although I recognize the importance of the read/unread distinction.

See also note 25 in chapter 6 of the present volume for my "media" explanation of why Columbus' trips to America had so much greater cultural impact than the Norse visits.

Chapter 6

The Double Entendre of Communications Media

On April 11, 1983, at 9:00 PM (Eastern Standard Time), some 400 million human beings in 73 countries sat down in their living rooms, shanties, thatched huts, and taverns to watch the 55th annual presentation of the Academy Awards originating from Los Angeles. The subject of this television spectacular—honoring the best work in cinema in 1982, along with several "life-time" achievement awards for film work during the past 50 years—was hardly spectacular, epistemologically speaking, or even of much real social import (though the awards given to the movie *Gandhi* that year gave this particular ceremony more social significance than usual). Certainly the world-wide television coverage of the funeral of Anwar el Sadat several years earlier was of far greater political consequence; and the live telecast to a smaller immediate audience of the first moonwalk in July 1969 communicated an event of immeasurably greater importance to the evolution of our species.

Nevertheless, the circumstances and construction of the Hollywood broadcast have a special relevance to the role of communications technologies in the growth of knowledge, for that one telecast captured almost the full gamut of current technological communication, epitomizing nearly everything that media now do and the long road these technologies have traveled since their inception in the speech and carvings of the earliest humans eons ago. To start with, Academy Awards programs are notable for their huge and immediate world audiences—like the wedding of Prince Charles and Lady Diana, the funeral of Sadat, and various sporting events, the proceedings in Los Angeles were instantly and equally available to virtually every part of the planet. As Marshall McLuhan often pointed out, communication at the speed of electricity or light obsolesces distance.[1] It mattered not whether those watching were 10 or 10,000 miles from the events in Los Angeles; they all saw the same thing at the same time. Indeed, the only distance that counted in this situation was the distance from the television screen to the viewer's eyes. But the Academy Awards, unlike the global broadcasts of news and sports, transcend not only distance but time, for they beam to the world film clips as old as half a century and more. To look at this ceremony on television was thus to peer across not only miles but years, to the moment in time when the film clip on the screen was first made. Furthermore, the broadcast was indicative of the high degree of realism (or lack of discrepancy between the representation and the original being represented) that increasingly has characterized media in the nineteenth and twentieth centuries. Film clips viewed in person are virtually indistinguishable from film clips viewed through television (other than the smaller size of most television screens), and the people who participated in the ceremonies in person were present on the viewers' television screens in most of the key attributes of observed life, including color, sound, and motion (exceptions are the missing third dimension—a lack now technologically, but not yet commercially, remediable via holography—and blindness of viewees to viewer).[2] In other words, the Academy Awards broadcast conveyed to its television viewers an experience in most respects the same as that perceived by those who physically attended the event—an accomplishment unique to modern photographic and electronic media and utterly unattainable in secondhand verbal description or faceless words on a page.

Extension across space and time, fidelity to the original—speed, permanency, and accuracy of information transfer—have been the goals of communication technology, achieved with unsteady but growing success since the origin of media in the grunts and scratches of our ancestors on the savannahs millions of years ago. The drive to send information beyond the biological limits of eyesight, earshot, and memory, across the physical and human-made boundaries of space and time, has been remarked upon not only by McLuhan, but by a variety of observers of the human condition including Emerson, Bergson, and Freud.[3] The complementary drive to preserve or retain elements of biological (pretechnological) perception such as sound, motion, and color, or to reclaim such natural elements sacrificed in early technological extensions across space and time (for example, telephone reclaims voice in the instantaneous transmission across distance made possible by the telegraph, and photography restores image to the permanency of communication made possible by writing), has not been noticed too often, though it has occupied a central position in the writings of film theorist André Bazin,[4] and has figured more recently in my own work.[5] The speed, permanency, and fidelity with which initial ideas and observations are spread from one to many persons sets the conditions for criticism and dissemination of knowledge—determining the number of minds, both at any one time and throughout history, that can help develop these ideas, and the eventual value or effectiveness of these social cognitive activities (the more distortion with which an idea is transmitted, the more difficult the idea is to constructively criticize). Thus, the rise and evolution of communications technology, in its shaping of criticism and dissemination, has been as profound a factor as technology in general and cognitive-intended technology in particular in the growth of human knowledge. We commence our consideration of media, and the swiftness, durability, and accuracy of their transmission of information, with the beginnings of these vehicles in spoken and written language.

THE AGONY AND THE ECSTASY OF ABSTRACTION

Of the numerous characteristics that distinguish human from animal knowledge, one of the most salient is this: most human knowledge is of things not physically present. All organisms, nonhuman and human alike, inherit structures selected in encounters of their ancestors with past environments, and in this biological sense all organisms have some knowledge of environments not physically present. Organisms also may learn through conditioning to respond to stimuli not physically at hand (in this way Pavlov's dog was taught to salivate—respond to the stimulus of food—at the sound of a bell) and thus display a knowledge, of sorts, of events or objects that are absent. But the ability to make an invisible, unhearable, nonsensible event the center of immediate perception and cognition, to quest for and be guided by realities entirely beyond the world at hand, seems to be largely if not exclusively the profession of human beings.

Events may be beyond our physical sensation in any of several ways. An object simply may be over the next hill, that is, absent from our immediate perception due to distance. More significantly, the object may have existed in the past, or be thought to exist in the future, in which case its presence in our perception entails a sense of past, present, and future, of invisible yet omnipresent time. Perhaps most intriguing of all, an object of our perception and thought may be physically absent because it in fact has no physical existence at all—like the concepts of goodness and badness or the concept of concept itself, such entities may be wholly products of our intellect, or of leaps our intellect makes that go much further than the usual contribution of cognition to the perception and knowledge of physically real objects and events.

In all of these cases, the perception of something not physically present is made possible by a process of abstraction—a selecting of features from an original perception of a physically real event, and a playing back of these features in the mind at a later date. Abstraction works by producing a representation that is incomplete yet sufficiently evocative of the original to serve as a stand-in for the original in our cognition. (Abstraction thus does for perception what perception does for touch—both serve as distortive though efficient substitutes for more primitive types of reality–contact, parts of the "hierarchy of vicariousness" described by Campbell and discussed in Chapters 2 and 5.)[6] To think about a lion that may be over the next hill or was in the area yesterday is usually to picture the beast and perhaps recall its roar—recreating characteristics clearly indicative of lion—yet rarely to reconstitute the lion so fully that we would run from the thought of the lion with the same speed we would run from its physical perception. In the case of higher-order abstractions such as goodness and badness, the thoughts are so removed from physical perception as to scarcely be reconstructions at all, though they are compatible enough with the real world to allow their human imposition on real events in a way that often makes these values seem intrinsic to the events themselves.[7]

Abstractions depend almost entirely on communication, and communication depends entirely on abstractions. Although private abstractions (totally internal to an individual) are possible in principle, such abstractions would be of little value to the species unless they were communicable to other individuals. In the case of direct, unabstracted perception of the environment, communication is not as necessary: everyone in a group of hunters chancing upon a lion presumably would see the same lion. However, if one member of the group happened to think that there might be a lion over the next hill, this thought would be of little use to the party until it was communicated to other members, for unlike the immediate perception of the lion, the abstraction most likely would not be in the minds of the other members of the party prior to the communication. Precisely because abstractions differ from the evidence at hand, they are infinite in possibility relative to that evidence, and thus they can be reliably shared only by a process of communication that goes beyond the perception of immediate evidence. The immediately perceivable lion can be seen by the group whether they talk or not; the lion over the next hill, differing from whatever is immediately perceivable, can be seen by the group only if they talk.

Meanwhile, the process of communication itself could not occur without abstraction. With the hypothetical exception of a complete duplicate of an original (in principle impossible or paradoxical, since a duplication is obviously a duplication—a second instance of an original and not the original itself),[8] all representations or transmissions of information entail the transfer of selected characteristics of the event or object being represented or communicated about. The representation may be more or less incomplete (a black-and-white photograph of a sunset is more incomplete than a color photograph of the same), and may relate to the original through a physical process (like the photochemical reaction of photography) or by cultural fiat (the meaning of words), but in some crucial respect the communicated representation will always differ from the original. The raison d'être of all communication lies in this discrepancy: whereas physical events or objects of any appreciable bulk cannot be moved very far without considerable difficulty (if at all), representations or abstractions of the events can be moved great distances (and often across time) with relative ease. Via abstraction, events become portable in our heads (in speech) and in our hands (in written documents or photographs), and durable in the records we keep of them (in stone and in more recent media).[9] Indeed, all communication processes can be characterized as commencing with an abstraction or rendering of an immediate perception or original into a form that can be transported, proceeding with the actual transportation of the abstraction, and concluding with a de-abstraction or reconversion of the transported form back into features that resemble the original. (The Shannon-Weaver model of communication describes these activities as encoding, transmission through a channel, and decoding.[10] I like to use the example of instant coffee to help explain what is going on here: Coffee companies would have great difficulty sending hot brewed cups of coffee directly to millions of consumers. Manufacturers thus convert or "abstract" the hot brewed coffee into a powdered form that is capable of being so transported. When consumers pour hot water over the powder, they reconvert or de-abstract the coffee back into its original hot brewed form.) In the case of our hunter and the lion, neither the lion itself nor the hunter's immediate perception of it may be easily transported in original form to anyone (direct transmission of a perception in original, unabstracted form from one person to another would require an act of mental telepathy, which as far as we know is not generally possible, or at least not usually practiced by the vast majority of human beings). When the hunter converts or abstracts the perception of the lion into the word "lion" (which bears no physical resemblance whatsoever to the physical lion), however, this abstraction may be easily transported through speech to anyone within earshot of the hunter (or within earshot of those who have heard and repeat the hunter's words)—one day or a thousand after the initial sighting—and if those who hear these words are able to de-abstract them (understand the meaning of the word "lion"), they will recover most of the hunter's original perception of the lion. In instances of more advanced technology such as the telephone, spoken words themselves are mechanically abstracted into electronic patterns transportable at the speed of light over long distances, and mechanically de-abstracted or reconverted into sounds highly similar to the original words at the other end of the phone line.

Speech and its offspring the alphabet are among the earliest and most abstract of media, trafficking in representations that are in every respect different from the original. The spoken word has no physical connection or resemblance at all to the events it describes (onomatopoeia is one, small exception), and the alphabet has no physical connection or resemblance to the sounds it records or the events described in this recording. (Since the sounds represented by the alphabet are parts of spoken words which themselves have no direct connection to the reality they describe, the alphabet may be considered doubly abstract, or removed by two levels of abstraction from direct perceptual reality. More recent artificial or purposely constructed languages such as Morse Code, which are abstractions of alphabetic writing, may for the same reason be considered three levels of abstraction away from the immediately perceivable world.) Alphabetic writing was preceded by many forms of pictographic writing which recorded not the sounds of speech but, as well as possible, the literal sights of the physical world, and thus operated on levels of abstraction far closer to the real world than the alphabet, and indeed far closer than speech itself. But with the partial exception of Far Eastern ideograms, the stylized suns and stick figures of primitive writing which captured elements of the literally perceived world failed to meet the needs of any long-enduring, literate civilization, and in time gave way to a variety of phonetic writing systems culminating in alphabets.[11] Thus in the first two great communication revolutions of humankind—speech and writing—very high degrees of abstraction carried the day.

Viewed in terms of the fundamental goals of communication—extension of information across space and time, and fidelity of the extended information to the original—the early media of speech and phonetic writing hence get good grades in the first category and poor grades in the second. This suggests that the primary purpose of communication is extension rather than fidelity, with transmission of less accurate representations to a larger audience preferable to transmission of more accurate representations to a smaller audience. Indeed, speech and phonetic writing attain the first objective only through sacrifice of the second: the transmission of information about trees, lions, and the physical world solely through the action of vocal chords on the atmosphere commits us to representations that necessarily have nothing to do with the originals (with the trivial exception, again, of onomatopoetic words), and the visual transcription of these vibrations into a series of letters compounds the abstraction. As a result, subsequent media have been directed not only toward furthering the extension of information across space and time, but to recapturing elements of the physical world lost in the abstract soup of speech and the alphabet. Photography, for example, provides the permanency of print without putting the visual world through a double wringer of abstraction.

Abstraction thus has a complex and two-sided relationship to communication and the growth of knowledge. On the one hand, abstraction (or discrepancy between original and representation) is the very lifeblood of communication—without such discrepancy there would be no communication, since, as explained above, neither events nor immediate perceptions can be transmitted in their fullness. On the other hand, one of the basic goals of communication is to reduce such discrepancies as far as possible in order to increase the likelihood that the recipient of the representation will perceive in it the original that is represented. But the situation is complicated even further by the fact that some originals can be communicated only by highly abstract representations, with the result that these extreme abstractions are the best or most accurate means of communication of such originals. Thus, while a color photograph of a sunset is clearly a more accurate representation of the scene than a verbal or written description (a picture is worth a thousand words), a photograph of "justice" would undoubtedly be less accurate than a verbal or written description, assuming that a photograph of such a concept were even possible.

When "justice" or some similar abstraction becomes the original or object of communication, abstraction has seeped from the process to the content of communication, and such seepage is of the profoundest epistemological import. In the transmission of information about lions and sunsets, or about molecules and galaxies, the proper function of abstraction is as the means of transmission—any other contribution constitutes a distortion of the original, to be factored out as much as possible by the recipient of the transmission.[12] In the transmission of information about objects which are themselves abstractions, however, lack of literal connection to physical reality is the very stuff of the communication, not to be factored out, but recovered as much as possible by the recipient of the transmission.

This situation of abstraction as the content of communication and possible object of cognition is remarkable on at least two counts. To begin with, such a role for abstractions contradicts Kant's insistence that the only appropriate objects of cognition—the only reality that we can ever hope to understand, in a scientific sense—are the data of our immediate sensory experience. While even our most immediate sensory experience is to some extent a product of our sensing structures and our intellect (Kant knew this), and the most abstract of abstractions is usually related in some distant way to some physical, perceivable reality, a chasm of difference separates the sunset from justice: the latter is a creature of our intellect to such an overwhelming degree that its subsequent investigation by our intellect runs the risk of our cognition chasing its own tail, or the paradox that results when a videocamera looks at a monitor that carries the image produced by the videocamera.[13] A second distinction of abstraction as the object of cognition, which gives Kant's problem biological expression, lies in the lessons such use of abstraction provide about the evolutionary derivation and subsequent development of our cognitive capacities. Since the environment that initially selected our cognitive structures presumably was nonabstract, consisting of immediate physical realities and perceptions, the application of our cognitive processes to abstract realms constitutes a novel use of these processes in an area in which they were not at first selected to perform. Indeed, the abstract originals communicated by speech and writing are an early manifestation of a more general transformation of the world from natural to technological, a shifting of the objects of our comprehension from the products of natural selection to the products of human construction, and a commensurate movement toward understanding not only the universe but also the consequences of our understanding of the universe. This study of the consequences of our understandings and actions is the stuff of ethics and the substance of abstract originals such as justice and freedom and good and bad—originals in which abstract human judgement, not sensory experience, is the primary level of experience, "preadapted" as cognitive data.

Whatever success we as a species have had with such a self-referential enterprise provides yet another argument against the view discussed in Chapter 3 that our cognitive capacities are necessarily limited in effectiveness to environments in which they were originally selected to perform—although a continued evolution of cognition after the appearance of abstract concepts would probably favor the selection of cognitive processes capable of making sense of abstractions, for example, understanding your enemy's morality. (The self as an object of its own comprehension, and the contribution of abstraction and technology to such self-understanding, are considered at great length in the next chapter.)

Thus the very process of spoken and written abstraction, which when used to transmit information about immediate physical perceptions serves as a handicap or sacrifice that makes such transmission possible, acts as an advantage or stimulant to the development of nonsensory objects of cognition that constitute a central part of the growing, humanly created environment. Further, inasmuch as our cognitive processing of even the most immediate realities into theories (or descriptions of classes) results in generalities or abstractions of specific perceptions, we are obliged to deal with abstractions as direct objects of our cognition whenever we communicate about physical events on the theoretical level. (Such abstractions are admittedly empirical in derivation—epistemological in Kant's sense—and thus not as intrinsically abstract as entities like goodness and justice.) But media such as speech may have done more than facilitate the communication and encourage the growth of abstract theories and originals—speech may be responsible for the very existence of these preternaturally cognitive creations in the first place.

Although humans may have had the mental ability to abstract prior to the development of speech, abstractions in a prelinguistic environment would have been of little use to the species since they necessarily would have been wholly private or noncommunicable. (Incidentally, modern gesture systems such as sign language for the deaf are postlinguistic or derivative of speech. Primitive prelinguistic gestures and expressions such as pointing or smiling are nonabstract in that they relate in a direct, physical way to an object or emotional state at hand.) Although every existing biological structure need not be of equal or immediate use to its species (indeed, preadaptation sometimes entails successful performance in a new environment by a structure whose apparent utility in its original environment was nil), the existence of such a specialized and communications-dependent capacity as abstraction for any length of time with no apparent communicability seems unlikely. Instead, a co-evolutionary scenario for speech and abstraction seems appropriate, and might read as follows: Speech emerged in a world in which there were no abstractions, serving the purpose of communicating perceptions of immediate physical events across distance and time. Abstraction was the process used by speech to achieve such extensions—a primitive "technological" process which communicated or extended sensory perceptions only through the complete transformation or extreme distortion of the object originally perceived. (Phonetic writing increased the extension by increasing the distortion. Later media such as photography and the telephone have been able to accomplish equivalent and even greater increments of extension with radically less abstraction or distortion.) Eventually, the distortive or extrasensory character of abstraction was turned to an advantage by our species: the abstract fat in the communication of immediate perceptions became the meat or content in the communication of general theories about these perceptions, and gave rise to a new class of cognitive constituents which were at base nonsensory or abstract. As these mental creations joined the immediate physical world as part of our selecting environment, the cognitive structures of our species began to be selected for their ability to make sense not only of the physical world, but of highly abstract concepts as well. Individuals not able to navigate a cultural environment in which abstractions played a growing role were at a distinct disadvantage for survival. The result of such self-generated evolutionary pressure toward abstraction is homo sapiens, a species whose every undamaged member is linguistic and abstracting.[14]

This scenario suggests that the most significant epistemological consequence of speech in the long run may have been its unintended fostering of a new, abstract realm of nonphysical conceptions rather than its primary task of extending perceptions of the immediate physical world. Nearly all human technology is to some degree a product of the primitive technology of speech which encourages the production of conceptions which, in turn, serve as the basis for the production of material embodiments of these conceptions.[15] (Note, however, that only those abstractions with some connection to physical reality can be directly embodied in a technology. The automobile, for example, is an embodiment of a concept (travel) which has some physical foundation; but no technological embodiments are directly available for a wholly abstract concept such as freedom. Such intrinsic abstractions can be actualized only through the application of technological embodiments of other, more concrete ideas. For example, freedom can be materialized on one basic level through the application of a blowtorch to iron bars—or application of an embodiment of an idea about cutting to an embodiment of an idea about containment. More political, philosophic notions of freedom are commensurately more difficult to technologically embody.) Moreover, the corequisite relationship of speech and abstraction has favored not only the growth of abstractions, but also the survival of speech and writing as media best suited to negotiate the increasing number of abstractions in our lives. Not only have the early media of speech and writing survived the advent of the industrial and electronic revolutions in communications, speech or writing has been an integral component of every new medium with the exception of the still photograph, including printing, telegraphy, motion photography, telephone, phonograph, radio, television, and most recently, computers. Speech is a parent whose presence continues in a literal way in virtually all of its media offspring.

From the perspective of communications technology, the upshot of this highly complex, mutually reinforcing relationship between speech and abstraction, in which speech does a better job communicating what it was not originally selected to communicate, is the development of two classes of media, more abstract and less abstract (for example, writing and photography), designed to transmit information about two classes of reality, the predominantly physical and the predominantly abstract. The shortcoming of highly abstract media in communicating about physical reality (for example, speaking or writing about a subtle color or flavor), and the far poorer showing of less abstract media in capturing the fundamentally conceptual realm (for example, taking a photograph of freedom), has created a fertile situation of conflicting and continuing evolutionary pressure for both improvement and retention of most media systems. As we shall see later in this chapter, current media such as video screens, which display images and words with musical and spoken accompaniment, are a logical response to this multidimensional environment set in motion by speech.

We have said little about the primary task of speech, which is the extension of perceptions and ideas of the real world from one individual to many people, and therein across space and time. Such a mechanism is crucial to the growth of knowledge even when operated with considerable loss of content (such as loss of color), for without it our knowledge would be noncommunicable and private, subject only to the suspect edit of internal criticism and to no dissemination at all. By virtue of even these basic extensional functions, then, speech is a full partner in the mutually reinforcing and overlapping triad of thought, language, and technology that makes us human.

But although the extensional abilities of speech, not to mention its encouragement of abstraction, help to make us human, these extensional abilities are not enough to make us civilized—that is, they cannot provide the type of dependable accumulation of knowledge needed for highly organized and differentiated societies. Speech is transpersonal in its communication, or transcending of any specific human individual, but is not transhuman, or transcending of all human beings: in an oral society, the only possible source and carrier of humanly produced knowledge is one or another human being, and we humans are neither objective nor permanent enough for the development, dissemination, and retrieval of the large amounts of variegated knowledge supportive of complex social structures. In the end, the extension of information across space and time via speech is only as reliable as word of mouth.

The remedy for this ephemerality of speech resides in writing, whose storage of information outside the human brain permits the accumulation of a theoretically infinite amount of knowledge. Unlike the spoken word, whose physical embodiment dies the instant it is created, the written word lasts as long as the material it is imprinted upon, which in the case of Marshack's fossilized ox-rib may be as long as 300,000 years and still going strong. Moreover, whereas spoken communication carries the bias or distortion of each person who passes the information along—each speaker in effect having to recreate the message, and in so doing adding a new dose of noise or distortion—the written word is subject to the production noise of only one person, the writer. (The inevitable exponential growth of noise in an oral chain is demonstrated in the children's game "telephone," in which a phrase hurriedly whispered through even a small line of children almost always reaches the end of the line in garbled form.) Although distortion also haunts the written system (as it does all systems) at the decoding or reading end, the confinement of encoding or production noise in writing to the one-time contribution of the writer is an immense improvement over the accuracy of spoken communication, in which noise is continuously compounded at the source, or each time a message is repeated. (The printing of written material adds an additional element of noise to the production. But the noise component of publication remains the same whether the book or newspaper is distributed to ten or ten million people.)

In this section we have examined writing only as it relates to the nexus of language and abstraction that originates with speech. We turn now to a consideration of the unique extensional capacities of writing—a consideration that takes us from the alphabet through the world of books to the printing press, and eventually to the appearance of printed words on computerized screens in our own day and age.

THE BOOK IN HISTORY

In a book in which self-reference has been such a prominent theme, we inevitably take up the subject of books. A book—any book—is a textbook example of the hierarchy of technologies within technologies that constitutes the operation of written communication systems. Thus, this book, itself a technology, consists of a more primary technology for the transmission of information, the alphabet; and this book is at the same time a component of a larger technological transmission medium such as a library or a bookstore. We thus can identify at least three distinct levels of technological transmission in written communication:

1. a primary encoding or abstraction of the object of the communication—in the case of pictographic systems, this encoding is a direct abstraction of external reality; in the case of phonetic systems such as the alphabet, this encoding is an abstraction of an already abstract rendering of reality, speech;
2. the placement of this primary technology in some physical medium suitable for dissemination at least across time and perhaps across space—typical conscripts have been cave walls, clay tablets, writ-

ten manuscripts, printed books, and, most recently, electronically encoded print;
3. the distribution of these secondary technologies through some larger system—for example, libraries, mail, computer networks—which becomes increasingly important as the secondary technologies become more portable (the immovable cave wall is in effect its own distribution system).

Until the invention of sound recording, there was no way (other than phonetic writing) of physically capturing the spoken word, and thus we encountered no placement and distribution technologies in our examination of speech. Indeed, unlike writing, speech may be considered a "unistructural" means of communication, with but one level of technology and content, or with all transmission and distribution functions encompassed in the initial technological act of spoken abstraction. In shifting our attention from speech to writing, we enter a multitiered world of structure within structure, whose uneven and interactive development has a long and fascinating history, rich in consequences for the growth of knowledge.

The earliest surviving examples of writing, in the form of pictures on cave walls, are in every sense the opposite of spoken communication: (a) speech is utterly abstract, whereas pictorial representations of visual reality were closer to the external world than the representations of any medium until the development of photography; (b) speech is as mobile as human feet, but cave paintings are utterly fixed; (c) speech evaporates the instant it is produced, while some cave paintings have endured more than 20,000 years. Although the high abstraction of speech and the low abstraction of pictures are, depending upon the object of the communication, both desirable and thus complementary, the second and third oppositions presented early communicators with an either/or situation, in which portability (desirable) was paid for by a complete loss of permanence (undesirable), or permanence (desirable) was attained through a complete sacrifice of portability (undesirable). Moreover, the subsequent streamlining of cave pictures into pictographic symbols on clay tablets was only marginally successful in disseminating permanent representations across space. Pictographic systems require a different symbol for every word or idea to be written; hence they are difficult to learn and cumbersome to operate, with the result that pictographic readers are few, and writers even fewer and very slow in output. This severe constraint on both consumption and production at the source prevented any widespread dissemination of pictographic communications, and would have done so even had the transmission vehicle been paperback books rather than clay tablets and papyrus. (Indeed, McLuhan suggests that although the printing press was invented in China hundreds of years before it appeared in the West it was never used there as a mass medium because of the unwieldy Chinese picture-based writing system. Imagine, in an age before computers, attempting to set even a line of type with 2,000 or more different characters from which to choose.)[16]

The development of pictographic and derivative writing systems therefore proved to be an evolutionary dead end. Not only did such an approach fail to substantially improve the distributability of literal pictures, but in transforming pictures into successively more stylized characters (for example, a human image becomes a stick figure, which in turn becomes a few intersecting lines) the pictographic system lost the literal picture's high degree of fidelity to the original. Thus, although a stick figure has more resemblance than either the spoken or the alphabetically written word "human" to the physical appearance of a human being, the stick figure nonetheless conveys very little of what a human actually looks like. A failure on both counts—dissemination and fidelity—pictographic or visually reflective writing systems became a curiosity of communications, lingering on in traffic signs around the world, fading ideographic systems in the East, and recently revived in personal computer "icon" systems.[17] Meanwhile, the high fidelity of literal pictures and paintings kept them in the mainstream of communication until the nineteenth century, when most of their practical (but not their artistic) functions such as military and medical sketching were usurped by photography.[18]

The decisive breakthrough in the development of a permanent vehicle that retained some of the disseminative power of speech arrived with phonetic or acoustically reflective systems—written symbols that represented spoken sounds rather than seen events—and in particular the deployment of completely phonetic systems of representation such as the alphabet. Although Egyptian hieroglyphics and Chinese ideograms incorporated a variety of phonetic devices (often in the form of phonetic symbols literally attached to pictograms), the continuing pictorial presence in these systems precluded any big improvement in dissemination, since the scribe in most cases was still obliged to use a different symbol for each word or event. (Few hieroglyphs or ideograms are entirely phonetic; a sentence in which each word is partially pictographic takes as long to write—and, more importantly, to learn to write and read—as a sentence in which each word is entirely pictographic.) A completely phonetic system, in contrast, drastically reduces the number of different symbols needed for writing, since the number of sounds that we use to describe events through speech is a minute fraction of the number of events themselves. Our alphabet arbitrarily divides the sounds that we speak into 26 different units which, when mastered along with rules for their combination into a theoretically infinite number of words and sentences, allows us to transcribe a world requiring thousands of different symbols to transcribe in Chinese. Lost in this revolution is the small degree of pictographic fidelity to the external visual world; gained is an exponential increase in communication economy, the ability to hold the world in a handful of letters—"to hold infinity in the palm of your hand."[19]

The impact of the alphabet on the walls of hieroglyphic ignorance hit like a one-two punch, with the second blow even more shattering than the first. The initial alphabetic volcano, explored by Harold Innis, Marshall McLuhan, Eric Havelock, Walter Ong, and others, disseminated knowledge on a scale that fostered the rise of science, mathematics, and logic in Ancient Greece, and the eventual triumph of monotheism in the West at large.[20] The alphabet's tsunami arrived nearly two millenia later in the form of the printing press, whose use of interchangeable type was facilitated by the alphabet, and whose outpouring of books, pamphlets, and newspapers swept away the medieval world in a wave of scientific revolutions, religious reformations, and national states whose effects are yet being felt.

Among the most interesting and instructive of the first series of consequences of the alphabet was the rise of monotheism. Elaborating on Innis' suggestion that Hebrew prophets were assisted by the flexibility and efficiency of alphabetic writing, one can see two crucial advantages of the alphabet over hieroglyphics in the dissemination of the monotheistic idea: first, the physically contradictory notion of an omnipresent yet invisible deity is inexpressible in pictures and picture-based symbols but easily represented in an abstract system like the alphabet which has no connection to visual reality; and second, the increase in written production and literacy fostered by the alphabet created an environment in which more and more potential missionaries and masses could be reached and converted. Historical support for just such an analysis is found in the striking example of the Egyptian Pharaoh Ikhnaton (Amenhotep IV), who 150 years before Moses attempted to convert his kingdom to the monotheistic worship of Aton, the sun god. The futility of Ikhnaton's monotheism, which was replaced by the older Egyptian polytheism several generations after his death, is usually attributed to the stubborn opposition and subversion of Egypt's priestly class, and to failure of the monotheistic idea to reach the Egyptian masses.[21] In terms of our media analysis, however, Ikhnaton's reformation can be seen as a prisoner and victim of the hieroglyphic system, in which the notion of a transvisual life-force was unintelligible and, even if it could have been comprehended, inaccessible to all except the hieroglyphic-polytheistic priests who had every reason to reject it. (Indeed, the near impossibility of describing a supraphysical monotheistic force in hieroglyphics has led some historians to conclude on the basis of the pictographic record that Ikhnaton's religion advocated the worship of the sun in a literal, physical, iconic sense, rather than as a symbol for a universal, life-giving energy. An unlikely interpretation at the other extreme is found in the 1954 movie "The Egyptian," which portrays the Pharaoh's claim for the sun god as really a call for the Son of God—that is, an early recognition of Christ.)[22] Thus it was that a monotheism championed by one of the most powerful people on Earth, the Pharaoh, ended as a personal lunacy. In contrast, the powerless Hebrews with an alphabet as their scepter eventually converted the whole Western world.

If the Ten Commandments—written in stone, not spoken in air—were the first expression of the alphabetically ushered rise of monotheism, the subsequent dissemination of the monotheistic idea is most associated with a medium whose physical flexibility was better matched to the cognitive flexibility of the alphabet: the book. Although the paper and binding technologies necessary for books had nothing to do with the alphabet (indeed the book, like the printing press, was first developed in ideographic China), a pictographic "Good Book" would have given comic-book treatment to much of the subtlety of the Old and New Testaments and the Koran, and again would have been inaccessible to all but a tiny portion of the population. In a nonalphabetic setting, the great portability of the book is wasted—a case of disseminative overkill. The principle seems to be that in a multistructure media system such as writing, the disseminative power of the entire system is only as great as the power of its weakest component; in the instance of hieroglyphics, this is not very great, with the result that the hieroglyphic book is not much more effective than the hieroglyphic tablet (perhaps less so, since the tablet is more durable). The coupling of the alphabet and the book, on the other hand, set loose the disseminative powers of both, and resulted in a means for propagation of knowledge that remained unrivalled until the twentieth century. The cognitive intercourse spurred by books created a population explosion of ideas, in which books were the sex cells and the alphabet was the DNA.

The fulfillment of the book's disseminative potential, however, awaited the Second Coming of the alphabet in the printing press, a device which at last began to bestow on the alphabet a physical accessibility commensurate to that of the spoken words it recorded—a device through which, in the words of contemporary Johannes Stradanus, "single writings cover a thousand sheets, just as one voice can be heard by a multitude of ears."[23] The audience for print was not only the equivalent of the audience for speech but immensely greater, because print is permanent. The shift from handwriting to print thus furthered an escape begun by the alphabet from the age-old either/or communication choice of permanence and lack of accessibility versus accessibility and lack of permanence: the press made information both accessible and lasting. What the press disseminated stuck; or, we might say that print moved mountains of information, or writings once recorded on mountains, to the masses. The millions upon millions of books—cells with the DNA of our acquired knowledge—turned out by the printing press thus stand both as the apex of written, mechanical communication, and a prototype for the even more accessible and permanent distribution of written information by electronics at the end of the twentieth century.

The impact of this intellectual dynamite, this explosive mixture of accessibility and permanence ("gunpowder of the mind," David Reisman aptly called it),[24] was comprehensive, and felt throughout the length and breadth of religion. A thousand or more years of heresies against the Roman Church, hissed into ears, amounted to very little; Luther's protest, taken less than a century after the development of the printing press in Europe, became the Protestant Reformation. More important even than the distribution of Luther's arguments was the printing of Bibles by the press—millions of them eventually—thus putting teeth in Luther's central thesis that people should read the Bible for themselves. Much the same occurred in the political sphere. For centuries, rumors of a New World had trickled into Europe in Norse myths and other orally transmitted tales with little effect; Columbus' voyages, made 40 years after Gutenberg's press, were read in printed books and treated as bankable fact, and became the basis for an Age of Discovery which midwived the rise of modern national states. (Innis overlooks this, suggesting that the press helped crystalize national identities by putting spoken vernaculars into visible print—a connection which I would regard as valid, but less significant to the rise of national states than printed stimulation of the Age of Discovery.)[25]

The decline of the Church and the rise of national states were mutually reinforcing catalysts; on their own they would have been powerful stimuli for the growth of science, but the press gave science a more direct assist as well. The golden age of monotheism which preceded the golden age of science was a handwritten or scribal age, not a printed one. Indeed, the press put an end to unmitigated belief: although the abstract arguments of monotheistic religion received a massive distribution in print, this distribution was in effect too much, creating too rich an information environment with too many fine points of view and options, to suit monotheistic creeds that preached the one, true way. Thus the Age of Faith, in the first place made possible by the disseminative powers of the alphabet, ultimately succumbed to too much of a good thing. Science, in contrast, was from the very beginning a continuingly revolutionary activity—seeking to destroy its own past, as Thomas Kuhn puts it,[26] or forever attempting to come up with better theories—and hence always able to benefit, to thrive in fact, from increases in dissemination. Thus the first sparks of science, nursed by phonetics but restrained in the West by the Age of Faith, leapt into flames when fanned by the books and pamphlets that flew from the press. The further fueling of the conflagration by telescopes and microscopes,[27] as well as improvements in the operation of the press itself, resulted in the Scientific Revolution of Newton, Darwin, Einstein, and Freud that continues unabated today. Throughout all of this, the increasing availability of printed theological, political, and scientific tracts created a pressing public need to learn how to read—a need which encouraged the rise of public education, which in turn further promoted reading and the escalating culture of print and knowledge.

The primary carriers of this wisdom, the cells which propagated this modern scientific culture, were individual books. As a vehicle of knowledge dissemination, the individual book (the printed pathbreaking treatise on planetary motion or evolution) was and continues to be a masterpiece of efficiency, combining all one could want of permanency and portability. The printed cell of knowledge, however, loses its portability when functioning as part of an aggregate or larger body of wisdom—for the library, unlike the individual volumes it comprises, cannot be readily bought, sold, borrowed, or carried. Indeed, the amalgamation of books into libraries recreates much of the ancient distribution system of writing, in which the reader was obliged to come to the mountain. Thus the book's revolution in portability finds its limitation in the library.[28]

In the parlance of media theory, this limitation of paper and print can be seen as a problem endemic to any medium in which hardware and software, or the conveyor of information and the information, are inextricably wed. Unlike our vocal chords (the hardware) which in principle are capable of transmitting an infinite number of words (the software), the printed page can transmit but one set of words—those which are printed on it. Although paper, to be sure, is a lot lighter than stone, with the result that books can be carried and writings on sides of mountains cannot, the words appearing on printed pages are as inseparable from the paper as carvings on the sides of mountains are from the stone, and hence in this sense are as inflexible and nontransposable. This means that the information in any given book is fixed and immutable (though of course subject to changing interpretations), which in turn means that we may add to the information that we carry in a book only by adding another book to our load—a procedure which, owing to the finite power of human muscles, rapidly neutralizes the book's mobility and leads to the library limitation. The situation has improved only now, late in the twentieth century, with the development of computer/video systems capable of displaying an infinite amount of printed material on a single screen.

ELECTRONIC FACILITATIONS OF PRINT

We have already met the computer as a powerful sorter of our sensory experience and thus an augmentor of our generation of knowledge; now we will find that the computer as a disseminator of printed information is as revolutionary for communication as it is for primary cognition. Computer terminals with the portability of individual books have been commercially available and in widespread use in the computing community for several years,[29] but the detachability of software from hardware in even cumbersome computer systems has at last liberated the printed word and thus us from bondage to the page. To possess a book is to possess only a book and those words frozen in it; to possess a computer is to possess a library, indeed a potential library of every book that has been written and will be written, provided they have been fed just once into a computer system. For computers, like vocal chords and blackboards, are stages upon which an endless series of word-plays may be performed. Moreover the computer, unlike these prior prolific media, takes good care of its performers, offering permanent abode to any word appearing on its screen. Thus, writing's age-old quest for a system that records information without sacrificing the flexibility of living speech—the grail of communication toward which the alphabet and then the printing press were mighty steps—is fulfilled in the computer.

The digital process with which computers perform this magic upon text is analogous to the alphabet's processing of external reality, and draws an equivalent (though electronically heightened) dividend: just as the alphabet's encoding of experience into a limited number of arbitrary symbols with no intrinsic connection to the experiences gives the alphabet enormous flexibility and facility, so the computer's encoding of letters into a limited series of simple on-and-off switches (usually eight, to be exact, and the "ons" and "offs" really correspond to above and below a voltage level) allows the computer word processor to record, display, alter, and reformat letters and words and paragraphs with lightning speed and consistent accuracy. Several immediate benefits to the creative, critical, and disseminative processes result.

To begin with, the actual production of text on a computer screen (once a word processing technique is mastered) is easier—takes less of an expenditure of physical energy—than just about any externalization of ideas beyond mumbling. Hands hovering over electronic keyboards in the middle of the night—at times when even electronic typing is too exhausting and too noisy for most environs—seem able to respond more directly and efficiently than any other amanuensis to the ideas in the brain. (I can testify to this effect personally—though, alas, most of the manuscript of this book was typed on a self-correcting typewriter, before I made the psychological investment in word processing.)[30]

Second, the ease with which text can be corrected once entered, through an instant resetting of electrical pattern in response to a keystroke or two, rescues the written word from the inevitable negotiation and compromise with error that placement on the fixed paper page entails. Such compromises take their toll both on the personal level, when an author is too tired to make yet another erasure or paste-over correction on a manuscript, and in the social arena, where writers may feel uncomfortable about asking their secretaries or assistants to do yet another revision. Further, the increasing delivery by authors of word-processed texts capable of being directly converted into printed books by the publisher without intermediary typesetting removes another level of likely error and accommodation, as authors who have fretted over missing and misspelled words and sentences in typeset manuscripts well know.[31]

Ultimately the electronic connection may make conversion of word-processed manuscripts into printed texts unnecessary altogether, as writers telecommunicate their texts to computers all over the world with an additional keystroke, and therein become publishers. Hundreds of thousands of people are already using personal computers and modems (a device which renders word-processed text into a form transmittable via conventional telephone connections) to send, receive, and organize texts of various lengths at the speed of light any place in the world.[32] Such "electronic mail," "electronic publishing," and "computer conferencing" will likely have as profound an impact upon the interactive globalization of knowledge as the jet plane. We will consider these developments in greater detail in Chapter 8.

The millions upon millions of written words already processed by computers quietly refutes the view that computers or electronic media in general are destructive of literacy. Indeed, from the very first harnessing of electrical energy in the telegraph, and the development of electric lighting which greatly facilitated reading, electricity has been more of a friend than a foe to the printed word.[33] The current placing of print directly on the electronic saddle now gives the written mode all the speed and exposure of presumably the fastest phenomenon in the universe: for the first time in history, a word can be read in virtually all parts of the world the instant it is written. This is destructive not of literacy, but of intellectual aristocracy.

Nor is the computer necessarily a threat to the ultimate survival of books, the traditional vehicle of print. Computer typesetting and video display terminals have greatly economized book and newspaper publishing,[34] and will certainly prolong the existence of these operations as active, mainstream media. Moreover, even when computer screens have replaced paper as the primary mode of printed communication,[35] we may for aesthetic or even sentimental reasons desire personal, permanent copies of certain treasured texts. Works like Kant's *Critique of Pure Reason* or McLuhan's *Understanding Media*, which we profit from reading over and over again, and enjoy turning the pages of by hand, may be best served by the physical format of the book. Thus, the ultimate victim of the computer is neither literacy nor books but the library: the immovable, inaccessible, nine-to-five, no talking, closed-on-Sundays library.

The image of a world of letters reeling under the sword of electronics is not, however, entirely unjustified: modern computers are descendants of electronic media which continue to flood our senses with literal sounds and images, at the very least competing for time and attention we might otherwise devote to reading. To the extent that the sudden development of electronic and photochemical media in the past 150 years has been a revolution in entertainment—a revolution of slapstick, melodrama, and rock music accessible with the flick of a finger—the immediacy and sensuality of electronic communication has certainly been a tempting distraction from the bloodless appeal of the book,[36] and indeed from the often difficult navigation of leisure in real-life situations (such as the cocktail party). But the casualty here is more likely *A Tale of Two Cities* than the *Critique of Pure Reason,* and if everyone who watched soap operas were to spend that time reading drug store novels, who would want to call this an improvement in our intellectual climate? (The "trashy" novel, incidentally, predates both television and cinema, and flourished as the "dime" novel that poured forth in the 1890s and earlier from newly improved printing presses.) Moreover, although the pursuit of pleasure has undoubtedly been served by electrochemical communications, the conveyance of literal images and sounds at the speed of light and otherwise has nonetheless been of assistance in the pursuit of knowledge. The photograph, for example, provides a copy of external reality free from the sort of personal distortion inevitable in any written or painted description or depiction; motion photography makes possible a speed up and reversal of recorded passages of time that reveal relationships and interconnections (for example, the unfolding of flowers) unapparent to the "real-time" eye; the telephone permits Socratic dialogue at a distance; and television brings people on the moon into living rooms. Such effects are far more than tricks or conveniences. They are fundamental departures from both the content and method of printed communication, and indeed they are the first reductions of the second-hand, retelling quality of communication that arises with speech itself. The stuff of our perceptions and cognition is highly perishable in transport; we may preserve our perceptions in print, but the preservative risks serious distortion of some of the delicate specimens. Think of electronic and photochemical media as "cognitive refrigerators."

THE CONQUEST OF IMMEDIACY AND ETERNITY

Like all living organisms and indeed all natural processes, the human organism is tied to the rhythms and cycles of independent, external mechanisms: the durations and tenuations of objective time. Unlike other organisms, however, humans are aware of time—Alexander Marshack defines our very humanity as an awareness of past and future[37]—and in a sense all technologies may be seen as human attempts to contest the ordinations of time. Agriculture and ranching try to make crops and livestock grow faster; medicine aims at delaying the onslaughts of age; transportation is designed to move us faster; and communication strives for both speed and preservation, or the movement and retention of information as rapidly and enduringly as possible. Whether the successes we have achieved in these and related areas are "victories" over time, or ingenious exploitations of properties already present in the physical, temporal universe, the result is the same. Human technology has increasingly manipulated the ebb and flow of natural processes to our own human, supranatural ends. In the history of communications, and indeed of all technology, the first decades of the nineteenth century rank high with the achievement of two of the most profound of these time-altering devices: the instantaneous transmission of information via the telegraph, and the attainment of literal eternity in the photograph.

Although the conquest of immediacy and eternity was, in the case of Samuel Morse, accomplished in part by the same inventor, the scientific bases of these devices are entirely different, and the representations of reality they produce are at opposite ends of the abstraction continuum. The telegraph utilizes the speed-of-light capabilities of electricity to transmit short and long signals (clicks and clacks) which, as abstractions of written words (which are themselves abstractions of spoken words, which are in turn abstractions of directly perceived reality) are three levels of abstraction removed from the perceptual world. (Computer programming operates on a similar triple level of abstraction, communicating through artificial "languages" which are abstractions of natural written forms, for example, the computer language BASIC as opposed to English.) The heliochemical process of photography, on the other hand, records images less abstract than the words of speech, and in fact as proximate to external reality as the images produced by natural vision. The difference between photographic and biological vision is thus not one of distance from external reality, but rather one of the number of external features accounted for, with early photography blind to such factors as motion, color, and the third dimension. Such deficiencies were eventually corrected, and x-ray and infrared technologies now make the naked eye seem blind by comparison. Meanwhile, the advent of video recording in our own time has combined the immediacy of the telegraph with the permanency and fidelity of the photograph, at last reconciling these divergent functions.

THE IMMACULATE CONCEPTION OF PHOTOGRAPHY

Painting, carving, sculpting, and the like have been providing permanent and accurate recordings of external reality since long before photography—as long ago, perhaps, as the origin of speech itself. However, even the most nonabstract or true-to-life painting conveys a representation which, like the abstract spoken word, is a product of an idiosyncratic, individual human nervous system. Painting and similar hand-made media are thus inevitably tainted, to paraphrase André Bazin, by the sin of subjectivity.[38] Although the photograph is also a product of human mentality, in that the camera taking the photograph is a product of human design, the human contribution here is more standardized and objective, that is, we can know in great and predictable detail the specific contribution of the camera to the photographic image (for example, the image is black-and-white because of the camera and film). Thus, although individual, subjective choices are obviously entailed in the photographer's decision as to where to point the camera, how much light to use, and so forth, a pristine, wholly objective reaction occurs between the light bouncing off external reality and the chemicals on the film at the instant the photograph is taken—a reaction in which no symbolism, human interpretation, or firing of neurons is involved. When the technological contributions to this process are subtracted or taken into account (again, for example, the image is black-and-white as a result of the camera), we recover a lasting image of external reality no different in epistemic principle than the image produced by the eye of the photographer or any human (or organic) eye.[39]

The image produced by the camera is thus equivalent to the image produced by the painter's eye, but vastly different from the image produced by the painter's hand.

The immaculate conception of photography is of considerable epistemological consequence, for unlike all prior communication systems, photography allows us to treat its representations as natural artifacts. In extending this effectively objective perception to objects and events no longer present, the photograph reduces or makes less relevant the impediments posed by time and distance to perception, and therein enlarges the realm of possible direct observation so crucial to the growth of knowledge. The botanist examining a photograph of a flower taken in 1853 sees all the aspects of the flower captured in the photograph exactly as they were in the original blossom long since shriveled and gone. The photographer determined the view, and the technology of the camera predetermined which aspects would be captured, but those aspects of the flower recorded in the photograph appear to the botanist in 1986 virtually the same as they appeared to the photographer's eye at the instant the photograph was taken in 1853. In this way, the photograph extends the scientist's field of firsthand vision across formerly inaccessible distance and utterly unviewable time. The whole world, past as well as present, becomes an oyster for our cognition, as objects existing at times proximate to the taking of the photograph become candidates for subsequent, quasi in-person scrutiny. Moreover, the development of sound recording in 1877, motion photography in the 1890s, motion photography synchronized with sound and color photography in the early part of the twentieth century, and holography (three-dimensional photography) since 1949 have all increased the extent or number of aspects of reality viewable through the technological window on the past. The results are pictures of the past nearly as comprehensive as they are direct.

The recording of literal images moving through time in motion photography has had scientific dividends beyond those obtained from the capturing of existence in its natural rhythms. To capture motion is to acquire the ability to manipulate it, and in manipulating motion we can alter the speed and order of events in ways that may make these events more humanly comprehensible. Time-lapse photography, for example, can speed up the images of unfolding flowers and developing crystals to the point where we may see patterns in these events invisible to the naked eye in natural time. Apropos Pascal,[40] our default disproportion to the cosmos is not only one of size but of speed and tempo—at our middling pace of life, even close-at-hand events like a raindrop's fall or a flower's bloom happen too quickly or slowly for us to adequately perceive—and motion representations such as film and video help us fine tune these events into a speed appreciable by our cognition. The artifice of motion photography thus humanizes time in the same way that computers humanize numerosity, and telescopes and microscopes make size and distance more humanly amenable[41]—although the cognitive as distinct from aesthetic benefits of such manipulation depend on the accuracy and entirety of the photographic images that comprise the motion picture.

Unlike the development of the telescope, microscope, or even the printing press, the growth of technologies that capture and distribute literal representations of external reality usually has been motivated only in part by a desire to learn more about the universe or help disseminate knowledge (the early development of motion photography is something of an exception in the first regard).[42] Yet, as I have argued in detail in my *Human Replay: A Theory of the Evolution of Media,* neither was the invention and deployment of successively more lifelike media an accident. We humans regularly perceive images in motion and synchronous with sounds, see in color rather than black-and-white and in three rather than two dimensions; can our systematic correction of each of these shortcomings of the initially still, silent, colorless, flat photograph have been mere coincidence? Eons of natural selection in the organic foundry have resulted in natural modes of perception which are highly efficient—thus the quest for efficiency in information transmission, whether overlaid with commercial, artistic, scientific, or no additional motives, leads quite logically (though likely unconsciously) to the natural.[43] The payoff for the pursuit of knowledge is a picture of reality communicable with increasing fidelity and fullness.

ELECTRONIC MERCURY

The lifelike character of contemporary technological communication pertains not only to the content or accuracy of the representation, but also to the manner in which the representation is delivered or disseminated. This returns us to the province of electricity.

In the natural, "pretechnological" human communication environment, information is readily exchanged between sender and receiver (though this exchange is circumscribed by the range of the speaker's voice and the receiver's hearing). The importance of such immediate exchanges of information to the intellectual process was appreciated by Socrates, who, as we have seen, decried the incapacity of the written word to respond to its reader's inquiries. Indeed, until the invention of the telegraph, all communication beyond proximate physical surroundings entailed the loss of interactive ability: the contribution of long-distance communication to cognition was thus restricted to the dissemination of knowledge and its criticism via a delayed-reaction, after-the-fact process. Not possible at the distances attained first by writing and then by printing were imaginative interplays of brainstorming and other social constituents (however minimal) of the generative phase of knowledge. More importantly, also missing was the give-and-take of immediate criticism prior to the preparation of knowledge for dissemination.

Because electricity travels at the speed of light, messages sent by electricity reach their destinations (however distant) in virtually the same time that they would reach destinations within shouting distance. Thus, electricity reinstates on a global level and beyond the conditions for immediate exchange of information characteristic of premedia, in-person communication environments (for example, two people sitting and talking across a table), and in this sense telegraphy trivializes or obviates the dimension of distance in communication in much the same way as photography renders irrelevant the dimension of time. Moreover, in an "anthropotropic" evolution (tropic as in "toward" and anthropo as in "human")[44] paralleling that of photography, the transmission of information via electricity has progressed from the sending and receiving of abstractions of written words by the telegraph, to delivery of spoken words by the telephone and images of the speakers' faces via videophone. The consequence of this increasingly lifelike instantaneous transmission of information is that the style as well as the substance of in-person conversation is obtainable at great distances, as parts of the world become corners of a single, grand Victorian parlor room.[45]

The telephone is arguably the most significant electronic medium in history, not only because it personalized the officious telegraph by bringing the human voice directly into the home,[46] but because unsuccessful attempts to improve the telephone contributed to the development of the two dominant electronic mass media of the twentieth century. First, Fessenden's dream of a telephone system without wires materialized into the quite different one-way mass medium of radio (in which one voice addresses many, rather than two voices interacting with each other), and then the attempt of numerous inventors to develop a telephone system with pictures begat a one-way radio system with pictures—television.[47] Economics were decisive in both telephone "failures": wireless receivers first of sounds and later of sounds and images were affordable to consumers whereas generator-transmitters and video cameras were not. (Video cameras became affordable as a consumer luxury product in the 1980s; generator–transmitter technologies, other than hobbyist "ham" radio devices, still remain entirely beyond the general public's economic reach.) Thus the sequential mass audience of print and the droplet mass audiences of motion picture theaters (sequential numbers of very small simultaneous audiences in movie houses) have been dwarfed by the huge (sometimes world-wide) literal simultaneous mass audiences of radio and television, brought into being by an initially unintended capacity to transmit sounds and pictures everywhere at once. The penetration of radio and now even television in many parts of the world has been thorough,[48] and in this sense the most striking achievements of twentieth-century communication have come from as yet unfulfilled visions of the telephone.

Of all the technologies discussed in this volume, however, the one-way transmission to multitudes first of spoken words through radio and then of words and moving images via television has had the least apparent impact on the cognitive process and the growth of knowledge. Radio and television have been entertainers first, political and social revolutionizers second, and vehicles of cognitive development only a distant third. Occasionally a social change fostered by one of these media has had a profound epistemological outcome. The intensification of central political authority facilitated by radio in the 1930s—in the most extreme case, permitting Hitler to relate to the German people as a father addressing children who sat at his feet—resulted in the emigration of German-speaking scientists to Great Britain and America, which had the positive epistemological consequence of placing leading thinkers in environments highly supportive of most scholarly and scientific pursuits.[49]

Television on the face of it has played even less of a direct and indirect role than radio in the furtherance of human knowledge. Educational television programs devoted to scientific and artistic themes are seen by relatively small numbers of the population, and the recent growth of nonbroadcast cable television has by and large increased only the quantity of movies available for home viewing.[50] Indeed, television has been the recipient of considerable criticism, unsubstantiated and even nonsensical, to the effect that its viewing scrambles, inhibits, or otherwise injures rational thought processes (especially their development in children), and thus is explicitly countercognitive in its consequences.[51] The milder form of this criticism, that television viewing of images offers an easy distraction from the rigors of abstract reading, may have some merit, but, as we already have seen, video systems can deliver written words as well as spoken words and pictures. Thus, in the long run, television technology may be of more help than hindrance to the learning and promoting of reading. The stronger claim that television viewing interdicts the logical process seems itself a product of impaired reasoning. Since I myself am a child of the television age—having watched an hour or two of television virtually daily since shortly after my birth in 1947—I am inclined to offer my own work (for example, this book) as evidence that television, though mosty banal, does not destroy the mind. Readers no doubt will find much to dispute in this book, but I trust that they will agree that the book's presentation and development follow some minimal form of rationality and logic, and thus count against the view that exposure to television damages one's rational faculty. (Of course, I may be so "far gone" from watching television that this book and my entire intellectual life may be a figment of my imagination. A third possibility, that I have some sort of unique mentality resistant to the ravages of the tube, is not without its appeal, but fidelity to the pursuit of truth about television bids that I reject it.) On a more serious note, although television generally is not on the cutting edge of any cognitive process—televised first encounters with planets and other celestial bodies would be a significant exception—television may yet make an important contribution to the growth of knowledge via the millions of minds, especially young ones, that it touches with its (too) occasional programs of scientific and philosophic interest. Who knows, for example, how many future astronomers and naturalists may have been hatched in the televised discourses on those subjects by Carl Sagan and David Attenborough, what excitement an unexpected glimpse of an animated galaxy or dinosaur may have kindled in a receptive young mind? Further, when the viewing of even a fictional "mini–series" encourages the reading of an associated novel, we see television serving rather than subverting the habit of literacy, albeit in a modest way.

The portability conferred by transistors for radio receivers in the 1950s, and by microelectronics for all receivers (including image screens) in the past decade, continues to progress toward one of the ultimate goals of electronic distribution of information: a system of infinite accessibility in which all data—printed, photographic (imaged), and acoustic—will be available in principle to everyone, any time, from any place on Earth and humanly inhabited outer space. Nonbroadcast media such as cable television and physically transported videotapes are already challenging commercial broadcast oligarchies in the West,[52] and although the content of these new channels has been if anything even less uplifting than usual commercial network offerings, the very opening of these new pathways increases the opportunity for the dissemination of wisdom. (Oligarchical control of broadcast media is predicated on the limited number of channels available for electromagnetic carrier wave broadcasting. No such constraints pertain to the number of cables that can be laid or videotapes that can be distributed. Uncontrollable "samizdat" videotapes are also beginning to make their presence felt in the totalitarian East.[53]) The development of cheaper, even more effective nonbroadcast carriers such as fiber optics will likely further increase the ability of individuals to select and construct information packages as opposed to merely taking or leaving the offerings of today's mass media programmers.

Economic and political factors initially may serve to limit universal accessibility, as they have in earlier media systems. But the radical convenience and inexpense of new microtechnologies should eventually undercut regressive social patterns in much the same way that the comparative cheapness of the first printed books once defeated the traditional authority of the Church. This time, however, the result will be more comprehensive, with a commensurately greater democratization of the knowledge process. Apart from the ethical and political desirability of such a development, the dissemination of knowledge to so many more potential critics and generators of knowledge may result in a new Scientific Revolution every bit as profound for our age as the Scientific Revolution spurred by books was for the age that preceded ours, and from whence we came.

The content of these universally accessible pipelines will encompass most past and present communication forms, and will vary according to the needs of the communication. Since speech and writing are far better suited to the communication of abstract ideas than is photography or any more literal media system, speech and writing should have a prominence in the universal pipeline equivalent to the centrality of abstraction in human life. (The ready availability of recorded spoken material may to some degree replace print in such a system, since speech conveys abstractions as effectively as writing and is easier to learn. However, the capacity of writing to be rapidly scanned, searched, and sorted electronically and by eye—not to mention our nostalgic attachment to it—will no doubt maintain a significant place for the printed word in the comprehensive, speed-of-light pipeline.) When we need to examine an object or event as our senses would perceive it, three-dimensional, motion holograms likely will do the job, although flat videos or even still photographs may be adequate, depending upon which aspect of the environment we wish to inspect or recall. In interactive situations in which two or more persons wish to converse, today's videophone will no doubt evolve into an "holographone" or three-dimensional interactive system, and this itself may someday be replaced by devices that instantly transmit not three-dimensional images, but complete, real, live people. (Universal interactive access—the ability to immediately converse with any other person—is circumscribed, however, by the availability of the intended partner. Thus noninteractive representations are the only form of communication that in principle can ever be accessible to everyone anytime.) Indeed, the evolution of transportation systems toward increasingly faster movement of people (on horses, trains, autos, and planes), and the corresponding evolution of communication systems since the telegraph toward instantaneous movement of ever more life-like representations (writing, speech, visual images), suggests that a convergence of transportation and communication at some future date in an instantaneous teleportation technology is not only possible, but even likely.[54] (Actually, such a development would be a reconvergence or reunion of the two functions, since the earliest type of transportation, walking, and the earliest communication, talking, co-occur in the same system: the human being. Moreover, in the prewritten and prepainted state, the only way that information can be moved any distance is by moving or transporting the speaker.) The conception of a thought, the miraculous emergence of more from less that happens somewhere in the synaptic interconnections of the human brain, will probably be the same as long as human beings are the same species. But the distribution of the products of this process seems about to undergo a technological metamorphosis of such fullness, diversity, and accessibility as to rival the very teeming, multidimensional, instant interchange of figments in the brain from which thought emerges.

Nearly a century and a half ago, Nathaniel Hawthorne's character Clifford in *The House of the Seven Gables* commented that the electric telegraph was turning the world into one huge brain.[55] The technological brain, the extended mind that we are now constructing, is actually far better than one huge, global brain, for unlike the cells that make up our individual brains, the minds—our minds—that make up the global brain are obdurately individual and possessed of free will. Any of the user/constituents of the global pipeline may pull out, in whole or in part, or attempt to improve it. As ease and diversity of selective communication increases—today a telephone in every home, later today an interactive computer terminal—so does our ability to focus our cognitive pursuits, to partake of a chosen piece of group wisdom, to withdraw for generative, sheltered reflection, to put the results back in the pipeline for criticism, context and distribution. Thus through technology we may gain the world, know of the universe and expand to its requirements and dimensions, without losing our souls.

RETURN TO SENDER

We have seen in this chapter that the technological distribution of information is as central to the growth of knowledge as is the technological enhancement of perception and mental calculation discussed in the previous chapter, and indeed as crucial as the material embodiment of knowledge that constitutes all technology, cognitive-intended and otherwise.

Without speech we would have little in the way of human mental characteristics, and without writing we would have had little accumulated abstract knowledge prior to the nineteenth century. These communications media thus have been more essential to the growth of knowledge to date than telescopes and microscopes. On the other hand, recent social media such as telephone and television thus far have played a comparatively minor and mostly unintended role in the construction of human knowledge.

If many of the cognitive consequences of modern communications technology and technology in general are unintended, we turn now to a highly significant cognitive consequence that not only is unintended in entertainment technologies such as motion pictures, but has also been unintended for the most part in the design of cognitive-intended devices such as computers: the knowledge that we may gain of our interior self, and the operation of our internal cognitive processes, when they become externalized in technologies designed to extend our experience across time and enhance the performance of our cognitive capacities. Knowledge of our knowledge processes is the goal of this book, and indeed has been sought by thinking people for as long as they have sought knowledge of the external world—a dual quest symbolized in the person of Thales, who was the West's first known epistemologist as well as scientist.[56] Despite, however, the perennial flare of attention given to the Socratic–Platonic plea to "know thyself," the thrust of Western intelligence and thus the scientific intelligence of the world has been directed toward knowing external reality (of which our bodies and even our brains are considered parts), with the internal something, the active knowing agent variously referred to as the soul, the self, and the mind, relegated to the sidelines of speculative (read "second rate") philosophy.

The occasion for this derogation of the mind is no doubt its intangible, transempirical character,[57] which not only makes it impossible to specify through the senses (extended or otherwise), but difficult to systematically investigate in a scientific climate which quite rightly sees empirical confrontation as a crucial control on ideas and theories. Indeed, the attempt to scientifically understand the mind in psychology often has pursued its subject by denying, ignoring, or otherwise trying to work around its existence (as in the behaviorist program in twentieth century psychology).[58] Moreover, the study of the mind or the knowing agent has been faulted even by "pure" philosophers, who point out seemingly insurmountable problems of self-reference in understanding the mind, or quandaries about how a knower can be an object of its own cognition or a subject and object at the same time.[59] In such a milieu, technologies purposely constructed to aid human cognition have quite naturally been turned to the external world, intended to assist either our perceptions or our processing of our perceptions of external reality.

At about the same time that the science of psychology was first stirring in the 1870s, however, a German by the name of Ernst Kapp, little known today outside of philosophy of technology circles, made the startling but sensible suggestion that aspects of human life and mentality unconscious or inaccessible to ordinary introspection become clarified when we observe the external products of our mentality, that is, the workings of our technologies.[60] This insight was later picked up by such giants as Freud and philosophers like Heidegger,[61] and most recently by the discipline of artificial intelligence, which has begun to formally investigate the knowledge we can obtain about the workings of human mentality through the programming and observation of the operation of our external mental surrogates, computers.[62] The common thread running from Kapp through artificial intelligence is that we inevitably, consciously and unconsciously, put something of ourselves in all our creations; more specifically in the case of computers, that we electronically project aspects of our cognitive functioning in these programmed devices, and that in so creating externally observable models of human cognition we at last put the mind—or its reflections—in a place where it may be amenable to empirical investigation. Computer programs may, like human minds, do surprising things; but unlike the structures of our minds, the workings of computers and all technologies are authored by us, and thus are presumably more readable. Further, once we grant that brains and computers function in somewhat similar realms, any differences between the two—ways in which human mentality seems markedly unlike computer operations—tell us almost as much about own cognition as do similarities between protein-based and microchip thinking. We proceed now to a fuller consideration of how technology, largely unintentionally, has sprung the mind from its cell of subjectivity, and set it loose as an object of study among the microbes and the stars.

NOTES

1. See, for example, McLuhan's "The Rise and Fall of Nature," *Journal of Communication,* Autumn 1977, pp. 80–81 (reprinted in *The Antigonish Review,* Summer-Fall 1985, pp. 98–99).

2. Reception and projection of 360-degree moving holograms on television requires fiber optic carriers (or some other method of increasing conveyed information above standard broadcast levels) and screen resolution of at least 900 lines. Both are technologically feasible today, but are not yet available for general commercial use. For the theoretical basis of holography, see George W. Stroke, *An Introduction to Coherent Optics and Holography,* 2nd

ed., New York: Academic, 1969. See "Observational Terminus: Holography and the Retrieval of the Third Dimension" in my *Human Replay* for a discussion of the significance of holography in the evolution of communications technology. (Note that one-way mass media such as radio and television cannot replicate the full interactive aspects of life—even in holographic projections, viewers cannot physically touch the person or object projected. See chapter 8 in the present volume for a discussion of the differences between one-way and interactive telecommunications, and their impact on the growth of knowledge.)

3. The following list is partial but representative of the scope and quality of thinkers who have addressed the extensional action of communications media: Samuel Butler, "Lucubratio Ebria," letter to the Press (Christchurch, New Zealand newspaper), July 29, 1865 (reprinted in *The Notebooks of Samuel Butler,* ed. H. F. Jones, New York: AMS, 1968, pp. 35–40); Ralph Waldo Emerson, "Works and Days" in *Society and Solitude,* Boston: Osgood, 1870, pp. 127–150; Ernst Kapp, *Grundlinien einer Philosophie der Technik,* Braunschweig: Westermann, 1877/1978; Henri Bergson, *Creative Evolution,* trans. A. Mitchell, New York: Holt, 1911; Hendrik Van Loon, *The Story of Invention,* New York: World, 1928; Stuart Chase, *Men and Machines,* New York: Macmillan, 1929; Sigmund Freud, *Civilization and its Discontents,* trans. Joan Riviere, New York: Cape and Smith, 1930; Lewis Mumford, *Technics and Civilization,* New York: Harcourt Brace, 1934; R. Buckminster Fuller, *Nine Chains to the Moon,* Carbondale, Ill.: Southern Illinois University Press, 1938; Harold A. Innis, *The Bias of Communication,* Toronto: University of Toronto Press, 1951; Edward T. Hall, *The Silent Language,* New York: Fawcett, 1959; and Marshall McLuhan, *Understanding Media,* New York: Mentor, 1964. Kapp, Innis, and McLuhan offer the most comprehensive analyses of media extensions. Summaries and comparison of some of these and related works can be found in William Kuhns, *The Post-Industrial Prophets,* New York: Harper Colophon, 1971; James M. Curtis, *Culture as Polyphony,* Columbia, Mo.: University of Missouri Press, 1978; and my *Human Replay.*

4. André Bazin, *What Is Cinema?,* trans. and ed. Hugh Gray, Berkeley: University of California Press, 1967 (1958–1965).

5. See my *Human Replay.*

6. See "The Biological Basis of Reason" in chapter 2, and "Telescopes, Microscopes, and the Human Equation to the Cosmos" in chapter 5 of the present volume.

7. On the intrinsic neutrality of specific technological objects, see my "Guns, Knives, and Pillows" (paper presented at the International Conference on Phenomenology and Technology, Polytechnic University, Brooklyn, NY, October 3, 1986), where I point out that a pillow, generally considered a beneficial or at least innocent technology, can be used to commit murder via suffocation; and a gun, generally considered bad, can be used to hunt food and insure our survival. (Similarly, we would likely judge a natural bad event such as cancer to be good if it killed someone we judged bad such as Adolf Hitler.)

On this issue of technologies and events acquiring goodness and badness only insofar as they have an impact upon human lives, I am in complete agreement with Dewey's approach—that is, our axiologies agree, whereas our epistemologies and ontologies differ a bit (see chapter 5, note 37 and chapter 4, note 22).

But see also my arguments in chapter 9 (and implicitly throughout this book) that the general technological enterprise—technology as a human activity—is intrinsically beneficial to our species.

8. A perfect duplication also transforms the original (duplication of a unique object renders the original no longer unique, for there now are two examples of this entity; duplication of a second or third member of a series similarly alters the character of the duplicated entity, though in a less profound way), and in this sense a perfect or complete duplication is similarly unattainable and self-defeating. The duplication attempts to capture

the original in all its qualities including its uniqueness, and in so doing renders both entities nonunique.

9. As Dewey puts it, "Where communication exists, things . . . acquire representatives, surrogates, signs, and implicates, which are infinitely more amenable to management, more permanent and more accommodating, than events in their first estate" (*Experience and Nature*, p. 167).

Dewey also suggests that the very separation of events from their "qualitative immediacies" (conditions that appeal to "sensations and passions") renders them more "capable of survey, contemplation, and . . . logical elaboration."

This is a vitally important consequence of abstraction, but we also need to identify the nature of the event being abstracted, and the purpose of the communication, before we conclude that abstraction is an unmitigated benefit. For example, in a situation in which the qualitative immediacy of color plays an important role—say, in a medical diagnosis based on skin tone—a written description or even a black-and-white photograph of the skin in question could be insufficient for a rational diagnosis of the symptom. The most effective use of communications in cognition thus requires a mechanism for conservation of physical attributes jeopardized by abstraction, or processes of abstraction that sacrifice as few of the attributes of the physical event as possible. See ensuing discussion of abstraction, and "The Conquest of Immediacy and Eternity" and "Electronic Mercury," in the present chapter. See also my *Human Replay*.

10. Claude Shannon and Warren Weaver, *The Mathematical Theory of Communication*, Urbana, Ill.: University of Illinois Press, 1949.

11. The exception is partial because Chinese ideograms employ nonalphabetic phonetic components (for example, use of modified pictographs to represent events whose spoken descriptions or words rhyme with or sound like the words for the events described in the original pictograph; see Colin A. Ronan and Joseph Needham's *The Shorter Science and Civilization in China:1,* New York: Cambridge University Press, 1978, pp. 8–12; a hypothetical example in English would be use of pictographs for "bee" and "leaf" to write the word "belief"). See "The Book in History" and note 16 in this chapter for difficulties with hieroglyphic (ideographic-pictographic) systems as compared to phonetic-alphabetic modes.

Many of the weaknesses of pictographs apply to current computer "icon" systems, offered as improvements to computer text systems. See note 17 below.

12. I am including the greater mental manipulability of abstractions as opposed to original events (see note 9 above for Dewey on this) as an intrinsic characteristic of the transmissibility of abstractions. At the same time, for reasons explained in note 9, I see the deliberate stripping of original qualities from the abstraction (for example, the sacrifice of color) for any purpose other than improvement of transmission to be counterproductive to the growth of knowledge.

Noise or distortion in transmission of course can sometimes have a serendipitous beneficial effect—for example, I order by phone McLuhan's *Understanding Media;* the clerk misunderstands and sends McLuhan's *The Gutenberg Galaxy;* I find this book of more relevance than *Understanding Media*. But the possibility of such happy endings cannot be cause enough for encouragement of systematic noise in telephone connections.

13. Kant's method of dealing with this problem was to develop a separate schema for ethics and total creations of intellect—one wholly distinct from his epistemology (see his *Critique of Practical Reason*). He continued to maintain, however, that treatment of entirely nonempirical entities such as the soul and the deity as possible objects of scientific or empirical knowledge was a fundamental error.

Dewey is in agreement on the pitfalls of nonempirical objects of cognition, and warns that abstraction can result in objects "hailed as modes of Being beyond and above spatial and temporal existence" (*Experience and Nature,* p. 167).

14. This scenario provides an evolutionary, naturally selective accounting of the innate, genetically transmitted "rules" for intelligible spoken communication (grammar or syntax) posited by Chomsky on psychological and philosophic grounds.

15. The existence of tools in nonlinguistic species such as beavers demonstrates that technology is an older biological activity than language, and was no doubt practiced by our prelinguistic hominid ancestors. Most prelinguistic technologies may be considered embodiments of genetically encoded knowledge, and this suggests that the defining characteristic of fully human technology is its embodiment of individually learned knowledge—that is, theories that are consciously derived and applied with the assistance of speech. Although our hominid ancestors likely used sticks to hunt termites in the same way as modern chimpanzees do, the human technological enterprise could not have gotten fully underway until it was talked about. (See chapter 5, note 1 on the biological antiquity of technology.) (On the likely existence of proto or extremely primitive human technologies prior to language, see the discussion of mimetic, nonabstract human thinking and its relation to tool making in *Human Replay,* pp. 256–260 ff., and Julian Jaynes' *The Origin of Consciousness in the Breakdown of the Bicameral Mind,* Boston: Houghton Mifflin, 1976, p. 130, which suggests that skills needed to make primitive tools could have been transmitted without language, "solely by imitation . . ." in the same way as "bicycle riding." See also E. H. Gombrich's dictum that "making comes before matching," that is, humans produce and fashion first, and compare and abstract second, *Art and Illusion,* Princeton, N.J.: Princeton University Press, 1960, p. 161.)

16. McLuhan, *The Gutenberg Galaxy,* p. 185. The figure of 2,000 is further increased by Chinese rhyming patterns that play a role in the written language (see Ronan and Needham, pp. 8–10). In contrast, the alphabet's 70 or so characters (26 letters in small and capital form, plus punctuation marks), can describe any object or abstraction, and are conducive to use in interchangeable type.

Even today, the difficulty of printing with so many characters is evidenced in the prevalence of one-page posters rather than newspapers in China. Takeshi Utsumi also reports that printers in Japan (where Chinese characters were imported in the sixth century) until recently suffered high rates of lead poisoning due to lead type carried in their mouths as a way of expediting the movement of large assortments of characters (from the computer teleconference "China Connection" on the Electronic Information Exchange System at the New Jersey Institute of Technology, Newark, New Jersey; comment #57, December 6, 1985, 7:17 PM; by Takeshi Utsumi).

An additional significant consequence of the large numbers of characters and the few competent writers of pictographic systems is that they foster communication elites and what Harold Innis termed "monopolies of knowledge"; see note 20 below for references on Innis, and notes 21–22 and associated text for discussion of how the Egyptian priestly class used its hieroglyphic monopoly of communication and knowledge to defeat pharaonic religious reform.

17. Chinese and derivative ideographic systems are the only survivors of an ancient age in which pictographic systems of various sorts were employed in places ranging from China to Egypt and the Middle East to the Mayan civilizations in America. However, as indicated in note 11 above, the Chinese system itself has been partially phonetic for two thousand years or more. Further example of the decline of even the current Chinese system is indicated by the Japanese introduction of an alphabetic writing system shortly after World War II. Meanwhile, computers have greatly increased the speed and facility with which Chinese

characters can be written (see R. W. Apple, Jr., "Two Britons Devise a Computer That Can Communicate in Chinese," *New York Times*, January 25, 1978, p. 2; Paula Chin with Bradley Martin, "Solving the Chinese Puzzle," *Newsweek*, August 18, 1986, p. 43); but these computers have little impact on the cumbersome process of learning how to read thousands of different characters.

Icon systems in personal computers have been promoted in the past few years as a means of making these devices more user friendly. In a typical non–icon system, for example, one removes or erases a text file (for example, a letter or paper stored in the computer) by typing "erase" or "delete" the name of the file. In an icon system, the user performs the same operation by moving a small picture (icon) of the text file (say, a drawing of a sheet for a letter file) across the computer screen to a small drawing of a garbage can; when the icon for the file is in proximity to the garbage can icon, a key on the terminal is pressed (or a lever on a "mouse"—detachable from the computer), and the file is erased. (See *Whole Earth Software Catalog for 1986*, ed. Stewart Brand, Garden City, N.Y.: Doubleday, 1985, pp. 217–219, for brief further descriptions of icon systems.) Such systems may ease the pain of entry into the computer world for some, but they are in no sense a substitute for the actual alphabetic text that is stored in files, and transmitted to readers. Indeed, even as a command system, icons sacrifice much of the subtlety of direction available in alphabetic commands. One can easily command a system to erase one file, two files, all files, temporarily erase a file, and so on via typed words from the keyboard; accomplishment of the same tasks via stylized garbage cans is more difficult.

Note that icon or picture-command systems are a computer function entirely distinct from the generation and manipulation of computer graphics of physical phenomena such as molecular relationships and weather patterns. Such computer modelling is of great assistance to scientific research on the generative, critical, and disseminative levels; see note 30 below.

Traffic signs and similar pictographic media seem unassailable in their advantage over written advisories when international readers are involved. Certainly an X drawn over a smoking cigarette is a more dependable way of conveying a "no smoking" advisory than writing such an advisory in any number of languages which the reader may or may not comprehend. Even when unfamiliar language is not at issue, a sign on a highway with stick figures crossing a street is likely easier to comprehend by someone driving 60 miles per hour than a written notice about pedestrians in the area. Such pictographic techniques are likely to continue in the foreseeable future.

18. Thus painter Paul Delaroche's exclamation "From today on, painting is dead!" on first sight of a daguerreotype (the earliest commercially successful form of photography) in 1838 was only partially sustained. Pressed to accomplish visual feats that objective photography could not, painting turned to capturing "pure" light rather than visual objects (Impressionism) and then to the highly subjective Expressionism and abstract art of Picasso and others in the twentieth century. (Delaroche later reconsidered and saw a photographic service to even portrait art in the quick copies of objects and scenes the painter could obtain for reference via the photograph. See Helmut and Alison Gernsheim, *L. J. M. Daguerre*, New York: Dover, 1956/1968, pp. 95 ff. for more on Delaroche.)

19. William Blake's "Auguries of Innocence."

Also lost in changes from pictographic to phonetic writing systems is a certain playfulness and even art dependent on manipulation of visually reflective characters. Perhaps the juxtaposition of visual realities in pictographs engenders an inventive attitude, advantageous to the initial generation of knowledge in playful speculation, whereas the alphabet is far superior as a vehicle of criticism and dissemination. This might account in part for the initial invention of many technologies in China (printing, the compass, gunpowder, rocketry) which in general were unimplemented on mass scales until their importation and development in the West. (See, however, William McNeill's *The Pursuit of Power*, Chicago: University of Chicago Press, 1982, p. 49, for a discussion of the Chinese failure to implement their technologies as a consequence of Buddhist and Confucian attitudes that punished autocatalytic capitalist growth processes. See also my "Cosmos Helps Those Who Help Themselves: Historical Patterns of Technological Fulfillment, and their Applicability to the Human Development of Space," paper presented at Conference on Space Agenda: Context and Opportunity, MIT, April 5, 1986, to be published in C. Mitcham, ed., *Research in Philosophy and Technology*, vol. 9, 1988.

The most sympathetic treatment of nonphonetic societies is offered by McLuhan, who throughout *The Gutenberg Galaxy* and *Understanding Media* and virtually all of his work suggests that the alphabet "flattens" and "neutralizes" the multidimensionality and "magic" of the oral society. (Phonetic systems puncture this magic because they seek to visualize the acoustic; pictographic systems pose no threat to the acoustic because they do not seek to coopt the acoustic in writing.)

20. Innis' works in this area are *Empire and Communications*, Toronto: University of Toronto Press, 1950, and *The Bias of Communication*, Toronto: University of Toronto Press, 1951. McLuhan's best works are *The Gutenberg Galaxy* and *Understanding Media*. Representative works by Eric Havelock are *Preface to Plato*, Cambridge: Harvard University Press, 1963 and *The Literate Revolution in Greece and its Cultural Consequences*, Princeton: Princeton University Press, 1982. Walter Ong's *The Presence of the Word*, Ithaca, N.Y.: Cornell University Press, 1967; *Interfaces of the Word*, Ithaca, N.Y.: Cornell University Press, 1977; and *Orality and Literacy*, New York: Methuen, 1982 are examples of his many works in this area.

21. See, for example, Hugh Thomas' *A History of the World*, New York: Harper & Row, 1979, pp. 133–134, where the defeat of Ikhnaton's creed is simply attributed to the "attitude of the well-entrenched priests," and characterized as the "earliest known clash between Church and State and won by the former." See also Sigmund Freud, *Moses and Monotheism*, trans. Katherine Jones, New York: Vintage, 1939/1967, pp. 25–26, who suggests that the "Aton [sun] religion had not appealed to the people," in particular running contrary to their taste for "myth, magic, and sorcery." Innis, *Empire*, p. 21, cites a variety of historical sources and concurs that Ikhnaton's religion "failed to gain the emotional support of the people." None of these accounts links Ikhnaton's failure with the masses to the limitations of hieroglyphic writing. (Innis' position on writing systems and ancient religions, then, is that hieroglyphic systems support polytheism and the alphabet facilitates monotheism—he makes no claim that failure of a specific monotheism is due to a hieroglyphic environment.)

22. A materialist rendition of Ikhnaton is found in Innis, *Empire*, for example, where Innis repeatedly describes the religion as worship of "the solar disk," and explains the attempted revolution in political and commercial terms (p. 21). See also Margaret Murray, *The Splendor That Was Egypt*, New York: Philosophical Library, 1949, p. 54: "The Aten [Aton] is the actual disk of the sun, the physical sun which emits heat and sends out visible rays. . . . Akhenaten [Ikhnaton] was therefore no heretic."

However, see also Freud, *Moses*, who cites Arthur Wiegall's conclusion in *The Life and Times of Akhnaton* (1923), that Ikhnaton's deity had "no form." (Wiegall also points out that no graven images were allowed of Aton—only symbolic disks.) Freud's thesis is that the Pharaoh's monotheistic religion was later picked up by Moses—an Egyptian—and became the real basis of Judaism. Were this the case, the "alphabetic" theory of monotheism would be even stronger, for then we would have literally the same religion failing with hieroglyphics and succeeding with phonetic distribution. I'm content to emphasize, however, that the failure of Pharaonic monotheism and the triumph of Hebraic religion is due primarily to differences in modes of dissemination, whatever the historical connections between the two.

23. Caption to Stradanus' engraving of a print shop in operation, reprinted in Joseph Agassi's *The Continuing Revolution,* New York: McGraw-Hill, 1968, p. 26.

24. Reisman quoted in Neil Postman, *Teaching as a Conserving Activity,* New York: Delacorte, 1979, p. 65.

25. See *The Bias of Communication,* pp. 55 ff., for Innis on the relationship of printed vernaculars and nationalism. McLuhan, *The Gutenberg Galaxy,* p. 239 sees a similarly nationalizing effect in printed vernaculars, but attributes this specifically to the transformation of vernacular into a closed system via print, and the consequent build-up of (nationalistic) energies in closed systems.

James R. Enterline provides excellent descriptions and analyses of Norse penetration of Greenland and North America in *Viking America,* Garden City, N.Y.: Doubleday, 1972. Archeological evidence shows at least two flourishing Norse communities in Greenland from about 950 to 1450 AD, and Norse presence in Canada in approximately 1000 AD. (See Gwyn Jones, *A History of the Vikings,* New York: Oxford University Press, 1984, p. 304, for a discussion of Viking artifacts at L'Anse Aux Meadows, Newfoundland.)

Viking knowledge and settlement of the New World thus was far better established than the knowledge that resulted from Columbus' initial voyages, and yet this extensive experience had little effect on Europe. Social, political, and economic conditions in Europe were likely unsuitable for an age of discovery in the year 1000 in any case, but the hard-copy descriptions of Columbus' discoveries seem nonetheless prerequisite to the flow of hard currency that financed the waves of European exploration some 500 years after the Vikings.

See Samuel Eliot Morison's *Admiral of the Ocean Sea,* Boston: Little, Brown, 1942, ch. 27 ("Spreading the News") for summary of the impact on Europe of the first printed descriptions of Columbus' voyage. Columbus' "Letter on the First Voyage" was "printed in Barcelona as early as April 1 [1493]. . . . The first Latin translation, dated April 29 . . . was printed in Rome . . . as a pamphlet of eight pages. This became a 'best seller'; it ran through three Roman editions in 1493, and six different editions were printed at Paris, Basle, and Antwerp in 1493–1494. [A Tuscan version] was printed in Rome on June 15, and at Florence twice in 1493. A German translation was printed at Strassburg in 1497, and the second Spanish edition appeared . . . at about the same time" (p. 379).

On the success of the post–Columbian European expeditions to the New World, Nigel Davies points out that "If, like the Viking expeditions, the Enterprise of the Indies had not been backed by gunpowder, its outcome would have been uncertain," *Voyages to the New World,* New York: Morrow, 1979, p. 239. Thus we might say that the Chinese inventions of print and gunpowder were the necessary stimulant and security respectively for the European settlement of the New World.

See also Elizabeth Eisenstein, *The Printing Press as an Agent of Change,* New York: Cambridge University Press, 1979, pp. 583ff., for arguments about the role of printed material in Columbus' preparation and execution of his voyages. Columbus' consultation of printed astronomical tables and treatises before and during his expeditions, however, is beside the point of the present thesis, which focuses on the role of the press in furthering the impact of the expeditions after they were made.

26. Thomas S. Kuhn, *The Essential Tension,* Chicago: University of Chicago Press, 1978, p. 345.

27. The cooperation of telescopes and printing in the pursuit of knowledge was captured beautifully nearly 100 years ago by Emily C. Pearson, who wrote in *From Cottage to Castle: The Boyhood, Youth, Manhood, Private and Public Career of Gutenberg,* Boston: Earle, 1888, p. iii: "Printing has been styled 'The telescope of the soul.' As the optical instrument brings near and magnifies objects remote and invisible, so printing puts us in

communication with minds of the past and present, and preserves the thoughts of this age for future generations."

28. This library limitation subsides as libraries and bookstores grow more numerous and accessible. For the scholarly researcher, however, sources of all but a handful of very popular and classic printed texts are few and far between.

29. Radio Shack's TRS-80 M100 "lap top" computer has telecommunication and word processing facilities, and weighs four pounds. It was introduced in 1983, and in 1986 sold for $399. Numerous other portables are currently marketed; see Richard Dalton, "Electronic Notebooks," *Whole Earth Software Review,* Summer 1984, pp. 87–95, for brief descriptions of a variety of models available then.

30. See Peter A. McWilliams, *The Word Processing Book,* Los Angeles: Prelude, 1982, and *The Whole Earth Software Catalog,* ed. Stewart Brand, pp. 46–63, for introductions to the techniques and benefits of word processing.

An area ripe for research is the difference between brain-tongue (or vocal chord) and brain-hand (or finger) connections in the human production and communication of information. On the basis of my own experience and the anecdotal reports of others, I'd conjecture that when fingers operate with essentially mechanical systems (that is, handwriting, self-correcting typewriters, or any system in which the immediate product is "hard copy" or relatively permanent), the brain-hand connection is less efficient than the brain-tongue; but when fingers operate with electronic or image systems (systems in which the printed words may be easily manipulated before conversion to hard copy), the brain-finger connection is at least as fluent as the brain-tongue. Note that the mere powering of typewriters by electricity does not create an electronic system in the word-processing sense: the electricity must work to make the immediate product electronically malleable. Greater effectiveness of word processing versus speech is also likely a function of individual cognitive styles.

An equivalent improvement in fluency results from the digitalization of graphics and music. Here the graphic artist or musician can express and manipulate visual or musical ideas directly, without the intervening physical graphic medium or musical instrument. In other words, an artist can produce pictures from a computer keyboard with no need of pen or pencil or brush; and a musician can create and play a beautiful melody entirely by typing on a keyboard (connected to a music synthesizer in which on-off sequences generate tones), never plucking, stroking, blowing into, or otherwise touching a traditional instrument. See Stanley Klein, "Computers That Draw Pictures," *The New York Times,* July 6, 1980, section 3, pp. 1, 5; Andries van Dam, "Computer Software for Graphics," *Scientific American,* September 1984, pp. 146–159; and *The Whole Earth Software Catalog,* pp. 122–137 and 198–199 for preliminary assessments of graphic and musical computer programs, and the impact of the former in engineering, architecture, and scientific research and dissemination.

Graphics programs have been of greater cognitive import than music programs, especially in the visual representation (and facility for rearrangement of such representation) of complex physical relationships such as those at the molecular level. Computer graphic projection and rearrangement of projections of physical relationships—known as computer modelling—assists not only in communication but also in fundamental understanding of the relationships, and in this second sense constitutes a hybrid of the primary cognitive technologies of laboratory experimentation and computer data processing discussed in chapter 5. See note 14 and "Computers and Tractable Immensity" in that chapter.

The main consequence of music digitalization is aesthetic, and renders the musician more a thinker and less a performer. Indeed, computerization of both graphics and music relocates art and its production from a dual residence in the brain and the hands to a position predominantly in the brain. Such "intellectualization" and movement away from physical dexterity in art may foster radically new types of art and artists—people with talents somewhat different from those of the traditional artist—and different conceptions and expectations of artists by society.

31. Douglas Hofstadter's *Gödel, Escher, Bach,* New York: Basic, 1979, is an early example of a text which was word-processed and typeset by the author. See p. xx in the preface of that book for the author's comments on this process.

On the reduction of error via word processing and author-controlled typesetting: Note that what is being reduced here is not the commission of error in the initial generative (thought) process—as we saw in chapter 3, such commission of error is impossible to completely eliminate and also seems a positive source of knowledge in the long run. What word processing reduces is rather the retention of error in a work after the error has been identified. An author who discovers that an argument can be more clearly stated after it has been written is in a better position to make such a clarification with word-processed as opposed to typewritten text.

Of course, increased facilities for revision could well result in a net increase of error introduced in the revised version. The assumption I am making, however, is that maximum opportunity for the effective application of the critical process is beneficial for the growth of knowledge, even though the critical process itself is ever liable to error.

Emphasis of error elimination in addition to error identification is an important part of Popper's philosophy. See, for example, *Objective Knowledge,* p. 126, where Popper contrasts Hegel's view of contradictions with Popper's: both see contradictions as important to uncover; but whereas Hegel revels in their existence as intrinsic to the fabric of reality— needed to propel the dialectic—Popper urges their removal via rational investigation as necessary for the human furtherance of knowledge. (In fairness to Hegel, the dialectic process in its own way leads to the removal or reconciliation of contradictions. See also chapter 9 of the present book, where I suggest that paradox may be the spur of evolutionary self-transcendence—a mechanism that bears some resemblance to Hegel's. In Hegel's dialectic, however, humans are much more the witness or conduit and much less the creative designer of progress than they are in Popper's and my cosmologies.)

32. McLuhan noticed that just as the printing press made everyone readers, so the Xerox machine makes everyone publishers ("Laws of the Media," *Et Cetera,* June 1977, p. 178). Electronic telecommunication of text furthers this development by allowing authors to publish without leaving their homes or places of business. ("Desktop publishing," or the increasing use of personal computer–printer combinations to produce newsletters and near-book quality materials in homes and offices, is yet another development in this area.)

See Warren J. Haas, "Computing in Documentation and Scholarly Research," *Science,* February 12, 1982, pp. 857–861, for an early assessment of the impact of computers on the dissemination of scholarly material.

Figures compiled by the Future Computing Company of Dallas, Texas, report about 19 million personal computers in use in the United States in 1984, with approximately 20 percent of these outfitted with telecommunicating modems.

See my "Impact of Personal Information Technologies on American Education, Interpersonal Relationships, and Business, 1985–2010," in Paul Durbin, ed., *Philosophy and Technology,* IV, Boston: Reidel, 1987 (in press), and "Communities of Representation" in chapter 8 of the present volume for further discussion of word processing and telecommunication in the production and distribution of scholarly texts.

33. David de Haan suggests that "electric lighting did more to improve the habit of reading books than anything before it," *Antique Household Gadgets and Appliances,* Woodbury, N.Y.: Barron's, 1977, p. 121. The flickering flames of gas and kerosene lamps may have been picturesque, but they were poorly suited for long bouts with the printed page.

34. See "The Times Enters a New Era of Electronic Publishing," *The New York Times,* July 3, 1978, p. 21, for a description of computerized typesetting and video display terminals in newspaper publishing. These have become commonplace since the late 1970s.

Similar procedures are now standard practice in book publishing. These generally require retyping of an author's manuscript by a publisher for entry into a computer typesetting system, and thus are no more under the author's control than traditional noncomputer typesetting. Direct author typesetting as described in note 31 above has yet to become an industry standard. See Ray Walters, "The Coming of the Computer," *The New York Times Book Review,* July 24, 1983, pp. 12–13, for a positive appraisal of the impact of computers on the writing, production, dissemination, and storage of books. Two-fifths of all published authors used word processors in 1983, according to Walters.

35. Not everyone agrees that computerized storage and dissemination of print will replace paper. See, for example, market researcher Steve Pytka's comment quoted in the *Wall Street Journal* that we will achieve "the paperless office at the same time we see the paperless bathroom" ("Paperless Future?" February 27, 1986, p. 1). Pytka's assessment followed a survey of 50,000 owners of computerized paper printers.

The analogy is inapt, however, in that the physical substance and texture of paper is more crucial to the tasks of bathroom tissue than to the distribution of printed information, which as we have seen can be accomplished by media such as wall carvings and computer screens, having little in common physically with paper. Nonpaper wash-and-dry technologies may yet replace bathroom tissue—but the point is that the intrinsic qualities of paper are more essential to the bathroom-eliminative than to the cognitive-disseminative function, Freud's anal-written connection notwithstanding.

Chemist Joseph Zuckerman makes a comparable observation in a discussion about lap-top (portable) computers versus books, in which he tells of a reported letter from nineteenth-century German composer Max Reger to a critic who had described one of Reger's works as sounding like "clanking chains," dissonant and unsingable. "I am sitting in the smallest room in my house," Reger is said to have responded. "I have your review before me. Soon it will be behind me." Zuckerman adds that such a response would not have been possible with a critique communicated to Reger on a computer screen (from the computer teleconference "Connect Ed Cafe" on the Electronic Information Exchange System, New Jersey Institute of Technology, Newark, N.J., comment #539, June 23, 1986, 10:25 PM). See Robert Hendrickson's *The Literary Life and Other Curiosities,* New York: Viking, 1981, p. 206, for a printed reference for this story.

See Walter Ong, *The Presence of the Word,* pp. 92–110, for a comparison of Freud's anal stage of human development and McLuhan's view of scribal-print communities.

36. Electronic media that communicate literal sounds and images thus run contrary to Dewey's conception of communication as a separation of events from their "qualitative immediacies." However, one need not accept Dewey's view that absence of such immediacies aids the process of cognition (see note 9 above).

For example, McLuhan in contrast argues that the abstraction of events from their surroundings, and the logical analysis and organization of these abstractions, is a consequence of one particular type of communication—writing and print—and results in a mode of cognition inferior to the gestalt sensory understanding that McLuhan sees promoted by oral and modern electronic communication. (See my "McLuhan and Rationality," *Journal of Communication,* Summer 1981, pp. 179–188, for a detailed consideration of McLuhan's views on media and cognition, and *McLuhan: Hot and Cool,* ed. Gerald E. Stearn, New York: Dial, 1967, p. 270, for an interview with McLuhan in which he provides a statement of these views. See also note 19 above.)

My position is that cognition is furthered by consideration of events both in and removed

from their immediate surroundings—that is, I see printed abstraction and electronic immediacy as contributing in indispensable and complementary ways to the growth of knowledge.
37. *The Roots of Civilization.*
38. *What Is Cinema?,* p. 12.
39. The comparability of photographic and biological vision is hermeneutic—equivalence in literalness or noninterpretativeness of the image produced—and not necessarily an equivalence in accuracy. Thus, an image resulting from a machine developed by humans in the past 150 years will likely provide a representation less accurate in many respects (and more accurate in a few regards) than that provided by a living structure refined through eons of evolution. Indeed, biological vision systems differ among themselves in degree of accuracy and aspects of reality they report, although they share the relatively unmediated relationship with reality characteristic of the photograph. (All systems—biological as well as photographic—inevitably bias or mold the visions they provide. Lack of mediation refers to the general absence of idiosyncratic human interpretation intrinsic to such communication modes as speaking, writing, and painting.)
40. See chapter 5, note 21.
41. The cognitive value of motion photography was discussed as early as 1895 by E. J. Marey, who wrote that "the [motion photo]graphic method, with its various developments, has been of immense service to every branch of science. . . . Almost all vital functions are accompanied by movement, but any attempt to investigate them is beset with difficulty, for the majority are very complicated or very rapid, but it occurred to us that many of these problems could be solved by chromo[motion]photography. . . . By means of chromophotography we were enabled to make direct examinations of the movement of the heart by a more subtle eye than our own and one that is capable of grasping in a movement the sum total of changes which take place in the cavity of the heart." From E. J. Marey's *Movement,* 1895, cited by Don Carlos Ellis and Laura Thornborough, *Motion Pictures in Education,* New York: Crowell, 1923, p. 13.
42. See Mast, *A Short History of the Movies,* New York: Pegasus, 1971, pp. 23–25, for brief descriptions of the work of Eadweard Muybridge and E. J. Marey, two scientists whose investigation of motion in living organisms created the first motion photographs in the 1870s (Muybridge) and 1880s (Marey). Muybridge's *Animal Locomotion* (1887) was republished by Da Cato Press, New York, 1969; see note 41 above for more on Marey.
Although most inventors have serious aspirations for their creations, inventions often make their initial entrance into society as novelties, gadgets, and playthings. Only later are their scientific and possible artistic applications realized. See my "Toy, Mirror, and Art: The Metamorphosis of Technological Culture" (1977) reprinted in L. Hickman, ed., *Technology, Philosophy, and Human Affairs,* College Station, Tex.: Ibis Press, 1985, pp. 162–175, for a detailed theory of this life-cycle of technologies.
43. This equation of the natural and the efficient is consistent with the cybernetic view that information transmission obeys similar principles whether in biological or technological contexts (see, for example, Norbert Wiener, *Cybernetics,* New York: MIT and John Wiley, 1948/1961). The equation runs contrary to views such as Jacques Ellul's (for example, *The Technological Society*) that pursuit of the efficient via technology engenders profoundly unnatural, artificial environments. See "The Burden of Rational Technology" in chapter 9 of the present volume for more on Ellul.
See also "Anthropotropic Media" in chapter 7 of this volume.
44. Term coined in my *Human Replay.*
45. Written exchanges conducted instantly via computer conferencing among participants situated in their homes also have much of the fluidity of in-person conversation. See "Communities of Representations" in chapter 8 of this volume and articles by Michael Spitzer, D. J. Pullinger, and myself (P. Levinson, "Marshall McLuhan and Computer Con-

ferencing") in the March 1986 *IEEE Transactions on Professional Communication* (special issue edited by Valarie Arms) for further discussion of the style of computer conferencing communications.

46. The direct contact between the telephone's earphone and the listener's ear brings the speaker's voice into intimate distance with the listener—only lovers enjoy such mouth-to-ear proximity. The telephone thus not only personalized but made intimate aspects of human communication.

See Man-Kong Lum, "A Study of the Impact of the Telephone on Human Sexuality," M.A. thesis, The New School for Social Research, May 1984, for consequences of this aspect of the telephone.

47. Marconi's wireless was intended and implemented as a radio-telegraph, that is, for the sending of Morse code via electromagnetic carrier waves. The Canadian Reginald Fessenden and the American Lee de Forest independently (and with mutual animosity) pioneered the sending of human voices via carrier waves. Fessenden's work (starting 1900–1902) predates de Forest's by a bit (1906), and eventually focused on the possibilities of "radio-telephony" to the extent of designing a system in which a voice originating on telephone was transmitted to a radio station via telephone line, broadcast, received at a distant radio station, transferred to telephone lines at the distant area, and then transmitted via those telephone lines to a person's home or office (1906–1907). De Forest is usually credited with being among the first to realize the public broadcast potential of radio (1909). See John V. L. Hogan, *The Outline of Radio,* Boston: Little, Brown, 1923, pp. 19–21, for a description of Fessenden's work, and G. Chester et al., *Television and Radio,* 4th ed., Englewood Cliffs, N.J.: Prentice-Hall, 1971, pp. 22–25, for a discussion of de Forest and Fessenden. Chester et al. point out that a serious drawback to radio as a wireless telephone was the lack of privacy of electromagnetic broadcast media. Key to the development of radio as a public mass medium was the recognition that "radio's very lack of secrecy was its great commercial strength" (p. 24).

AT&T's attempt to define both radio and television as forms of "telephony"—so as to exclude these under AT&T's government-endorsed monopoly from development by other companies—also contributed to the public's early perception of radio and television as wireless telephones and picture telephones. AT&T relented in 1926, but the picture telephone transmission of Herbert Hoover from Washington to New York in 1927 via a device developed by Herbert Ives of AT&T's Bell Laboratories was one of the most dramatic accomplishments in the early development of television. Ives' mechanical scanners were soon replaced by the electron gun (iconoscope) of Zworykin and his associates, who worked for AT&T's early rival, RCA. See Eric Barnouw, *Tube of Plenty,* New York: Oxford University Press, 1975, pp. 40–50, for AT&T's early work in radio and television, and David Phillips et al., *Introduction to Radio and Television,* New York: Ronald Press, 1954, p. 22, for discussion of the Hoover two-way television broadcast.

48. Jonathan Miller, "The Global Picture," *Channels 1986 Field Guide,* 1985, p. 16, states that 99 percent of homes in the United States and Japan and more than 90 percent of homes in Western Europe receive television. Steve Behrens, senior editor of *Channels,* indicated in a telephone interview that data on television penetration in non-Western parts of the world is less reliable. However, Behrens says a variety of sources provide the following figures: 75 percent of homes receiving television in Brazil, and approximately 10 million and 6 million television sets in the People's Republic of China and India, respectively. Although the last two figures are only a fraction of the total number of homes in these two countries, the habit of viewing television in public places rather than in private homes in much of the Third World suggests that these televisions reach greater numbers of people than equivalent numbers of television sets in the West.

49. The strengthening of political power in the age of radio was evidenced not only in

Nazi Germany, but also in the Western democracies of Britain and the United States. Churchill and Roosevelt were both masters of the radio address, and used this capacity to become the dominant political leaders of the twentieth century.

The visual accessibility of television apparently defuses much of the mystique of radio, and political leaders since the rise of television in the 1950s have in general not had the same sway with constituents as did their radio elders. See Joshua Meyrowitz, *No Sense of Place*, New York: Oxford University Press, 1985, chapter 14, for a discussion of television and the decline of the political hero. See also McLuhan, *Understanding Media*, chapter 30, for the view that Hitler's "political existence" is "directly owing to radio and public address systems." McLuhan also speculates here on the "cooling" effect of television upon political images.

See Jarrell Jackman and Carla Borden, eds., *The Muses Flee Hitler*, Washington, DC: Smithsonian Institute Press, 1983, for accounts of the cultural impact on England and America of artistic and scholarly emigration from Nazi Germany.

The cognitive impact of social transformations—including the difference between one-way media such as television and two-way technologies such as telephone and computer conferencing—is examined in detail in chapter 8 of the present volume.

50. The A. C. Nielsen Company of Chicago provides the following statistics for May 1986 American television viewing: for noncable homes: average of 32.2 hours watching commercial network television, 10 hours commercial independent television, 1.2 hours public-educational television; for pay-cable homes: 27.7 hours commercial network, 12.4 commercial independent, 8.3 cable, 1.3 public-educational.

51. Jerry Mander's *Four Arguments for the Elimination of Television*, New York: Morrow, 1978, pp. 348–349, contends that television "inhibits cognitive processes . . . is a form of sensory deprivation . . . disorients a sense of time, place, history, and nature . . . suppresses and replaces human creativity . . . limits and confines human knowledge," and so forth. Neil Postman has presented a somewhat more moderate line of argument that television is disruptive of traditional contemplative rationality—see his *Teaching as a Conserving Activity*, New York: Delacorte, 1979; *The Disappearance of Childhood*, New York: Delacorte, 1982; and *Amusing Ourselves to Death*, New York: Viking, 1985.

Typical of the "evidence" Mander offers in support of his claims is "Ten kids were asked to watch their favorite television programs. . . . They just sat back. They stayed all the time in alpha [a quiescent brain state]. This meant that while they were watching they were not reacting. . . ." (p. 210). But who were these "ten kids"? How old were they? Was there an experimental control group? Did the presumed evidence of mental impairment (the alpha state) continue after the children had finished watching television? Mander provides not a clue. See my review of Mander's book in *The Structurist*, 1979/1980, pp. 107–114.

Postman takes a deductive approach. He starts with the premise that people seem to have trouble thinking clearly these days, and seeks to demonstrate that conditions of television viewing are responsible for this sad turn of events. Nowhere, however, does he present evidence that capacity for rational thought (and quality of life) was or is palpably superior in nontelevised societies. (To the contrary, the absence of television in the Union of South Africa as late as the 1960s and the quality of thought and life in that society at that time offers one of numerous examples of the "crimes" of television committed in places where television never trod.)

See also Gene Maeroff, "Reading Achievement of Children in Indiana Found as Good as in '44," *New York Times*, April 15, 1978, p. 10, for evidence that contradicts the widely-held view that reading ability has declined since the advent of television. See Flora Schreiber, "The Battle Against Print," *The Freeman*, April 20, 1953, for an anticipation of the widely-held view.

52. As of the end of 1985, total audience share of American network television had

dropped 13 percent since the 1970s, reports Ben Brown, "Bracing for the Aftershock," *Channels 1986 Field Guide,* p. 22. Cable television in the same period penetrated 46 percent of the 86 million American homes with television (Theresa Izzillo and Jeffrey Wolf, "Banking on a Windfall," *Channels,* p. 36), and nearly 30 percent of the 86 million television homes were equipped with home video recorders (David Lachenbruch, "From Gizmo to Household Word," *Channels,* p. 74).

53. According to Jonathan Miller, "The Global Picture," *Channels 1986 Field Guide,* p. 16, "Moscow has a lively market in 'samizdat' videos." Decentralized photocopying technologies have for many years been a primary vehicle of Soviet dissentniks, and officially restricted. (The new "glasnost" or openness policy of Mikhail Gorbachev may change this.) See Andrew Nagorski et al., "Moscow Faces the New Age," *Newsweek,* August 18, 1986, pp. 20–22, for discussion of Soviet attempts to cope with personal computers, Xerox machines, satellite television transmissions, and other decentralizing media. See also Annette E. Dumbach and Jud Newborn, *Shattering the German Night,* Boston: Little, Brown, 1986, for a sensitive account of how six heroic people ("The White Rose") worked against the Third Reich from within with the help of a photocopying machine.

54. See my *Human Replay,* pp. 290–294 ("The Reunion of Talking and Walking") and the sources cited there for discussion of possible electromagnetic propulsion systems that could transport people from New York to Los Angeles in seven minutes. See also note 19 in chapter 8 of the current volume.

55. Nathaniel Hawthorne, *The House of the Seven Gables,* New York: Collier, 1962 (1851), p. 239; said by Lewis Mumford in *The Pentagon of Power,* New York: Harcourt Brace Jovanovich, 1970, p. 314, to be an early recognition of the totalitarian effects of electronic communication.

56. The designation of what we now call physics as "natural philosophy" into the nineteenth century shows the continuing close connection of science and philosophy until almost the present day. The connection goes back to Thales, who as far as we know was the first philosopher to seek support for his theory of existence—that water is the basic element of all existence—by logical and empirical rather than mystical means.

57. By "transempirical" I mean that each human mind is a function of an individual human empirical brain, but transcends this specific empirical basis by producing ideas embodied in empirical technologies outside the originating human brain, and comprehensible by other empirical human brains. Thus mind is never nonempirical—it always has some sort of empirical association—but it intrinsically goes further than its origins in a given human brain.

58. The converse explanation of physical reality as wholly a function of mind or sensation by some philosophers and by variants of quantum mechanics in the twentieth century is equally regrettable, and has been termed a "Dark Age" of philosophy by psychologist R. L. Gregory (cited in Bartley, "Philosophy of Biology versus Philosophy of Physics," p. 58). See also chapter 1, note 7 and chapter 4, note 2 in the present volume.

59. Kant concluded that scientific knowledge or even objective perception of the mind was not possible, since "we cannot obtain the least representation of a thinking being [the self] by means of external experience, but only through self-consciousness," *Critique of Pure Reason,* p. 236.

60. Kapp, *Philosophie der Technik.*

61. Freud writes in *Civilization and Its Discontents,* pp. 37–38, that technologies not only extend motor and sensory functions, but are materializations of fundamental psychological processes (for example, photographs and phonographs are expressions of our powers of recollection and memory). See my *Human Replay,* pp. 38–46, for further discussion of this aspect of Freud's thinking.

Heidegger holds that technological activity "reveals" otherwise recondite workings of the human mind and its relation to nature. See chapter 4, note 20 for Heidegger's writings on technology and principal discussions of these. See also chapter 9 note 5.

See also Colin Cherry (in Edmondson, p. 111), who agrees with Giambattista Vico that "in order for us to really understand anything . . . it is necessary that we first should have made it" (see note 4 in Chapter 8 of the present volume for more on Vico). I agree with this view insofar as "making" includes observation in the active Kantian sense—that is, although our production of technologies facilitates our understanding of them and ourselves, we may nonetheless understand phenomena that are natural or not of human origin via observation. (Because this observation is inevitably filtered through language and often through more tangible tools, we may say that the observation is "technological" in this sense. See "Productive Knowledge" in chapter 4 of the present volume.)

62. Margaret Boden's *Artificial Intelligence and Natural Man,* New York: Basic, 1977, provides a comprehensive survey of the many areas of human cognition investigable via computer programs. Wiener's *Cybernetics* was one of the first to offer some intriguing examples in this area, such as a comparison of "neurotic" programs and human symptoms.

Chapter 7

Socratic Technology: Media as Mirrors of the Mind

In the light of day the external world presents us with all manner of shapes, sizes, structures, and colors to engage our senses and our intellect. In hot pursuit of knowledge about this external array, bathed in its wonders and beckoned by its intricacies, we are prone to pay little attention to the nature of the endowment that permits us, presumably alone among the creatures of the cosmos and certainly of Planet Earth, to so appreciate and reflect upon such visions. Kant's teaching that everything we find in external reality is as much a projection of ourselves as the world pales beside the vivid testimony of our senses, as indeed any immaterial idea, however powerful, will likely do.

In the evening, however, the blackness offers far fewer rewards to inquiring eyes, and our gazes at the heavens quite naturally tend to be reflected back on their sources. "The stars do not shine, our minds do," Henryk Skolimowski has recently written in support of Kant,[1] and the location of such an arresting statement in the nighttime, when the relative simplicity of perception facilitates perception's self-analysis, is no accident.

The darkness and standard star patterns of nighttime skies have no doubt served as blank screens on which humans have sought a clearer image of their internal selves since the time that humans first began thinking of themselves. The technological transformation of night into day, first through fire and most recently through electric light, in one sense has brashly distracted from such nocturnal contemplation, greatly extending our exposure to the viewable external world and its delectations. And yet the motion picture, television, and computer screens at which we gaze bear an unintended epistemological affinity to the screens of the evening sky, for unlike any natural event viewed under natural or artificial light, the media screens and technological processes by which they operate are material as well as perceptual products of human mentality, and thus repositories of the human selves that made them. In fact, we may find that the combined human origin and tangibility of these media offer representations of the self and tracings of our cognitive processes far clearer than what we may see in evening skies or when we close our eyes, and that the detachability and externality of these technologies make their clues to the mind more amenable to empirical study. In this chapter we consider such possible technological contributions to the age-old Socratic goal of knowledge of the self.

ESCAPING THE CELL OF SUBJECTIVITY

Because each and every technology, as we have seen throughout this volume, is a material embodiment of some human idea or ideas, all technologies are externalizations of human thoughts in which and through which we may read the thoughts and ideas that the technologies embody. A chasm of difference, however, persists between thoughts and ideas on the one hand, and the minds and cognitive processes that produced these thoughts on the other; in order to read the minds and cognitive processes which produce thoughts, we thus must encounter as our "book" a very special type of technology. This is a technology which embodies not only an idea but an idea that extends, reflects, replicates, or replaces mental functions and cognitive processes, bringing such capacities into the light of day where they may be more objectively examined.

The kind of technology suitable for such cognitive lessons is actually quite rare, even among technologies that extend or contribute to the knowledge processes. Telescopes, for example, extend our perception without externalizing the cognitive processes which make sense out of these perceptions; photographs extend our perception in a way that externalizes our memory, but not our active thinking processes. The lenses of the telescope and especially the camera, to be sure, replicate and thus externalize aspects of our visual system, and therein may provide valuable insights into the workings of human (and other organic) sight. But the eye and most of its attendant neurophysiology are already part of external reality vis-à-vis our internal cognition, and thus, unlike our internal cognition, are in no drastic need of technological externalization in order to be objectively observable.

In what sort of media, then, are we to find projections of the knowing agencies? The difference between still and motion photography provides an instructive example. The still photograph is a simple replication or recording of an image as the eye sees it; motion photography, in contrast, is actually a series or sequence of still photographs, presented to the eye so rapidly that the human mind unconsciously weaves them together into a single, continuous, connected moving sequence. Perception of the motion picture thus requires not only the straightforward act of vision necessary for perception of still photography, but also demands a further act of mental connection, an act which goes beyond the eye to draw upon processes we would commonly associate with the province of the mind. The mind is hence a constituent of motion photography in a way that is far more active and prominent than the necessary mental foundation of still photography (and, of course, all other modes of perception). In a manner profoundly unlike the snapshot, recorded voice, and virtually all perceptual media, the motion picture and its video equivalent are incomplete and underconstituted on their own, kits for perception in need of assemblage by the human mind.[2] An examination of how we see motion pictures may therefore uncover elements of the mental processes which make such perceptions possible.

Although motion photography was in fact first invented in attempts to scientifically understand the nature of motion and its perception, the subsequent development of motion pictures has been mostly devoted to the purposes of entertainment;[3] the use of motion photography to elucidate the workings of the mind is thus another classic case of unintended cognitive benefits deriving from a technology. The situation may be put as follows: motion pictures replicate motion to entertain (and communicate factual information); in order to replicate motion on the screen, motion pictures must call upon metaperceptual, cognitive processes which we use in the perception of motion events in real life; motion pictures thus provide an artificially constructed and hence more knowable external forum for the study of cognitive processes which might otherwise be hidden in our perceptions of real life. In short, motion pictures in pursuit of noncognitive goals replicate and externalize in a controlled and specifiable way mental processes which are so infused in everyday life as to be all but invisible in their natural habitat.

At the other end of the intentional spectrum we have computers, which from their inception and throughout their use have been conspicuously developed to enhance and externalize our cognition. Although expanding cognition and understanding cognition are two quite distinct processes, the difference is far less than the gap between entertainment and self-understanding via film: thus the computer is a likely place to look for a technological exposition of the soul. This indeed is just what has been happening in the past 30 years, as the field of artificial intelligence has asked with increasing perspicacity what the operation of electronic circuits can teach us about the mysteries of the protein-based brains that created them. Since a computer once constructed and programmed functions autonomously in many ways, computers are an extreme of externality and detachment from the human mind among extensional media, and thus are unrivalled as a source for the study of cognitive, information-sorting processes outside the human mind. Moreover, because computers and their programs, unlike motion pictures, are explicitly designed to perform cognitive tasks (for example, provide conclusions logically derived or chosen from given data), dissimilarities between computer and human cognitive processing may be as instructive about the workings of human mentality as similarities. Not only does a dissimilarity tell us what human cognition is not (in the Popperian falsificationist tradition), differences between computer and human processing raise intriguing questions about whether the discrepancy is due to the relatively early state of computer development, or to enduring, fundamental differences between the transmission of information via electronic switchery and biochemical tissue.

In what follows we consider both the increasingly studied area of the computer's lessons for the mind, and the scarcely mapped territory of motion pictures as a playground for mental functions. We start with motion pictures, in deference to the historical priority of their cultural impact. (Devices which incorporated principles later used in motion picture and computer technology were first developed in ancient times, and refined in numerous real, attempted, and hypothetical inventions through the nineteenth century; motion pictures attained cultural notice in the twentieth century nearly 50 years prior to computers, however.[4]) We conclude with further discussion of the question raised in Chapter 6 as to why we seem to replicate so much of ourselves in our technologies.

LIGHTS, CAMERA, COGNITION!

Persistence of vision—the perception of a quick succession of images as one continuous sequence—has been described by at least one film historian as a fortunate defect or idiosyncrasy of human vision upon which the earliest film makers chanced to light.[5] Although the process of preadaptation discussed in Chapter 3 suggests that many originally nonadaptive and even maladaptive traits prove useful in later circumstances, persistence of vision seems instead to belong to a different sort of evolutionary pattern in which a trait of apparently no value or even counteradaptive general value is discerned upon closer scrutiny to be part of a behavioral repertoire of long-standing benefit to the species. (See, for example, the recent suggestion that motion sickness is a side effect of an organic mechanism for the expulsion of ingested poison.)[6] Thus, persistence of vision allows us the illusion of continuous visual contact with external reality in a world in which innumerable necessary blinkings and split-second closings of the eyes for cleansing would otherwise cut our perception into a million pieces: what seems in the laboratory to be a mere trick of the mind or incapacity to see discrete units of reality as they actually are presented turns out in the real world to be a saving illusion which in fact improves our perception of the external world. Just as our inability to see at extremely close proximity to the eye enables us to see the rest of the world (capillaries on the surface of the eye would otherwise obscure our vision), our inability to see brief gaps between images gives us an uninterrupted view of the world despite the intrusion of blinking eyelids.

The question occurs as to whether what we see with our eyes wide open is actually an illusion—whether the world, like the motion picture, actually consists of discrete, motionless units which come upon the eye with such rapidity that we spin them into an apparently seamless web of experience. Such issues are not thoroughly decidable by technology (or any empirical or logical means), since, as Kant so clearly saw, knowledge of the external world can never go beyond the knowing system and its specifications. Thus, were our cognitive faculty on its deepest level incapable of computing stillness, insistently reading motion instead into everything it perceived, then no amount of extracorporeal technological detection, no slight of hand of slow motion photography, could ever possibly show us stillness in the world. Whatever data in favor of stillness such devices might uncover, our cognition would process this data as motion. (We nonetheless might imagine or in some abstract sense logically deduce stillness on the basis of such data.) However, since in fact we perceive stillness in the world as often as we perceive motion, we have no reason to suppose that our cognition is blind to stillness; indeed, the very fact that we can pose the question of whether motion is an illusion of stillness, and recognize the motion picture as an illusory presentation of still images, shows that we are not immune to the perception of stillness. Consequently, in principle we should not be blocked from discovering through some procedure that the everyday world we see in continuous motion was really a rapidly flipping sequence of unconnected, unmoving images. (An even less likely possibility is that motion and stillness are both fabrications of the mind imposed on whatever is actually out there. As has been suggested throughout this volume, however, evolutionary theory presupposes some minimal correspondence between external reality before humans appeared on the planet and what our cognition tells us about the external world—though the recalcitrant solipsist may consistently insist that evolution and reality are also illusions.)

In any case, the invocation of Kant in a discussion of motion pictures (or of motion pictures in a discussion of Kant) points to the profound issues put up in lights by this former back-door, penny arcade form of amusement. Interestingly, the Kantian–motion picture connection was first drawn when the "flick" was scarcely out of its penny stage, in Hugo Münsterberg's 1916 *The Photoplay: A Psychological Study,* generally neglected even by film theorists until its republication by Dover a decade ago.[7] Schooled in nineteenth-century German psychology and thus in Kantian epistemology, Münsterberg recognized that the motion picture, a product of human mentality to a far greater extent than our perceptions of the external world, offered a wonderfully pure and controlled environment in which to discern the operation and impact of our cognitive faculties of space and time. This insight can be explicated as follows: Meaning both in film and the external world is a product of the agent of perception (the human perceiver) and the objects perceived, but in the case of film the object of perception is itself a product of a human mind, or a filmmaker's product, before its perception by audiences. The seashore we perceive in a scene of a seashore in a film looks the way it does in large part due to human decision, whereas the seashore we perceive offscreen is usually a consequence of nonhuman causes. (Of course a seashore in a documentary will be by human intention more faithful to the literal seashore than a seascape used as background in a Hitchcock mystery. Note again also that a still photograph of a seashore will almost always be a more direct reflection of external reality than a motion picture of the same subject, except insofar as the movie captures the seashore's motion.)

The unencumbered expression in film of our cognitive capacities or organization of perception is especially the case for the filmmaker, who from scraps and snippets and sequences of images stolen from the real world creates a wholly new, self-contained, spatiotemporal world upon the screen. The filmmaker's perception of the world off the screen is also a product of our intrinsic cognitive powers of organization, but, because we have much less control over the images we encounter in external reality, our and the filmmaker's perception of the outside world is to a far greater extent determined by the nature of external reality, or at least determined in such a way that the operation of our internal cognitive structures is less clear than in the deliberate construction of a filmic world. Moreover, contrary to a strict interpretation of Kant, our cognitive imposition of relationships among external events must to some degree accommodate relationships that already exist among external events independently of the human mind, for how else could we discover different relationships among events with the same cognitive apparatus? (If events were objectively relationless, and relationships entirely a product of mind, we should expect to find the same relationships and properties in all events. Certainly in local perception of everyday events this is not the case, though we continue to search for fundamental principles that underly all local relationships.) In contrast, the filmmaker not only selects which events will be woven into the filmic narrative, but determines almost entirely the relationship among these events that constitutes the narrative, that is, which image will follow which other image, how rapidly, in what manner (such as through cuts or dissolves) and so on. Indeed, although some juxtapositions of film frames or "shots" can destroy a narrative illusion or be unamenable to narrative integration (such as a succession of abruptly different angles or shots of the same person talking, as in an interview), the early history of motion pictures is to a large degree a story of filmmakers weaving frames of greater and greater diversity—close-ups, long shots, side and above shots—into cohesive narrations.[8]

The arbiter of the filmmaker's vision is the film-viewer, who on the one hand has no choice in the images presented by the filmmaker, and far less input than the filmmaker regarding the relationships of events on the screen, but on the other hand nonetheless must ultimately make sense out of the filmmaker's projected relationships if the film is to be a successful vehicle of the filmmaker's ideas. The question of where the ultimate source of the film narrative resides has been hotly disputed in the years following Münsterberg, with montage theorists (usually in the majority) citing the filmmaker, and mise-en-scène theorists citing the viewer.[9] Münsterberg himself was generally silent on this issue, often blurring or ignoring the distinction between maker and consumer.

In terms of the epistemological issues which are the concern of this volume as well as Münsterberg's—as distinct from the ideological and social concerns which have been the usual agenda of film theorists—the difference between filmmaker and viewer is to some extent not even important, for the filmmaker and viewer alike have human minds, and the projection of either or both mentalities on the screen should expose elements of a same human cognition. Similarly, the question of whether the computer reveals the mind of the programmer or the user is not highly relevant to the central theme of this volume, since in either case fundamental aspects of human cognition presumably pertinent to programmer and user (and to all humans) are at work. Nevertheless, the question of what we might call "determination of the narration" specific to film has bearing in a very clear-cut way to the issue of external versus internal governance of perception and cognition, and thus is worthy of a bit of further consideration.

The dispute between montage and mise-en-scène advocates has been far-reaching and multi-leveled, encompassing practitioners as well as theorists of film, and entailing questions both of what film is and what it ought to be.[10] For the montagist, the essence of filmic experience is not the image or captured sequence of images, but the juxtaposition of images often unrelated to each other in reality, and the ideas and feelings that these manufactured relationships call forth in the viewer's mind. The action here is at the intersection of images—in the interstices rather than the images themselves—and the narrative comes from the viewer's negotiation of the intersections and inevitable attempts to fill in the gaps or draw conclusions from the juxtapositions.[11] But although the viewer thus may seem to have an active role in such an engagement with film, the montage filmmaker expects the viewer's negotiations to be largely unconscious, and predetermined or at least strongly suggested by the juxtapositions which the filmmaker has carefully designed. The successful montage film, then, is overwhelmingly an expression of the filmmaker's rather than the viewer's mentality, as the viewer creates the story that the filmmaker intended the viewer to create.

The mise-en-scène film is of course no less a paean to the filmmaker's ability to tell the viewer a story, but here the maker treads more lightly on the rhythms of space and time as they are found in the external world, and indeed tries to weave a story as much as possible with the threads of these external rhythms, through the capturing and mimicking of patterns of motion in the real world rather their reconfiguration. For the mise-en-scène devotee, the glory of film and essence of cinema lies in its capacity to simulate on the screen the seemingly uninterrupted, contiguous flow of images in the external world. Unlike the sudden cuts and rapid dissolves of montage, the mise-en-scène film advances the action through fluid turns, shifts in focus, lingering dissolves, and other techniques that enhance the viewer's illusion of seeing a single, continuous frame. When successful, the result for the viewer is a panorama somewhat like external reality, or, more modestly, like the experience of live theater in which the viewer is given a far greater choice of what to attend to, and hence more leeway and initiative in the construction of a narrative, than is the case in the dictated experience of montage.

However, an unqualified equation of mise-en-scène with external reality/viewer choice and montage with manufactured vision/filmmaker's expression would not be correct. The mise-en-scène viewer is given more to choose from than the recipient of montage, but these choices are nonetheless presented by the filmmaker, as indeed they must be if the film is to convey a story or even a point of view. Thus the mise-en-scène viewer picks out a story from an external reality purposely stacked by the filmmaker. At the same time, neither are the conventions of montage entirely at odds with perception in the external world, where quick turns of the head, for example, often yield visual experiences more akin to the rapid cutting of montage than the smooth uninterrupted flow of motion aspired to by mise-en-scène.[12]

Nonetheless, despite the occurrence of montage-like modes of perception in the natural world, and the inevitable element of filmmaker predetermination in mise-en-scène, the difference between these two types of filmmaking remains: the flow of information or arrow of the narrative travels from the outside in or top down in montage, and from the inside out or bottom up in mise-en-scène. Since what we see in the external world is presumably not a product of any deliberate external agency, the more realistic or natural of the two modes of filmmaking is mise-en-scène. Moreover, mise-en-scène accords well with the Kantian recognition that perception is largely self-determined, that is, preconfigured by the perceiving agent from the inside out, rather than imposed by an active environment on a passive mind as supposed by both the British empiricist and Marxist materialist traditions.[13]

Yet montage quite obviously works—indeed, rapid cutting techniques are used far more typically than lingering pans in Hollywood movies—and this success must be accounted for in terms of human modes of cognition. Ironically, the success of montage spotlights the active role of the human mind in our perception of nonfilmic external reality, for when we leap to conclusions from a sequence of two or more disparate, unconnected, rapidly presented images, our minds demonstrate the ability to see rich stories in the flimsiest, most incoherent of evidence. The montage filmmaker capitalizes on this effervescent quality of the mind—its penchant for telling us stories on the most minimal of promptings—and the ease with which such narratives are elicited suggests that the mind regularly makes a major generative contribution to perception outside as well as inside the movie house.

An enumeration of the mental activities displayed in the viewing of both montage and mise-en-scène film would constitute a whole discipline of filmic psychology, or at least of perception and cognition. Through the juxtaposition of various sorts of images (rapidly in montage, gradually in mise-en-scène), the viewer is moved from the present to the past and back again in the context of the narrative, that is, transported across time with the same facility with which film images transport us across space or places; through split-screen and similar techniques, the viewer is provided with a simultaneous perception of events occurring in different places (and/or times) within the narrative; through lingering superimpositions, the viewer is given to understand that he or she is witnessing some internal mental action of a character on the screen, for example, a character's recollection, imagination, or daydream. None of these viewer perceptions—seeing the past, looking at disparate events simultaneously, directly viewing another's mental experience—are literally possible in the real world, and yet the legerdemain of film somehow triggers such perceptions in the viewer, in a manner that nonetheless leaves the viewer aware on some level that what is being perceived is a film, not external reality.[14] A full investigation of precisely how and why certain sequences of images on a screen elicit these and numerous other sorts of perception would, I think, tell us epics about the workings of our perception and cognition. Thus far, however, this territory has gone mostly unmined, with little study beyond the initial claims made by Münsterberg.

As suggested earlier, the general neglect of film as a source of self-knowledge is probably due to the primary role of film in our society as a vehicle of entertainment, a role that renders any knowledge of cognition that we may derive from film radically unintended. In the case of computers, however, we encounter a technology which, from the start of its recent spurt of development after World War II, has been cited by some as a guidebook to human mentality—sometimes, as we shall see below, to an excessive degree. In shifting our search for knowledge of the self from film to computers, we thus change from a scrutiny of unintended consequences to what might be called the overintended consequences of a metacognitive technology.

ARTIFICIAL INTELLIGENCE AND REAL LIFE

I have so far written quite glowingly about the performance and promise of computers in both the primary generation of knowledge and its dissemination. For those unaware of the profound contributions made by computers to the processes of human cognition, or, worse, for those who view computers as a detriment to the cognitive enterprise, I would reiterate the conclusions of the past two chapters with full force and effect. In light of this optimism, however, the reader may now be taken aback by what I have to say about computers as models or replicators of human cognition: In a nutshell, I think that computers, as they have presently been designed, bear only slightly more resemblance to the autonomous, generative functioning of human mentality than a parrot's speech does to a human's. I think that the notion that computers "think," manifest "intelligence," or even communicate or are communicated to through "language" entails a fundamental distortion of these concepts as they are used to describe human experience, and that their application to computers is thus a misapplication or an illness of metaphor. And yet I find this very gulf between computer and human processing a wellspring for self-discovery—for in inquiring why current computers are by and large hands, not brains; extensions, not originators; assistants, not masters of cognition; we inevitably elucidate aspects of uniquely human mentality.

Let us begin with an examination of what seems to be one of the most indisputable capacities of computers, their use of language. On one level, nothing seems more obvious than that computers employ languages. Although computer lexicons such as BASIC, PASCAL, FORTRAN and operating systems such as CP/M, MS-DOS, UNIX and the like use varying amounts of mathematical symbols, they consist primarily of letters and even words taken from English; indeed these words often are used in a manner strikingly like their use in natural English (for example, when one types "safety" before turning off a Kaypro 10 CP/M personal computer to prevent accidental erasure of information). Moreover, mathematical symbols such as 1, 2, and 3 also play important roles, albeit not central, in natural languages. So why, then, does not the use of these words and symbols in a computer machine constitute a language—if not a complete linguistic alternative to English, then at least a new dialect in the sense of Cockney or Brooklynese?

Let us see exactly what happens when we type the word "safety" into a Kaypro CP/M computer. To start with, this command has meaning for the computer, that is, it can evoke the desired response, only when the command is typed into a computer that has been previously set to a prompt mode appropriate to receipt of such a command. Typing the word "safety" in the absence of such a prompt may result in responses of question marks, "unknown command," or most likely no response at all from the computer. Similarly, at the other end of the communication chain, once the computer has accepted the safety command, the machine can have but one response: its drives will be placed in a position such that the sudden loss of power resulting from the shutdown of the computer will not erase any stored information. Barring a malfunction, no other response, no equivocation, no disobedience is possible. (In more sophisticated systems, a computer may initially respond by flashing something like "Are you sure?" on the screen, to which a response of "yes" or "y" will result in the safety maneuver. These computer queries, however, are in no sense spontaneous, and are as inevitable in certain systems and circumstances as the simple safety response.) Such constraints on the acceptance and execution of commands demonstrate what really goes on when we type "safety" into our computer: a sequence of six electronic signals (produced by pushing the keys S-A-F-E-T-Y) react on one specific electronic state (the appropriate prompt mode) to produce another specific electronic state which causes the drive heads to change position.

In what sense can a computer operating in this fashion be said to have understood the word "safety," or responded to it in the way a human being might? If simple behavioral response were the sole criterion of understanding, then we might indeed reasonably attribute human-like understanding to a computer that moves its heads in response to a safety command, for such an action is clearly analogous to the response of a person who jumps back on the sidewalk on being advised of the approach of a speeding automobile. But human behavior is in fact far more complex than such analogies suggest. To begin with, humans need not be tuned to a specific prompt mode in order to receive or understand a safety command or any other command or communication. The human receiver of course must possess some capacity for hearing or vision as well as knowledge of the language in which the command is communicated, but such preconditions are analogous to the computer's general condition of being unimpaired and programmed for a specific "language," and are in no way approximate to the highly precise and particular prompt mode which is also an absolute precondition for the computer's acceptance of any command. Furthermore, once a command is understood by a human being, he or she has an infinitely greater latitude of response, including all manner of query, ridicule, refusal, and other modes of partial or noncompliance in addition to the full compliance which is the unimpaired computer's only possible response. Thus, both input and output ends of human language processing are characterized by a multidimensionality and flexibility radically at variance with the one-dimensionality and irresistibility of computer communication: once humans know a language, we may be totally surprised by a communication and still understand it; and having understood the communication, we may just as easily ignore or sneer at it as respond in the manner the communicator intended.

Nor is this difference reducible on the human side to an interesting but insignificant quirk of human language, or on the computer side via an increase of electronic circuits. To the contrary, the multiplicity or ambiguity of human linguistic reception and response, including the possibilities not only of deliberate disobedience mentioned above but an infinite variety of misunderstandings, is the very tissue of linguistic experience, arising from, among other things, the inevitable discrepancy between representation and original discussed in Chapter 6, and the endemic imperfection of evolutionary organisms both as perceivers and as performers discussed in Chapter 3. Whether one takes from these facts of language and life the nihilism of deconstructionism, the Peircean/Popperian optimism-from-fallibilism that I advocate and elaborate on in Chapter 3, or some intermediate position, one cannot ignore the ubiquity of noise, the undershooting and overshooting that nonetheless often touches part of the target, the error and room for growth that is the bane and hope of living and cognitive experience, and more specifically, a necessary precondition for rational and self-directed thought and action.

Examples of programmed, prompted behavior (as opposed to autonomous or self-willed thought and activity) abound in noncomputerized aspects of human experience, and serve to clarify the persistent difference between all computer activity and the self-directed core of human life. I was party to an especially vivid example in March 1986, when I had occasion to give a talk to a Japanese delegation at The New School for Social Research. The group of 15 spoke not a word of English, and every word that I spoke was conveyed to my guests by a translator. In my usual fashion, I could not resist sprinkling my talk with a goodly number of strained jokes, and I was delighted to see my audience chuckle and in some instances roar after listening intently to the translations. Indeed, so pleased was I at this demonstration of the transcultural quality of at least my humor, that I related the incident to a student of mine (whom I was instructing via computer network connection) in a phone call to Tokyo several weeks later. "I wouldn't be so pleased, if I were you," he offered after hearing my tale. "And why not?" I demanded. He proceeded to tell me of a talk he had attended in Tokyo by an American professor several months earlier. This professor also had a penchant for humor, and his sallies like mine were rewarded by laughter from the audience. But not, in this case, because of the translation ability of the translator: my student informed me that each time the American in Tokyo delivered a gem in English, the translator simply instructed the audience (in Japanese) that a joke had been made, and that either slight, moderate, or hearty laughter was called for.

The poignance of these incidents (my wounded pride aside—although I maintain that my translator indeed translated my jokes) stems from our conviction, likely shared by all humans, that however much laughter in its ordinary expression may be a product of prior experience and conditioning, it is nonetheless authentic in a way that laughter in response to being told to laugh surely is not. The observable behaviors in real and prompted laughters are presumably identical, but the behavioral criterion is clearly not enough. We require an evidence of self-expression in our judgements of authenticity, and although the nature of such expression is poorly understood, we immediately perceive its absence in the prompted laughter and the computer's programmed activities.[15] Painting by numbers is not art, and "thinking by numbers" is not thought.

But cannot a computer be programmed to make errors, operate ambiguously, produce unintended consequences, and function of its own accord? In terms of the current conception of AI programming—the "expert systems" approach, where for example a medical doctor's decisions are reduced to a series of rules which are in turn embedded in a computer program[16]—the answer is that a project to program for unprogrammed consequences is in principle self-defeating, for much the same reasons that a request to "be spontaneous" is inherently unfulfillable. Any errors that the computer made in accordance with such a program would not be errors, since in fact they were made in accordance with the program; ambiguous performances and unintended responses would necessarily be chosen from some predetermined menu of possible responses (embedded in the rules), and thus would not be fully unintended; and a computer programmed to operate on its own accord would be operating in accordance to the rules of the program, and thus in accordance to the programmer's intentions, and not the computer's. The problem here is that the ex post facto results of human intelligence, the skimming off the top of the consequences of human expert action, constitute the results of intelligence, not the intelligence itself: thus the construction of these results in a computer program at best rehashes intelligent action rather than creates the circumstances necessary for its generation. To achieve human intelligence, artificial intelligence will likely have to aim much lower—not at the consequences of intelligent action, not even at a literal rehearsal of the error–making constituents of intelligent action, but at a recreation of these constituents in a special environment that allows them to interact in a genuinely undirected and potentially self-transcending manner. In short, the only known suitable context for the emergence of intelligence—to date natural, perhaps tomorrow artificial—is the self-transcendent environment of life.

To be sure, electronic transactions of computers suffer no less than any other aspect of the physical universe from the tithe of entropy, and thus they break down, malfunction, and therein commit all sorts of unintended errors. Further, computers can be easily programmed to produce random results, and in this sense yield unintended consequences. Nevertheless, the reversal of entropy and noise in life that engenders intelligence entails more than mere entropy and noise: evolution is not random but blind, as Donald Campbell puts it,[17] meaning that emergence and self-transcendence require structures and organization that take advantage of the unintended possibilities of noise, and make the possibilities themselves something less (or more) than totally random. Mere noise, then, is not enough. Living organisms and especially humans have the capacity to generate and then exploit possibilities that while unintended are not entirely inapt (successful mutations and preadaptations in organisms; rationally developed theories in humans), and this shift from noise (entropy, randomness) to "noise-plus" (unintended structure) makes possible the process of evolution and progress from errors.

My thesis is that as a minimum requirement computers will need to be programmed with this "noise-plus" ability—poorly understood at present even in its natural living situation—if they are to achieve the boldest hope of artificial intelligence and function as autonomous, creative "thinkers." Today's computers work at the leading edge of the great human-living reversal of entropy when they operate in tandem with human intelligence, integrated as assistants and extensions into the human cognitive enterprise. On their own, however, these same complex wind-up toys collapse into the entropy that awaits most nonliving systems—a lack of generative capacity relieved only by the evolutionary emergence of life.[18]

The observation that current computer performance is in no meaningful sense the equivalent of human linguistic and cognitive activity is not new, and has been made in varying detail by Fred Dretske, Hubert Dreyfus, Douglas Hofstadter, and others in recent years. Some of these critics of artificial intelligence recognize that its limitations arise from its lack of situation in a living, naturally selective context—that human mentality is not an addition to but rather an outgrowth of living organisms, and that we think as we do because we are and were, phylogenetically, first alive.[19] These critics presumably would agree with Maurice Merleau-Ponty that a living body is a necessary precondition of mind.[20] In what follows, I will expand this view that computers cannot be intelligent because they are not alive by relating it to the theory of the general evolutionary origin of the mind which has served as the framework for this entire book; in so doing, I will part company with some of the more extreme critics of computers who claim that artificial intelligence can never be achieved, and will argue instead that we cannot have artificial intelligence until we have created artificial life, or created from the constituents with which we wish to obtain artificial intelligence (for example, electronic circuits) the mechanisms and processes we associate with living organisms.

In Chapter 4 we explored the strengths and weaknesses of Popper's three-part division of existence into "World 1" of nonliving and living material and energy, "World 2" of subjective human thought (the act of thinking, feeling, and so forth), and "World 3" of the objective products of human thought such as ideas. In particular, we saw that this schema is inadequate for the classification of human technology (which has an uncomfortable citizenship in two of these worlds, 3 and 1). I suggested an adjustment of Popper's divisions that would better accommodate technology: a "T-World 1" of natural non-living and living material and energy,

"T-World 2" of human beings (the active or decisive criterion of this world being the human brain or mind, which not only thinks but also contains uncommunicated and unmaterialized ideas), and "T-World 3" of human technology or material productions (including spoken words, which are highly abstract representations of ideas transmitted in the material of air). In addition to cleanly categorizing technology, such a revised schema has the advantage of working without the troublesome metaphysical concepts of unembodied mentality and unembodied ideas (ideas are either embedded in the brain of T-World 2 or embodied or carried in the material—again, including air—of T-World 3),[21] and of emphasizing the crucial roles of communication and production as liberators of T-World 2 brain-bound denizens into a T-World 3 inhabitance of the universe at large. A disadvantage of the "T" schema, however, lies in its crowding of all nonthinking natural entities (living as well as nonliving) into World 1 (a problem also present in Popper's original schema), with the result that a monkey resides much closer to a rock than to a human in this division.

The mixing of living and nonliving entities in one realm causes special problems for an assessment of artificial intelligence. Computers are obviously part of T-World 3; but are they, like automobiles, citizens of T-World 3 that function like the components of T-World 1 (nonliving material), or, as some artificial intelligence enthusiasts suggest, human productions which replicate the performance of T-World 2, the human brain/mind? The question of whether computers have made the jump from artificial material to artificial intelligence is difficult to consider from the standpoint of a model that fails to distinguish between living and nonliving material, for the lack of a clear category for natural life (in contrast to natural material) obscures the possible role that a category of "artificial life" might play in the development of artificial intelligence. The revised three-world schema thus could benefit from unpacking to a four-part model of existence as follows: natural, nonliving material and energy; natural living material (including many aspects of the human organism); natural thinking material (the human brain); and artificial existence (products of natural thinking material: human technology).[22] Posed in terms of this four-part schema, the question about artificial intelligence amounts to this: how many of the first three worlds have we humans been able to reproduce in the fourth realm of our artificial creations?[23]

In addressing this question, we need to be aware that the creation of artificial entities—be they material, living, or intelligent—entails a level of technology far more complex than the harnessing of preexisting material and life for human purposes. The opportunistic use of already existing forms no doubt constitutes a profound technology of sorts, as in the employment of horses for transport and even more so in their deliberate breeding for such ends, but such activities are no more a creation of artificial or technological life than the borrowing and hoarding of a preexisting flame is the creation of fire. As the movie *Quest for Fire* graphically demonstrates,[24] the human creation of a technology such as fire from scratch requires a proficiency and degree of civilization far greater than that attendant to the borrowing and preservation of a naturally created flame.

Bearing this distinction in mind, we may ask how much of the natural order of material, life, and intelligence we have succeeded in recreating through our technology. Although we have yet to create matter from absolute void, our bombardment of atoms with subatomic particles to create new elements and release huge quantities of energy, indeed our liberation of energy from matter in the ancient process of starting a fire, attest to our ability to create artificial material, or at least manipulate preexisting matter to the point where a new form has been fashioned. The situation with respect to artificial life is not as clear. The purposeful breeding of organisms for desired ends (what Darwin called "artificial selection" and Luther Burbank called "training plants to work for man"[25]), and, more significantly, the recent development of genetic engineering, certainly seem headed toward the creation of artificial life, especially when such techniques result in the production of viable original species; but even the most sophisticated modes of gene splicing, as well as the older forms of deliberate breeding and hybridization, thus far seem better described as a grafting or borrowing from a preexisting flame of life rather than creation of an independent spark.[26] A genuine creation of artificial life, I think, would entail creation of a living organism from nonliving material, or at least creation through gene splicing of an organism which most likely would not (and perhaps could not) have arisen on its own (that is, via nonhumanly intended causes) in the natural world. This last criterion, in accord with the view that artificial material has been created in the production of new elements which would not have arisen outside of special laboratory conditions (for example, high-speed particle chambers), is just beginning to be satisfied with the development through gene splicing of bacteria that manufacture human insulin.

Difficulties in the creation of artificial life underscore the immensity of the challenge of creating artificial intelligence, and the hyperbole of those who claim that this challenge has in any sense been met by today's computers. No human technology has as yet created life from the preceding realm of nonlife or called forth matter from nonexistence—the best that we have done is the creation of new forms of life within the realm of life and new forms of physical existence within the realm of the already existent—and yet some computer experts claim that we have created or are in the process of creating an intelligence, not from the preceding realm of life, but from a material, nonliving realm two steps removed from intelligence in the evolutionary order.[27] As far as we know, intelligence in the natural world has arisen neither as a disembodied entity nor a property of matter, but as a special characteristic of life, that is, a unique living strategy for coping with the environment. This strategy transcends its living origins by actively understanding its environment, understanding itself, and, through technology, dramatically changing its environment in deliberate and unintended ways, but this strategy is inextricably living nonetheless. Life in the natural scheme of things is thus a necessary substrate of intelligence in the same way as matter is a necessary substrate of life. The claim for intelligence in current computers thus amounts to a claim of having skipped or dispensed with the substrate of life, of situating an intelligence directly in a nonliving, material medium. In view of the fact that we have not yet fully succeeded in creating life from nonliving material, the claim that we have created intelligence from nonliving material is doubly dubious—an egregious case of putting Descartes before the horse.

Of course we cannot discount totally the possibility that somehow we have or will put together through electronic circuits a basis of intelligence—that protein complexes and flesh and blood are not necessary prerequisites of reason. Nor can we discount the less likely possibility that in building a presumably intelligent computer we have in fact created a living form—that protein complexes are not necessary for life.[28] These possibilities remain open to the extent that we lack complete and perfectly clear knowledge of life and intelligence; and as the fallibilism discussed throughout this volume suggests, these deficiencies will likely never be entirely remedied. A flat denial forever and anon of AI possibilities in nonliving circuits thus amounts to a counterfallibilistic and unbecoming protein chauvinism. But lack of complete understanding is by no means the same as no understanding, and in the case of life and intelligence we surely know enough of the real thing—such as the ubiquity of unpredictability and spontaneity mentioned above—to conclude that today's computers are neither intelligent nor alive. Current computer programs no doubt replicate and indeed greatly improve upon many aspects of human intelligence or cognitive functioning, but even an infinite series of discrete cognitive functions or electronically embodied perfect solutions to computational problems does not add up to the sloppy performance of the strange capacity with which I am now writing and you are now reading this book: the capacity which we call intelligence.

We know so very little of ourselves and our mentalities, but, assuming the general accuracy of the evolutionary perspective, one thing we do know is that intelligence is the centerpiece of our struggle for survival as a biological species. Like the living processes which it both transcends and is constituted of, intelligence is a product of millions of years of exposure to shifting environments, of countless command performances in life-and-death situations, of innumerable events, subtle and large, expected and not, in the funhouse of life on Earth. Like the human languages through which it is usually expressed, intelligence deals with these problems by saying both less and more than it means to convey, by inevitably failing in its intended tasks yet somehow succeeding beyond our wildest expectations. Computers know none of this. They neither die with their failures nor necessarily flourish due to their success. True, humans may "kill" an ineffective computer program or species of computer, or "propagate" a useful program or machine, but the decisive factor here is not the computer's performance but a human evaluation of it—evaluations which, for example, in the case of a computer company with a monopoly, might well result in the propagation of an ineffective or unfit machine. Were human life and intelligence ultimately governed by some supernatural authority in the way that humans dictate the survival of computers, then computer "intelligence" might indeed be analogous to the human kind. We humans, however, seem governed by the dictates of natural selection and its products (such as human intelligence), and this unpredictable, life-and-death, self-reflexive performance of human intelligence in the whirlwind of natural selection in no sense seems comparable to the performance of computers. Instead, computers are bred to work dependably, accurately, predictably; when functioning properly, they do exactly what was intended, never more or less (we may get more or less from the computer than we expected—for example, improving our writing style through use of a word processor—but this is a case not of the computer producing more but of our receiving more than we expected); even when computers produce unpredicted results, this is due to an oversight or glitch in the program, in effect an unnoticed aspect of the program which made the unpredicted result inevitable. In sum, today's computers are incredibly and brilliantly designed artifacts (not living or intelligent products) of human intelligence. A friend of mine, a new father, recently called me and quipped that he had just purchased a personal computer as a "high-tech" balance to his new-born daughter, who could do little more than cry. My friend had it wrong. Next to a new-born infant, even the highest of high-tech mega computers is unbreachably low-tech and stupid, immutably frozen in its ways. Even the lowly amoeba and its capacity for evolutionary if not personal self-transcendence tops the current high-tech computer in many respects.

Computers, then, with all their awesomely rapid processing of huge and tangled amounts of data, are no more thinking and intelligent than the intricately moving mechanical ducks that ingested and evacuated food in the eighteenth-century courts of Europe were eating and alive.[29] Notice that I am not claiming that humans will never create artificial intelligences—only that current computers are not intelligent; that this is so most likely because they are not alive; and that in order to create an artificial intelligence we must first create an artificial life in which the artificial intelligence can be situated. If we take protein as a metaphor for the self-replicative, self-transcendent diversity of life, we might say that we must make proteins from protons before neurons from neutrons. (Protein compounds have in fact already been created from nonliving molecules, but we have yet to create any sort of functioning life from these inorganic origins.) I suppose that those who shudder at the prospect of humans losing their uniqueness as intelligent beings to humanly created technologies may derive some small comfort from these arguments, for this critique of current artificial intelligence certainly suggests that any notion of benign or malign computers soon taking over or replacing us as a species is utterly absurd. Such optimism, however, is a trivial consequence of my thesis, whose thrust is not that artificial intelligence is impossible, but rather the far more optimistic view that artificial intelligence awaits the invention of appropriate preconditions. I thus have not argued against artificial intelligence, but against the trappings and superficial resemblances of intelligence in today's computers, in the hope that the identification of such false starts will help the development of a genuine, artificial life-based, artificial intelligence.[30]

A more immediate result of current attempts at artificial intelligence is this: in goading us to explain why these machines which seem to think in some ways like us also elicit in us a deep-seated, almost Cartesian conviction that these machines are nonetheless not like us, artificial intelligence obliges us to come face-to-face with the organic, evolutionary nature of our intelligence and humanity which our intelligence might otherwise take for granted. The ironic but happy consequence of programs designed to think like us is that they serve as mirrors of our mentalities even though they fail in their intended task.

But even the failure is by no means utter, and the trappings of intelligence in current computers are quite impressive in a variety of ways in their resemblance to aspects of human cognition. Consider, for example, the distinction between "ROM" (read-only memory) and "RAM" (random access memory) in the personal computer (whose cultural ubiquity gives thinkers an increasing opportunity to notice fruitful analogies between electronic and living data processing). ROM comes pre- or hardwired in computers; that is, it is built into the machine. It can instruct or control the operation of the computer (or at any rate set limits on what the machine can and cannot do), but cannot be altered or amended except by the transplantation of a new ROM component. RAM, on the other hand, is a blank slate. It can be written on, programmed, and replaced indefinitely—but its contents vanish when the power is turned off (and thus can be saved only by recording to a disk or printing). Does not the ROM-RAM contrast provide an apt analogy to Chomsky's distinction between genetically embedded rules of grammar or language organization (which are activated or not, but are not learned or acquired) and the learned vocabulary through which these embedded grammatical structures gain expression? Furthermore, are not the two Rs of computer cognition reminiscent of Kant's distinction between a priori mental categories and the knowledge they produce? At very least, such possibilities show that, contrary to denunciations of computers as dehumanizing of our self-image and understanding,[31] these cognitive adjuncts have the capacity to put us in closer touch with aspects of our own inner workings. Once freed of the misapprehension that computers as they now are constructed are the equivalents or competitors of human intelligence—and prepared for the lessons that this central negative comparison holds for an understanding of our own modes of intelligence—we can begin to explore the many positive analogies and their meaning long predicted by the field of cybernetics.[32]

The question remains of why the electronic circuits of computers should in any respect be so much like the living connections of human mentality. The resemblance in the performance of the products of human perception and cognition to human perception and cognition themselves is found not only in computers but in film, and in the increasingly lifelike performance of most communications media as discussed in the previous chapter. The explanation of this reconvergence of technology with its human parents lies as we have seen in the coincidence of the efficient (the general object of technology) with the natural (which is intrinsically efficient owing to the exigencies of natural selection), and taps into the underlying current of cosmic evolution in which the human mind and its technological emissaries are involved.

"ANTHROPOTROPIC" MEDIA

In inquiring into why technologies perform so similarly to their makers, we first must ask whether these technologies are indeed replicating some of the functions of living human systems, or whether we for some reason are merely interpreting the operation of these cold machines in human terms. Here as always we encounter the Kantian truth that all we ever can know of anything is what we perceive of it; in this sense what we see in technology is as much a reflection of our preconceptions as what we see in the natural world—indeed probably more so, since technologies are material products of human cognition whereas natural objects are not. Nonetheless, just as the subjective basis of our perception of the natural world does not prevent us from distinguishing reality from illusion (as in investigating whether a perception of one sense such as vision is corroborated by other senses), so the subjective quality of our perception of technologies need not prevent us from making meaningful distinctions about less and more humanly performing technologies. Thus, in the evolution of photography from black-and-white to color, still to moving pictures, and two- to three-dimensional images (holography), we have in the same humanly produced, humanly perceived domain a shift of machines from less to more correspondence with human modes of perception, which entail color, motion, and the third dimension.

Since we have no reason to believe that our perception of these technologies has changed for any cause other than that the technologies themselves have changed, we are entitled to conclude that the shift from black-and-white to color and so on is a real, objective, external increase in the human-like performance of photography. Similarly, were we to construct computers which, in addition to answering yes or no or maybe, sometimes answered no but meant yes, or, better yet, sometimes answered no—but just as a joke!—we could say that such computers were more human in their performance than today's models.

Granting, then, that our technologies are indeed performing more like we do—in a real, ontological sense that transcends our mere interpretation of technological performance—we return to the question of why. On the psychological level, the answer may be simple narcissism: we invent technologies that mirror our modes of perception and cognition because we so very much enjoy seeing ourselves, or aspects of ourselves, whenever we can. However, just as the psychological pleasure of sex serves the deeper biological function of procreation, so the pleasure of figuratively seeing ourselves on television, of starring in our own movies, may work on behalf of a deeper survival mechanism. When we design a technology such as color photography that performs in a human, lifelike way, we are borrowing a design from nature—in this case, sensitivity to various wavelengths which permit us to make fine discriminations of objects, that is, perception of color—which has worked well for millions of years and probably contributed to our survival as a species in at least some small way.[33]

Most visual and acoustic commercial entertainment and consumer recording media are purposely designed to be as lifelike as possible, but technological development in general usually has little to do with narcissism, and is directed instead toward performance of tasks as efficiently or effectively as possible without regard to how lifelike such performance may be. Yet precisely because our natural modes of functioning are necessarily quite efficient—were they anything less, neither they nor we would have survived—our pursuit of the efficient in technology often amounts to a pursuit of the natural. Indeed, on occasion the most sophisticated of metacognitive technologies have been suggested by the mundane mechanics of nature, as in the case of a multimirrored x-ray telescope inspired by the honeycombed reflectors of the lobster's eye.[34] Thus, whether the ostensible goal of information-processing technologies is the narcissistic pleasure of seeing ourselves perform in our own extensions, or the functional agenda of getting a job done as effectively as possible, the result often replicates a pattern of information processing pieced together during organic millennia—some ancient mode of interacting with the environment which has played a part in the development of life and human life on Earth.[35]

Attainment of the ancient and natural through pursuit of the new and artificial may have deeper meaning still. Existence as far as we know it seems to have developed in a sequence of void, matter (and energy), life, intelligence (presumably only human), and now the tangible expression of human intelligence in technology, with each level unstable in the sense that it gave rise to the next, yet stable in that subsequent levels have built upon and incorporated rather than replaced or obliterated prior levels. The placement of human technology at the current summit of this cosmic sequence reflects our technological injection of a deliberate materialized structure into a heretofore blindly evolving universe; the cosmic provenance of technology is meanwhile attested to by the presence of prototypical technologies at the most rudimentary, prehuman levels of life. Seen in this widest of possible perspectives, the replication of human patterns of communication in technology has a significance that transcends the human realm: for if humans are indeed the stuff of the cosmos examining itself, as Carl Sagan has so aptly put it,[36] then one would expect technology, as the stuff through which most of this self-examination is accomplished, to perform in a way that befits its ultimate origins. In scrutinizing our technological productions for traces of ourselves, we thus peer into a fissure that goes straight back to the origin of the universe; we find that the discovery of ourselves in our technologies is a piece of a much larger and older cosmic process of self-discovery and growth in which the universe discovers itself through life and through us. These and other cosmic implications of technology will be the subject of the volume's final chapter.

First, however, we turn to a realm halfway between the individual and the cosmos: the social dimension and the group contribution to knowledge. Thus far in this book, we have for the most part considered knowledge as a product of individual minds, each a consequence of eons of cosmic and biological evolution and each a potential for changing the cosmos and the course of evolution through the material implementation of ideas in technology. This emphasis on the individual is due to the fact that the proximate source of any and all knowledge, the instant of an idea's conception, is always an individual mind. Nonetheless, the social conditions in which the individual mind functions can facilitate, hinder, or otherwise condition the mind's production of knowledge; and as we already have seen, the criticism and dissemination of knowledge is inevitably a group activity. The technological contribution to the communication of information, and thus to the criticism and dissemination of knowledge, was examined in detail in Chapter 6. We turn now to a brief examination of a broader level of technological/social interaction, and consider how the technological impact on the formation and function of groups has in turn influenced the generation, criticism, and dissemination of knowledge. In assessing the technological shaping of what might be called the social preconditions of knowledge production, we will frequent the haunts of restaurants, ride in automobiles and jet planes, and call again upon our old friends the telephone and the computer.

NOTES

1. Henryk Skolimowski, "Evolutionary Illuminations," *Alternative Futures*, Fall 1980, p. 34.

2. The video process is the more radically underconstituted, for unlike the motion picture's series of individual still images punctuated by blankness, the video image consists of a series of constantly moving dots or points of light (electrons shot from the cathode ray tube), and thus can never really be reduced to a sequence of complete constituent images. (Hence the greater difficulty of obtaining a video "freeze" frame.) Although these differences have led to divergences in editing technology (physical splicing in film versus electronic editing in video) and consequent distinctions in production styles and values, video for the unschooled viewer is for all intents and purposes a filmic experience, and thus we can include video in the following general discussion of the cognitive dynamics of film in this chapter. (Production values may further differ in films made for theatrical release as opposed to television broadcast—as in the degree of explicit sexual activity portrayed—but the basic issues of recording versus rearrangement of reality, producer versus viewer "creation" of the story, and so on, remain the same. These are considered in detail in the next section.)

On the cognitive difference between still and motion photography (including video), compare Stanley Milgram's observation that we "take" a still photograph ("The Image-Freezing Machine," *Psychology Today,* January, 1977, p. 52) with the idiom that we "make" a motion picture.

3. See chapter 6, notes 41 and 42 above, and Mast, *A Short History of the Movies,* New York: Pegasus, 1971. Commercial documentaries and, to a lesser extent, historical dramas are exceptions whose pursuit of entertainment has been accompanied by varying degrees of educational motive.

4. Simulation of motion in painted and drawn figures presented to the eye in rapid succession dates to Heron of Alexandria (ca 200 BC), and appeared in the Zoetrope, Phenakistiscope, Praxinoscope, and hundreds of other exotically named "wheel of life" devices throughout the nineteenth century. See Robert S. Brumbaugh, *Ancient Greek Gadgets and Machines,* New York: Crowell, 1966, for details of Heron's device; and Mast, *A Short History of the Movies,* pp. 19–21, for brief descriptions of the nineteenth-century precursors of motion pictures.

See also Edward Wachtel, "The First Picture Show: Cinematic Aspects of Cave Art," paper presented at the Fifth Conference on Culture and Communication, Temple University, Philadelphia, March 24, 1983, for a daring assessment of 10,000 year-old cave art in Europe as motion pictures. (Wachtel bases this hypothesis on the light conditions in the interior of the caves, which he suggests could create a flickering environment suitable for the illusion of motion in the painted and carved images.) See also Siegfried Giedion, "Space Conception in Prehistoric Art," in E. Carpenter and M. McLuhan, eds., *Explorations in Communication,* Boston: Beacon, 1960, p. 79, for a similar though less explicit suggestion along these lines. Also see Jean Charnot, *Art from the Mayans to Disney,* New York: Sheed and Ward, 1939, ch. 25, "From Altamira to Disney" for a discussion of ancient animation.

Computers are often traced to the Chinese abacus (ca 3000 BC). Mechanical adding machines and other nonelectronic calculation devices date to the work of Pascal and Leibniz in the seventeenth century, and to Charles Babbage's unrealized plans for a computer-like "analytical" engine in the nineteenth century. As J. David Bolter points out in *Turing's Man,* Chapel Hill, N.C.: University of North Carolina Press, 1984, pp. 32–33, Babbage's vision required an electronic facility not possible in the mechanical gears of his age. See Archibald Williams, *The Wonders of Mechanical Ingenuity,* Philadelphia: Lippincott, 1910, ch. 2, "Calculating Machines" for details on mechanical computing devices including cash registers through the nineteenth century; and Herman Goldstine, *The Computer: From Pascal to von Neumann,* Princeton: Princeton University Press, 1972, for the computer's development in modern times.

The need for supportive related technology in the actualization of inventions—Babbage's missing baggage—is explored in depth by Roger Burlingame, who explains Leonardo's failure to realize most of his inventions as due to a lack of "collateral" technology ("The Hardware of Culture," *Technology and Culture,* 1, 1959, p. 16). See chapter 6, note 16 of the present volume for a similar example in the Chinese failure to implement the printing press as a mass medium. We might say that the lack of a phonetic alphabet was to the Chinese printing press in the tenth and subsequent centuries as the absence of electronics was to Babbage's analytic machine in the nineteenth century (at the time of Babbage's work, the relatively simple telegraph was the only extant electronic application). The lack of a placeholder in the Roman numeral system may have handicapped the development of technology two thousand years ago in a similar way. See the brief discussion in "Computers and Tractable Immensity" in chapter 5 of the present volume.

Bolter's observation that "Babbage's blueprint and disassembled parts could not change the world" (p. 33) is also consistent with the distinction between unimplemented ideas and material technologies made by Marx, Dewey, and the thrust of the present volume (see especially chapter 4).

5. Talk given at The New School for Social Research by Gene Stavis, Director of the New York Cinemathèque, 1975.

6. Michel Treisman, "Motion Sickness: An Evolutionary Hypothesis?" *Science*, July 29, 1977, pp. 493–495.

7. *The Film: A Psychological Study*, New York: Dover, 1970 (1916).

8. Early (silent) modes of film narration are in many respects epitomized in the works of D. W. Griffith. See Mast, *A Short History of the Movies*, chs. 3–4 for descriptions of film pioneers Méliès, Porter, and Griffith.

9. Sergei Eisenstein's *The Film Sense* and *Film Form* (collections originally published in the 1940s containing previously written essays; republished as *Film Form and The Film Sense*, ed. and trans. Jay Leyda, New York: Meridian, 1957) contain masterful arguments for montage from montage's master filmmaker. Film theoretician André Bazin's *What Is Cinema?* makes an equally masterful case for mise-en-scène. See G. Mast and M. Cohen, eds., *Film Theory and Criticism*, New York: Oxford University Press, 1974, for snippets from these and related species of film theory.

10. Alfred Hitchcock's works embody the best of both traditions. The last scene of his *North by Northwest* (1959) cuts from a shot of Cary Grant pulling Eva Marie-Saint up into the berth compartment of a train to a long shot of a train roaring into a big dark tunnel. For non-pornographic-film audiences of the 1950s, this was the most explicit mode permissible for the depiction of sexual intercourse. Hitchcock's 1941 *Suspicion*, however, deliberately avoids fast cuts in a sequence of Cary Grant (a suspected wife-murderer) carrying a possibly poisoned glass of milk up a flight of stairs to a bed-bound Joan Fontaine. Instead of shifting frames, Hitchcock keeps the camera mercilessly focused on Grant climbing up the stairs with the glass of milk (really a glass painted white into which Hitchcock had placed a glowing candle to draw viewer attention). The unbroken mise-en-scène image creates far more tension than would have resulted from quick cutting between, say, Grant's face, the glass, and Joan Fontaine. See François Truffaut, *Hitchcock*, New York: Touchstone, 1967, for further discussion by Truffaut and Hitchcock of these and numerous other examples of montage and mise-en-scène in Hitchcock's films.

11. As McLuhan might have said, "the action is in the intersection." He did often make the point that "resonant intervals" or "grounds" were more important than the pieces or figures of an environment. See, for example, McLuhan and Barrington Nevitt, *Take Today: The Executive as Dropout*, New York: Harcourt Brace Jovanovich, 1972, pp. 10 ff. McLuhan here also makes the "point" (a metaphor he disliked) that modern physics focuses on "fields" not particles.

Taken to extremes, such exaltation of intervals approaches the derogation of material urged by quantum mechanics and (for different reasons) the Platonic primacy of forms. (McLuhan approvingly references Heisenberg, and McLuhan's Catholicism is consistent with a Platonic/Christian emphasis of spirit.)

Ironically, such "de-materialism" runs counter to the dialectic materialism (Marxism) that was Eisenstein's political basis and message (see note 13 below).

See chapter 1, note 7 and chapter 5, note 16 and associated text for references of various quantum mechanical interpretations, and how these relate to the evolutionary materialism urged by the present volume.

12. Joshua Meyrowitz points out in "Television and Interpersonal Behavior: Codes of Perception and Response" in R. Cathcart and G. Gumpert, eds., *Inter/Media: Interpersonal Communications in a Media World*, 3rd ed., New York: Oxford University Press, 1986, pp. 253–272, that sudden cuts are as much a part of everyday offscreen perception (as when we quickly turn our heads and thus change our field of vision) as slow pans. Further, as I argue in *Human Replay*, rapid cuts from long shots to close-ups on the screen may reflect

similar psychological shifts in focus in real life—as when we "zoom in" from a "long shot" of everyone in a room we have just entered to a "close-up" of a person's face.

This suggests that montage works because it too is in accordance with a natural mode of perception—in this case, the "quick change" mode. The difference between montage and mise-en-scène would therefore not simply be that montage is violative of natural rhythms and mise-en-scène faithful, but that each is more or less faithful to a different natural pattern. Montage is usually the less natural of the two because its quick "turns of the head" are created by the film maker rather than the viewer, and the total ambience of juxtaposed images in montage is less likely to occur in the real world than is the sum total of mise-en-scène images.

13. As seen in chapter 4, Marx held the external world to be more primary in human affairs than the mental or cognitive (contra Kant's a priori interactionism and Hegel's metaphysical dialectic), and the centrally determining factor in achievement of both human understanding and satisfaction. These expectations of Marxist epistemology are fulfilled in the explicit external direction of the viewer in montage, and suggest that the development of montage in the Soviet Union was no mere coincidence of geography.

See Mast, *A Short History of the Movies,* ch. 8, for a discussion of Soviet montage, and James Monaco, *How To Read A Film,* New York: Oxford University Press, 1977, pp. 308–314, for an assessment of Eisenstein's dialectic of film making.

But see also note 11 above for a countermaterialist (and presumably counterMarxist) quality of montage.

14. The capacity to enjoy a mediated or fictive experience as such—responding to depicted events as if they were real, yet at the same time knowing they are not actually happening—is captured in Coleridge's comment that "willing suspension of disbelief . . . constitutes poetic faith." The "willing" quality of the suspension suggests its dual character: emotional involvement in an illusion accompanied by logical awareness of illusion. This in turn demonstrates both the centrality of rationality in human activity, and the ability of reason to happily coexist with and indeed enhance emotion. See chapter 2, note 7 in the present volume.

15. Our untutored ability to distinguish between behavior inevitably conditioned by a lifetime of experience on the one hand, and a specifically prompted or elicited response on the other, has been been labelled "superstition" by behavioral psychologists such as B. F. Skinner, and presumably would be belittled by AI champions such as Marvin Minsky who describe human beings as "meat machines" (unintentionally programmed biological mechanisms, albeit far more complex than current electronic circuits). My response is that humans may indeed be the product of genetic and a myriad of experiential conditions, but that the sheer quantity and unpredictable interactiveness of these preconditions and stimuli makes their molding of human behavior nondeterminable, and utterly unlike the controlled conditioning of rats in mazes or mathematical modelling of today's computers.

For a clear statement of Minsky's position, see his "Why People Think Computers Can't Think," *The AI Magazine* (Fall 1982), pp. 3–15.

16. Edward Feigenbaum and Pamela McCorduck summarize the accomplishments and aspirations of expert systems in *The Fifth Generation: Artificial Intelligence and Japan's Computer Challenge to the World,* Reading, Mass.: Addison-Wesley, 1983. Expert systems that function with very high degrees of accuracy include spectroscopic analysis of chemical compounds and blood analysis yielding blood disease diagnosis.

Hubert L. Dreyfus and Stuart E. Dreyfus, "From Socrates to Expert Systems" in C. Mitcham and A. Huning, eds., *Philosophy and Technology II,* pp. 111–130, critique Feigenbaum and McCorduck's claim that expert systems are equivalents of human thinking on the grounds that human expert behavior is intrinsically irreducible to sets of logically arranged rules. Expert systems work well in areas with highly restricted sets of possibilities, Dreyfus and Dreyfus argue, by application of rules which replicate only a sliver of human reasoning dimensions.

I would add that even were human expert behavior thoroughly programmable in machines, the result would nonetheless differ radically from human thinking. Even the most cloistered human expert can easily perform a myriad of cognitive tasks (most medical doctors can strategize about tennis, and a few are gifted writers), whereas the AI expert system can accomplish but one task. Computer expert systems are thus more like human idiot savants—humans who can perform highly complex mathematical calculations in their heads but little else of adult human thinking—than human experts.

My critique, however, applies only to current AI expert systems programs, and is in no sense a metaphysical claim that human intelligence can never be replicated outside of human brains. See chapter 3, note 4 for the incompatibility of "never" dogmatism and fallibilism, and the ensuing discussion in the present chapter for positive suggestions about preconditions for AI development.

17. Donald T. Campbell, "Blind Variation and Selective Retention in Creative Thought as in Other Knowledge Processes," *Psychological Review,* 67, 1960, pp. 152–182. See also Campbell's "Evolutionary Epistemology," 1974, pp. 421–422.

However, see Campbell's "Unjustified Variation," 1974, where he says he prefers "unjustified" to "blind" as an appellation for nonforesighted but not nonstructured generation of organic possibilities.

I prefer "blind" due to its commonsense applicability to both biological (organic) and human cognitive realms. "Unjustified" seems alien to the biological sphere, where conscious attempts to validate knowledge—the process of "justification" in science, philosophy, and human pursuit of knowledge—never arise in the first place (except insofar as the biological gives rise to the human).

18. John R. Searle, "Minds, Brains, and Programs" (1980) in D. R. Hofstadter and D. C. Dennett, eds., *The Mind's I,* New York: Basic, 1981, pp. 353–373, distinguishes between "weak" versus "strong" claims for artificial intelligence (computer programs as appendages or extensions of human cognition versus computer programs as equivalents of human mentality). The distinction is crucial, but I find the terms "auxiliary" versus "autonomous" AI more descriptive of the difference between the computer as a glorified pencil (discussed in detail in "Computers and Tractable Immensity" in chapter 5 and "Electronic Facilitations of Print" in chapter 6 of the present volume, and to be considered in an additional context in "Communities of Representations" in chapter 8) and the computer as in any sense a "thinking" entity (the theme of the present chapter).

Donald B. Straus' distinction between IA (intelligence assist or amplification) and AI (artificial intelligence) is also apt, although the widespread use of the term artificial intelligence to describe both kinds of computer applications makes the reservation of "artificial intelligence" for autonomous-only activities somewhat impractical now. Straus discussed IA versus AI in "Information Technologies as Vehicles of Evolution," an executive-level seminar that I conducted for the Western Behavioral Sciences Institute via computer teleconferencing on the Electronic Information Exchange System, New Jersey Institute of Technology, Newark, New Jersey (Conference 349, Comment 1637, September 19, 1984, 9:54 PM). See also Straus, "Artificial Intelligence in Maintenance," *Applied Artificial Intelligence Report,* November 1984, pp. 11, 18, 20.

19. Fred Dretske is the most explicit in his linking of thought to living systems, arguing in his "Minds, Machines, and Meaning" in C. Mitcham and A. Huning, eds., *Philosophy and Technology II,* pp. 97–109, that intelligence entails pursuit of one's needs and interests—that is, survival. Against the cybernetic equation of information processing in technological and

living states, Dretske points out that "information is irrelevant to the operation of a machine in a way it is not irrelevant to the operation of sea snails or bacteria. If the sea snail doesn't get information about the turbulence in the water, it risks being dashed to pieces. . . ." Of course, a machine with incorrect information in these circumstances will also risk being dashed to pieces, but its destruction would frustrate our self-interest—that of the makers— not the machine's. The key to creation of machines with autonomous intelligence, according to Dretske, thus lies in building them so that they "not only get the information they require to satisfy [survival] needs, but [have] the capacity of using this information for this purpose . . . Only then will we be in a real competition with machines . . ." (See "On the Nonrationality of First Causes" in chapter 2 of the present volume for a discussion of how the nondesigned origins of life give living organisms an authenticity not found in products of deliberate design.)

Douglas Hofstadter argues along more general systems theory lines that current expert systems approaches fail because intelligence cannot be created in a "top-down" manner, or by attempting to reproduce the skimmed cream of intelligent action. Hofstadter recommends instead that technological intelligence can best be created in a "bottom–up" direction that begins with disorganized, unpredictable, error-producing systems—much like the processes through which life and intelligence arose in the natural world, I would add. See Hofstadter's unpublished paper "Artificial Intelligence: Subcognition as Computation" and James Gleick's essay on Hofstadter, "Exploring the Labyrinth of the Mind," *New York Times Magazine*, August 21, 1983, pp. 23 ff.

Hubert Dreyfus and John Searle are more uncompromising in their critique of artificial intelligence possibilities, but both leave doors slightly ajar in the direction of living systems. In "Minds, Brains, and Programs," Searle says that although no quantity and complexity of mere "programs" could ever attain thinking intelligence, in principle we could create an artificial intelligence by creation of artificial human beings ("a machine with a nervous system, neurons with axons and dendrites, and all the rest of it, sufficiently like ours," p. 368). And although Dreyfus, *What Computers Can't Do*, New York: Harper, 1972/1979, pp. 282, 290, flatly states that the phenomenological (nonobjective) nature of human intelligence means that it "in principle" could "never be fully realized" by artificial intelligence, he admits that his critique of "current computer techniques" does not consider the possibilities of programming computers "to behave like children and bootstrap their way to intelligence."

I associate myself with these critiques of artificial intelligence because none denies the possibility of AI on eternal metaphysical principle. Rather, their invocation of self-interest, bootstrap and bottom-up development, and axons and dendrites, points to a failure of current artificial intelligence efforts to take into account the properties of living systems in which the only known form of intelligence—the natural human kind—in fact arose.

Note that such criticism is quite different from Weizenbaum's—for example, *Computer Power and Human Reason*—which is directed against both autonomous and auxiliary artificial intelligence (see note 18 above), and attacks not only the feasibility but also the ethical advisability of most forms of computer functioning. See also note 20 below for differences between Dreyfus' and my perspectives.

20. "The body is our general medium for having the world," Merleau-Ponty observes in *Phenomenology of Perception,* trans. Colin Smith, Atlantic Highlands, N.J.: Humanities, 1962/1981 (excerpted in L. Hickman, ed., *Philosophy, Technology, and Human Affairs*, College Station, Tex.: Ibis, 1985, pp. 211–216), referring to the ways our bodily interactions with the world create environments of sensations and objects later perceived by our intellect. One could say here that such physiological relationships are the foundation upon which Kant's intellectual categories subsequently operate. Dreyfus explicitly employs Merleau-

Ponty's insistence on the inextricable contribution of living body to cognition in chapter 7 ("The Role of the Body in Intelligent Behavior") of *What Computers Can't Do*. See also Larry Hickman's treatment of Merleau-Ponty's "metaphysical structure of our flesh" in "Filosofia Hoy: El Filosofo y Techne," paper presented at conference on "Para Que Filosofia Hoy," University of Navarra, Spain, 1981.

Dreyfus' position differs somewhat from mine on this issue in that Dreyfus argues (a) that mind and body are intrinsically inseparable, (b) therefore mind cannot be fully explained by objective scientific-philosophic approaches, and (c) therefore current scientific attempts at building "minds" in machines must fail. I accept (a) and (c) but not (b)—I see no reason that science cannot eventually satisfactorily account for the mind-body mix—and I therefore suggest (d) that artificial intelligence may indeed be eventually constructed on a successful resolution of (b).

Dreyfus' perspective is in accord with such diverse claims of limited capacity for self-understanding as Kant's, who argued that scientific knowledge of the cognizing agent is not possible because "we cannot obtain the least representation of a thinking being by means of external experience, but solely through self-consciousness"—that is, mentality is not in external experience, and is therefore beyond scientific knowledge (*Critique of Pure Reason*, p. 236). Dreyfus' view is also in accord with Friedrich Hayek's observation that we cannot ever explain the function of the brain in detail because "any apparatus . . . must possess a structure of a higher degree of complexity than is possessed of the objects" it is attempting to explain (*The Sensory Order*, Chicago: University of Chicago Press, 1952, p. 185).

But see my "Evolutionary Epistemology Without Limits" for refutations of both these claims. Hayek's is disposed of by the recognition that the brain/mind is not a single apparatus but a multiplicity of uneven structures, which allows for the possibility of structures interacting in ways that would afford them an understanding of each other. (For example: structure A comprehends structure B, structure B comprehends structure C, and structure C comprehends both structure A and the interaction of A and B, and so on.) Kant's objection is undermined by the technological enterprise in general and the possibilities of artificial intelligence in particular—both of which externalize aspects of human mentality in a variety of ways (as we have seen, for example, in the case of film discussed in the present chapter).

See also Larry Hickman, "Philosophy, Techne and the Body," paper presented at the Fifth International Conference on Culture and Communication, Temple University, Philadelphia, Pennsylvania, March 26, 1983, for a brief elaboration of Dreyfus' views in this area.

A third type of claim about limits to self-understanding comes from Stent ("Limits") and Chomsky (*Reflections*, pp. 24–25): humans cannot understand their own deepest workings (such as will) because understanding of our interior environments was not germane to our evolutionary history. The general "evolutionary limitations" claim—the view that we cannot understand areas not pertinent to our evolutionary past—was refuted in chapter 3 of the present volume. Stent's and Chomsky's argument would be untenable even were evolutionary relevance a necessary factor for understanding, for surely knowledge of the inner mental workings of one's opponent would confer survival advantages. As Bertrand Russell observes, "the savage deceived by false friendship is likely to pay for this mistake with his life" (*Mysticism and Logic*, p. 15); see chapter 1, note 5 above.

21. See chapter 4, "Unideated Material and Immaterial Ideas" for more on the distinction between an idea's embedding or containment in the brain and its embodiment or expression in material technology. Embodied ideas have greater immediate impact in the material world than do embedded ideas, since the embodiment of an idea changes the material to express the content of the idea, whereas the embedding or lodging of an idea in the brain does not (thinking about an airplane does not give the brain wings). Representation of an idea in a communications medium would be a third type of material connection—much

like containment of an idea in the brain, since the representation of an idea in a material communications medium does not make the medium become that idea (a book about airplanes does not fly).

22. Prehuman organic manipulation ranges from reshaping of external material for organic ends (for example, bird nests) to secretion of organic substances for environment manipulation (as in spider webs). (The latter type of manipulation goes back to the very origins of life, in the molecular production of catalysts that encourage molecular reproduction.) All of these prehuman technologies may be considered prototypes or ancestors of the humanly created realm of artificial existence. See also chapter 5, note 1.

23. This fourth realm of artificial creations bears resemblance to Dessauer's fourth realm of technology. The two differ, however, in that mine is wholly created by humans (from the material and structures of the first two realms), whereas Dessauer's is called forth by humans from a preexisting Platonic world of ideal forms. My model coincides with Dessauer's to the extent that human technologies incorporate relationships (natural laws) already inherent in nature. But unlike Dessauer, I emphasize the ways that technologies may not only reveal but even bend natural laws (Heidegger focuses on the revealing, Dewey and I on the revealing and the bending, Marx on the bending), and I recognize—as per Kant—the important role of human cognition in shaping these laws as they are revealed to us. See chapter 4, note 21 for references on Dessauer; and see section "Can Technology Change Natural Laws?" in my "What Technology Can Teach Philosophy" in *In Pursuit of Truth* for more on the relationship of technology and natural laws.

24. *Quest for Fire,* motion picture released in the United States by Twentieth Century Fox, 1981 (first released in England in 1980).

25. Darwin, *The Origin of Species,* p. 153. Darwin distinguishes between two types of "human" selection of organisms: "methodical" selection as in the scientific, painstaking breeding of specific sorts of dogs, horses, and birds, and "unconscious" selection as when people favor attractive plants with better care (and thus better reproductive likelihood). The plasticity and improvement of species demonstrated under this human guidance—traced to ancient times by Darwin—is repeatedly cited by Darwin as evidence of the mutability and perfectibility of organisms via selection in nature: that is, natural selection. See chapter 1 of *The Origin of Species* ("Variation Under Domestication") and Darwin's *The Variation of Animals and Plants Under Domestication,* London: Murray, 1868.

Luther Burbank's development of numerous new varieties of fruits, vegetables, and flowers at the turn of the century is a high-water mark of human "methodical" selection, as yet unequalled even via genetic engineering. His principal works are *How Plants Are Trained to Work for Man,* New York: Collier, 1921, and *Partner of Nature,* New York: Appleton-Century, 1939, which was published 13 years after his death.

26. As Burbank (*Partners,* p. 61) puts it about the state of breeding in an era prior to genetic engineering, "We have seen that we can put nothing into a plant that is not contained somewhere in its bundle of family tendencies, habits and qualities that we call heredity."

At the same time, Burbank repeatedly emphasizes that the mere knowing of nature is inferior to use of this knowledge to actively improve on nature for the benefit of humankind ("while the counting of stamens and the classification of plants are important, they bear little relation to the vital thing, which is the improvement of plants—making them better to serve and delight mankind," p. 83). Burbank thus is in accord with Marx, Dewey, and the present volume on the primacy of material implementation. See last paragraph of note 4 above, and chapter 4 of the present volume.

27. See, for example, Feigenbaum and McCorduck's view that "machine intelligence" will soon be "faster, deeper, better than human intelligence," *The Fifth Generation,* p. 236.

28. Polytechnic University (Brooklyn, NY) chemist Frederick Eirich writes in private

correspondence that he is seeking to mathematically demonstrate that minimum necessary functions of "an intelligent brain" are supportable only via the specific chemical configurations of protein molecules. Success in such a project would foreclose the possibility of both natural (for example, extraterrestrial) and artificial non-protein-based intelligence. (Artificial life would need to be created via proteins or protein equivalents if the artificial life were to serve as a basis for artificial intelligence.) Eirich, however, is presumably a long way from realization of his goal.

29. See Siegfried Giedion, *Mechanization Takes Command,* New York: Norton, 1969 (1948), pp. 32–35, for a brief account of the defecating mechanical ducks.

30. I stop short of saying that such a development would be unconditionally advisable, since genuinely autonomous artificial intelligences would be by definition somewhat beyond our control, and thus potentially not always in the interests of our species.

Isaac Asimov's robot quadrilogy (*The Caves of Steel, The Naked Sun, The Robots of Dawn, Robots and Empire,* New York: Doubleday, 1953, 1956, 1983, 1985) provides a sustained and penetrating analysis of what can go awry even when otherwise autonomous systems are explicitly programmed to further human welfare. Asimov postulates as the first and presumably unbreakable law of his Three Laws of Robotics that a robot can never by action or inaction allow harm to befall a human—yet the flexibility and room for interpretation in even this definite commandment (apropos fallibilism) is precisely what gives Asimov's plots their punch. The story in *Robots and Empire,* for example, hinges on a robotical analysis of whether the projected future good of many humans should take precedence over the immediate well-being of individual humans.

Robots are an appropriate locus for a discussion of genuine autonomous artificial intelligence, for the robotic ability to move gives them a capacity to interact with external environments in a fuller, more tactile, and more lifelike manner than stationary computer processing units. Thus, if lifelike processes are a precondition of AI, we can expect genuine AI to arise first in robots.

See R. K. Miller, *Intelligent Robots,* Fort Lee, N.J.: Technical Insights, 1983, for a survey of real current developments in robotics.

[Note added in page proofs: the capacity for pattern recognition in new "neural-network" computers may be a promising development in artificial intelligences that function in a more life-like fashion. See Michael Rogers, "Mimicking the Human Mind," *Newsweek,* July 20, 1987, pp. 52–53.]

31. For example, Weizenbaum, *Computer Power and Human Reason;* and Jerry Mander, "Six Grave Doubts About Computers" in *Computers as Poison,* special issue of *The Whole Earth Review,* January 1985, pp. 10–20.

32. Wiener's *Cybernetics* (1948) is the pioneer in this field. Margaret Boden's *Artificial Intelligence and Natural Man* (1977) is the most comprehensive assessment in the past decade.

33. We might articulate a principle which holds that technologies survive to the extent that they replicate natural patterns. Silent movies, for example, were eradicated by the advent of talking motion pictures, but radio survived the rise of television: our principle explains this on the grounds that seeing without hearing rarely if ever occurs in the natural human world (we cannot close our ears), whereas hearing without seeing is quite common (we close our eyes, the world grows dark every night but not often silent, and so on). Similarly, still photography survived the introduction of motion photography, but black-and-white has all but commercially vanished in an age of color pictures: stillness is an important part of everyday perception, whereas black-and-white environments appear only at twilight, dusk, and in unusual shadow situations. See "Principles of Media Evolution: Survival of the Closest Fit" in my *Human Replay* for more.

A corollary of this principle would be that in general we develop technologies in order of the importance to us of the natural patterns the technologies intentionally or unintentionally replicate. Thus, motion is a more crucial perception than the third dimension—which wanes at long distances—and consequently we developed motion photography nearly a century earlier than holography. This corollary might explain the unconscious motive for technological development; the success of an invention also depends on the level of technical proficiency required to realize its completion, availability of necessary collateral technologies, and so on. See note 4 above.

34. "Methods of making high quality mirrors have changed," Michael F. Land writes in "Animal Eyes with Mirror Optics," *Scientific American,* December 1978, pp. 126–134. "Instead of consisting of a single layer of silver or aluminum they are made up of multilayered stacks of very thin films of alternating high and low refractive index. This turns out to be the way living organisms have made mirrors all along!" See also Beverly K. Hartline, "Lobster-eye X-ray Telescope Envisioned," *Science,* January 4, 1980, p. 47.

35. Darwin makes an analogous point about the relationship of artificial (human-directed) breeding and natural selection of organisms: "Can we wonder, then, that nature's productions should be far 'truer' in character than man's productions; that they should be infinitely better adapted to the most complex conditions of life, and should plainly bear the stamp of far higher workmanship? It may be said that natural selection is daily and hourly scrutinizing . . . rejecting that which is bad, preserving and adding up all that is good; silently and insensibly working, whenever and wherever the opportunity offers, at the improvement of each organic being . . ." (*Origin of Species,* p. 133).

36. Stuart Baur, "Kneedeep in the Cosmic Overwhelm with Carl Sagan," *New York,* September 1, 1975, p. 30.

Chapter 8

Technology and the Globalization of Intellectual Circles

The history of ideas is inextricably a history of places. To speak of the birth of democracy is to speak of Athens; to trace the growth of Western religion is to travel from Jerusalem to Rome to Mecca; more recent and less profound developments of thought are peopled with members of the Bloomsbury Group, the Vienna Circle, and the Frankfurt School. Such placements are indicative not of some mystical geographic determination of ideas, but of the fact that if an idea is to leap from an individual's conception to a role in intellectual affairs, it often must be nurtured or incubated (which includes being criticized) by a small group of minds with easy access to one another—an access best made possible by the sharing of a locale. Moreover, although the larger social structures in which these small groups must operate undoubtedly influence the texture of the nurturing (for example, whether the locale is democratic or totalitarian), more mundane factors such as the efficiency of transportation and the hours kept by restaurants may play a more significant role by directly though unintentionally regulating the interaction of individuals.

Up until the past few decades, the nature of technology was such as to promote exchanges among people within small areas rather than across large distances. The university, for example, was a self-contained unit for scholars, a sort of medieval manor for thinkers designed as much to keep away distractions from the outside as to facilitate discourse within. As we saw in Chapter 6, the inventions of writing and especially of printing provided the first technological bases for intellectual exchanges at a distance, but the time consumed in the conversion of ideas into writing and print rendered these media unsuitable for the rapid, free give-and-take of information characteristic of communication among individuals in close physical proximity. The application of electricity to communicative tasks in the nineteenth century finally did provide the means for an immediate exchange of ideas across vast distances, but the first use of speed-of-light communication in the telegraph was much too clipped and formal, and the second use in the telephone not formal and structured enough, for the service of intellectual ends. The abrupt messages of telegrams and the intimate gossip of telephone calls were thus both in their opposite ways unencouraging of scholarly discourse. As the twentieth century began, the primary place of intellectual exchange remained, as it always had been, the physical locale.

In addition to these psychological impediments to intellectual traffic, the economics of telegraphy and the telephone worked against the easy flow of knowledge across great distances until very recently. International telephone rates, for example, were until the 1980s so expensive as to make instant exchanges of ideas between continents impractical, with the result that when the telephone eventually did get drawn into intellectual affairs in the twentieth century, it tended to reinforce communication within existing local circles rather than create new and wider realms of discourse. More important than the recent lowering of long-distance telephone rates may be the growing deployment of telephone lines for the transmission of computer-created written communications, whose mixture of ease and formality recreates across great distances much of the comfortable yet careful style of in-person scholarly exchanges.[1]

What exactly is this in-person intellectual circle, whose magic defies capture by the fixation of paper and the fluency of the purely acoustic medium alike? Is it ineluctably a function of face-to-face physical presence, of access to entire, flesh-and-blood human beings rather than just their ideas, however precisely and vividly such ideas may be presented? Although every medium since and including the most primitive forms of carving and writing has easily communicated ideas without the physical presence of the human creators of these ideas—such indeed has been the very purpose of virtually every medium invented[2]—the essentially pre-media intellectual circle with its foundation in physical locality has continued fundamentally unchanged into the present century.

The relationship between most communications media and the in-person intellectual circle has been analogous to the relationship of communications media (especially computers) and the brain discussed in the previous chapter: technologies have been extraordinarily good in distributing the products of intellectual circles across long distances and time (and this distribution has, in the ways described in Chapter 6, contributed greatly to the growth of knowledge), but, much like the computer vis-à-vis the brain, communications technology generally has not been able to replicate or extend the special conditions of intellectual incubation that constitute the initial or initiating thinking community. Like the workings of the individual generative mind, the mechanisms of what might be called the localized collective mind—the first-line social shaping or generation of knowledge—have till now been mostly beyond the touch of technological reconstitution.

Two types of machines have lately begun to provide the means for what may be a genuine extending of the localized collective mind, or a recreation on a global scale of the access and moderation characteristic of the intellectual locale. One sort of technology, alluded to above, is exemplified by computer teleconferencing and its ability to create a socially and intellectually satisfying substitute for in-person environments in its instant though permanent conveyance of both abstract and personal information. (The face-to-face immediate imagery of videophone may have a compatible effect, contributing nonverbal aspects of physical presence utterly missing in computer and audio teleconferencing.[3]) A second sort of technology, with perhaps even a better chance of widening the nature of intellectual circles in the imminent future, is the jet plane or any transportation device which allows people to travel across continents as easily and economically as across town.

The impact of such technologies on the growth of knowledge is potentially enormous, for if they succeed in stretching the functional intellectual circle to global proportions, the number and variety of minds that can be brought to bear on the nursing of an idea will increase exponentially. On the other hand, the delicate nature of the local intellectual community may be incompatible with such magnification, even via supersonic jet planes operating at low cost. The very scarcity of great minds in any localized physical community, for instance, may well be one of the essential conditions for the successful functioning of such a community, thus rendering a jet-set Bloomsbury Group or an electronic Vienna Circle a contradiction in terms. Of course, the membership of global communities could be limited by deliberate regulation of the easy access provided by the enabling technologies, but such control would sacrifice the serendipitous membership that plays a part in organic or localized communities.

Assessment of these possibilities will be the main agenda of this chapter. However, before scrutinizing the technological impact on the social production of knowledge, I would like to say a few more words about social versus individual constructions of knowledge—in particular, about the many social accounts of knowledge that have been offered in the past hundred years, especially in the past few decades, and about why I have by and large given short shrift to such accounts in this book. In examining the pros and cons of the sociology of knowledge, I will attempt to identify those aspects of the social condition that are of genuine epistemologic import. These relate to the evolutionary formation of our species and knowledge, and unfortunately have been generally unaddressed in the prevalent approaches.

A REALISTIC SOCIOLOGY OF KNOWLEDGE

In the history of the sociology of knowledge, Kant's philosophy is a turning point, much as it has been in so much else of our attempt to understand our capacity for understanding. Before Kant, philosophers ranging from Aristotle to Descartes and Bacon saw the world as directly comprehensible by some combination of human thought and experience. Since Kant, most philosophers and indeed most students of the human condition have tended to see the world as comprehensible by humans only through an inevitably distortive lens of social attitudes, symbols, and like products and equipment of the human species. (Vico would be a notable exception to the first group; Mill and the Vienna Circle to the more modern tendency.)[4]

How did Kant, a champion of reason, realism, and progress, engender a profusion of epistemological relativisms and in some cases a cynicism that knowledge of any sort is a chimera? Kant of course was not only a champion but a critic of reason, and most especially a critic of the view that reason or any human faculty could reveal to us the world exactly as it is. Indeed, Kant emphasized that to perceive or understand the world is inescapably to bend the world (or our perception of it) to the contours of our perception and understanding, thus making our vision of the world an irreparably human vision—filtered and distorted—and in no sense a vision of the world as it is literally or completely. From this human-biased, cognitive structure-relative epistemology, so alien to the classic notions of unfettered interactions of mind and environment, the small but portentous jump was made to the knower and the known as mediated and distorted by, and ergo constituted of, economic conditions (Marx), unconscious, irrational symbols (Freud), or language (Wittgenstein).

For the evolutionary epistemologist who, like myself, views our ways of knowing as naturally selected adaptations to our proximate environment, Kant's original formulation (refined by Popper's fallibilism) is more adequate than most of its predecessors and successors as a basis for explaining how we know the world. The claim that reality may be perfectly copied through the senses (Aristotle, Bacon) or perfectly represented in innate ideas (Descartes) is far too optimistic for a naturalistic view which holds that cognition, like other products of evolution, is necessarily unclear, inexact, and partial—possessed of traits that imperfectly impose themselves on the objects of its transactions. But since our evolutionarily derived cognitive structures initially are ontologically prior to any humanly created social structure, the evolutionary epistemologist is also on firm ground in granting our biological faculties a privileged position among sociocultural factors. Of course, as humans evolve in response to environments of an increasingly humanly created, social and technological nature, given biological faculties may well be selected by social conditions, and thus appear subsequent to, or as a result of, such social factors. Even in this case, however, the selecting social environment is a product of prior biological structures.

The biological primacy of our mental capacities suggested by evolutionary epistemology accords well with the logical primacy of our mental capacities argued for by Kant—in both cases the relationship between cognitive structure and external reality is a favored one (albeit eternally imperfect for evolutionists) that transcends and indeed informs social conditions. From the standpoint of the evolutionist, a genetically derived cognitive structure (namely some momentously complex combination of protein molecules in the brain, formed according to some DNA-transmitted instruction, and stimulated, elicited, and otherwise acted on by an enormous variety of environmental influences) is a survivor of eons of encounters with the environment, an old hand at negotiations with innumerable external realities, and thus privy to an external world that surpasses any social or cultural condition much as Kant's a priori categories of thought are panhuman or cultureless. (Critics of Kant point out that his analysis is very much a product of "Western" style rationality, yet such criticism is itself of a traditional rational form. One therefore might well ask these critics what form of serious criticism would not be?) Cultural conditions, to be sure, may and do intervene in the cognition/reality relationship, but a thesis like that of Thomas Kuhn, for example, which states that knowledge is primarily a creature of cultural construction or social reality, is simply incompatible with evolutionary theory, for it ignores the physical reality that is the ultimate editor of all life and its productions.[5] Other social constructivists such as Marx and Wittgenstein may be gauged by the evolutionist on the basis of how much of a role, if any, they assign to physical or natural reality. (Marx would come out far better on such an assessment than would Wittgenstein.)

What part, then, does the social realm properly play in knowledge construed as an imperfect, nonetheless adaptive (that is, accurate to some degree) representation of a culturally transcendent external reality?[6]

Among those who view our mental capacities as enjoying an evolutionarily advantageous relationship with the external world, social conditions are quite often regarded as limiting, intrusive, or even destructive of the cognitive process. Stephen Jay Gould, for example, cites social factors in the form of predisposing metaphors and unexamined cultural suppositions as the most serious impediment to the search for truth conducted by fundamentally flexible, biologically derived modes of cognition.[7] Other social obstacles such as regressive political institutions and insensitive bureaucracies are often similarly cited.[8] Although such assessments of social reality as a limiting factor in the growth of knowledge are, from the evolutionary perspective, far superior to treatments that take the social condition as constitutive of knowledge in a primary way (to the belittlement or virtual exclusion of Kantian/biological factors), these assessments nonetheless address only half the story, and indeed give scant attention to the ways in which social factors may combine with biological and technological capacities (the latter often a significant form of social activity) to enhance our cognitive operations. The social restriction of knowledge is undoubtedly worthy of further investigation, but in a volume generally concerned not with the least but rather with the most that humans can know and do, we will focus instead on the even less explored area of how social milieus—themselves the products of communication and transportation technologies—may work to support, augment, and catapult the technological cognitive enterprise.

One of the few to consider the social matrix as a positive transformer of evolutionarily derived cognitive capacities is Aharon Kantorovich, who argues that many of our most important scientific theories such as quantum mechanics are products of "collective world pictures" or pooled mental abilities that go beyond the capacity of any individual's cognitive structures to create (and, one might add, to adequately understand or conceptualize).[9] Looking beyond any Jungian-type mysticism that we might be tempted to impute to such collective images of reality, we can readily see that these collective operations are for the most part not truly universal, globalized, or inclusive of all human thinkers, but instead are functions of small numbers of thinkers, working in limited, locale-based groups or intellectual circles. In the twentieth century, such groups of course have made varying use of telephone, air travel, and similar technologies to draw on worldwide sources for their membership. But the result thus far has been a partial and idiosyncratic tapping of worldwide roots rather than a comprehensive worldwide participation in Kantorovich's knowledge "collective," and the question remains open as to whether the small group works best for knowledge production or is merely the best that technology till now has been able to facilitate. To the extent that emerging technologies foster social environments in which knowledge can be an outgrowth not only of selected global sources but also of extensive global operations—globally interacting communities in which minds can be brought to bear on problems regardless of geographic location and numbers of minds available—the ultimate dimensions and potentials of the collective vision will at last be tested. In such circumstances, knowledge would be not only a world picture but a world production, though no one can say with assurance that these productions will consequently be more accurate representations of reality than their predecessors.

In this chapter, then, we accept as a premise Kantorovich's thesis that certain kinds of knowledge of objective reality may be attainable and even comprehensible (and certainly actionable) only by groups, and inquire into the extent that technologies of communication and transportation may globalize or otherwise sustain and shape such groups, and therein contribute to the growth of knowledge. The social contribution to knowledge considered here is therefore neither the social world as the prime or sole constituent of knowledge or the only reality that knowledge can discover (for this approach is wrong, or at least lethally incompatible with evolutionary theory), nor the social fabric as a wet blanket on the quest for truth (an area warranting of further study). We look instead at the potentials of technology for the knowledge process—more specifically, the epistemological potentials of technologically engendered community that may go beyond the possibilities of technology utilized by individual mentalities—on the grounds that, whatever obstacles social factors may put in the way of knowledge, we gain an advantage from an awareness of the potentials the obstructing social factor may have impeded. (Consideration of potentials makes our discussion future-oriented, but the questions could be applied by others historically as well, in researching for example the extent to which Darwinian and Freudian circles were products of the technologies of their times, and the extent to which the unique intellectual achievements of these groups were therefore a result in some sense of the contours of their technological networks.)

COMMUNITIES OF REPRESENTATIONS

In the paradisiacal climate of La Jolla, California, a new community of worldwide intellectual connections is under construction. Since 1982, the School of Management and Strategic Studies of the Western Behavioral Sciences Institute has been conducting a series of seminars through "computer conferencing." Participants meet in person for a week in La Jolla, then disperse to their homes and businesses across the United States (and, to a small but growing extent, across the world). Each participant is given a personal microcomputer and modem with word processing and telecommunications capabilities,[10] through which he or she receives lectures, comments, questions, and assignments from the seminar leader via a central computing system, and is expected to respond in kind.

Although the hookup is such that one or all of the seminar members (usually not more than 10 to 25 active participants) may be "on-line" or connected to the central computer system at the same time, the preferred and more effective use of computer conferencing is asynchronous, and works as follows:

1. The seminar leader composes a comment "off-line" on a word processor, and then transmits this comment to a central computer system via modem and telephone.
2. The comment is stored in a file or "conference" on the central system, where it is accessible by seminar members anywhere in the world (with telephone connections) any time night or day.
3. Seminar members access the central system via their personal computers and modems (connected to telephone lines).
4. They read the comments in the conference and enter their own (usually composed off-line) at times of their choosing.
5. The seminar leader in turn reads the participants' comments, and enters responses and new material at appropriate times.

All exchanges are conducted via this electronically written procedure, and the result in a few weeks is usually a rich, multifaceted, permanently recorded and often spontaneous dialogue produced by ten or more minds.[11]

In contrast to the in-person seminar or classroom where questions are often unasked or unresponded to, the continued access to stored comments in the computer conference gives participants the opportunity to raise issues or redirect focus to earlier topics any time such questions may occur to the participants throughout the course of the conference. (Conferences generally run from one to several months, though occasionally they are much longer in duration.) Further, the asynchronous mode of exchange allows members to participate—to read and write—at times when they are at their best (in a typical synchronous in-person or audio conference meeting, some members—perhaps even the leader—will likely not be fully prepared or in a mood most propitious for intellectual exchange). These and related factors heighten the quality of the computer conferencing dialogue.

Seminars of the Western Behavioral Sciences Institute (WBSI) do not carry formal academic credit; participants are for the most part leaders of business, government, and the not-for-profit sector. Use of computer conferencing for formal academic education began in 1984 with The New York Institute of Technology's program of on-line courses for undergraduate credit. Since 1985 Connected Education, Inc., an independent not-for-profit corporation, has been offering a program of courses via computer conferencing for Master's degree credit in conjunction with The New School for Social Research's Media Studies Program. In the first year of operation, "Connect Ed" attracted students from Japan, Singapore, the Middle East, England, South and Central America, and across the United States and Canada—several of whom matriculated for the M.A. degree entirely on-line. "EIES"—the Electronic Information Exchange System developed by Murray Turoff at the New Jersey Institute of Technology—provides central computing facilities for WBSI and Connect Ed, and has been host to other computer conferences on a variety of scientific and academic issues since 1975.[12] In all these cases, participants access the central system via personal computers and modems located in their homes and/or places of business.

To what extent do these electronically constituted communities qualify as global, or at least fully participational subglobal intellectual circles? Unlike McLuhan's "global village," a halfway, metaphoric community in which inhabitants receive common information from television and satellites but are not usually in a position to create or exchange it, the computer conferencing network establishes on a geographically irrelevant basis the possibilities for reception and initiation and exchange of information that heretofore have existed only in in-person, localized situations. The result is that members of these electronic communities know each other much as members of a real-life local intellectual community such as a university do, but as members of a global television audience do not.[13]

Telephone users have long enjoyed the interactive familiarity of computer conferees, and recently across distances equal to those attainable via computer conferencing (CC) networks. But people connected solely by telephone—still a predominantly one-to-one instrument—seem to lack the minimal cohesion and continuing group presence necessary for achievement and maintenance of community[14] Thus the telephone group, though readily interactive, is in its own way ultimately as disintegrative and noncommunal as the television audience.

Computer conferencing, then, seems to be a genuine advance beyond earlier electronic systems such as television and telephone in the creation of global, nonlocalized, intellectual circles. Unlike the television community, the CC group is interactive; and unlike telephone talkers, members of the CC seminar work from agendas that are literally visible and persistent both on the immediate computer screen and in central and local computer storage (and printouts on paper).

Moreover, the very lack of physical presence in CC, combined with its removal of the need to meet at appointed times and places, reduces many of the anxieties that occur in face-to-face groups, and may make computer conferencing superior in these respects to the in-person circle as an intellectual tool. Fingers fluttering lightly on keyboards in the privacy and convenience of homes may be more faithful servants of the intellect than tongues wagging in smoke-filled restaurants and crowded halls.

Can a communion of ideas without the physical presence of their creators perform as effectively as the in-person group, where bonds of community come from handshakes and smiles and work to sustain the group process regardless of the specific ideas produced? Although the ultimate answer for certain tasks is likely negative, many of the shortcomings of electronic communities are consequences of current technological capabilities and applications; enduring differences between electronic and in-person communities thus may be significant but few.

One obvious limitation of current computer teleconferencing is the absence of nonverbal cues in the communication process. Although a respondant's or audience's facial expressions may on occasion inhibit a speaker, numerous studies and common sense tell us that lack of facial cues, body postures, and the like can impair communication, and result in a reduction or distortion of the information conveyed.[15] The problem can be remedied by video conferencing, and indeed a video/audio image community may be a beneficial adjunct or even outrightly preferable to a community of texts for some intellectual endeavors. (Text, however, is amenable to searching and sorting in a way that video and audio recordings currently are not—thus giving computer text conferencing important advantages.)

A more serious limitation of both computer and audio-visual conferencing is the technological proficiency and/or equipment required to join and participate in the group. Although in-person intellectual groups often impose strict and sometimes secret entrance requirements (for example, the Cambridge Apostles),[16] and even the most accessible public universities impose some minimal criteria for selection of entrants (such as possession of a high school diploma or equivalent), the lack of any special technology necessary for participation in physical communities makes even the most restricted more accessible in principle than a group whose membership requires technological mediation.

A related impediment arising from technologically mediated membership is the systematic exclusion of the unexpected or uninvited serendipitous participant in a conference or seminar. Computer "hackers" and the movie "War Games" notwithstanding,[17] the in-person meeting or classroom is more vulnerable to the uncredentialed, late-arriving member than a computer system for which complex access codes may be required. The extent to which this vulnerability and unpredictability of in-person groups improves their performance is unknown, but presumably the unexpected member or student who strolls into a meeting or class offers a potential benefit, if only on the evolutionary ground of providing a larger, more varied population of ideas from which to choose. (Such unexpected membership is also in accord with a naturally selective, nonintentional, model of evolution.) However, as personal computers become increasingly commonplace in our society and our world, their possession and use will be scarcely more unusual than possession and use of vocal chords (the telephone has already attained this popularity), and thus we can expect most if not all of the admission limitations of electronic communities to pass in good time too.

So what, then, are the irreducible differences between electronic and in-person communities—differences which no amount of progress in communications technology will relieve? These differences are real, and relate to aspects of the group production of knowledge which are noninformational, or cannot be reduced to an exchange of information (even emotional information). If love making, for example, were intrinsic to the group knowledge process, then no amount of representational communication, however multidimensional and accurate (for example, holography), could replicate or provide for all the in-person aspects of this "knowledge" process. The physical substrates of in-person knowledge groups are less dramatic than the physical necessities of love making, but no less profound. How, for example, could any electronic community recreate the atmosphere of colleagues over lunch, dinner, or even coffee—environments in which tastes, shared aromas, and casual talk mix with more serious conversation in innumerable ways which are often more productive of knowledge than the formal in-person seminars which the lunch breaks are supposed to punctuate?

The inability to break bread via electronic circuits is metaphor for a host of privations in food for thought that occur in electronic substitutes for in-person social engagements—a loss of the myriad unnoticed, minor ways that shared physical presence acts as a stimulant to good thinking. Development and dissemination of current technologies will no doubt allow accessibility and comfort in global communications very much the equivalent of—and even greater in some ways—than the in-person cafe. We may well figuratively share lunch with colleagues in an electronic meeting spanning several continents, but no conceivable electronic representation will allow us to pour milk in our colleague's cup of coffee or pat him or her on the back.

Of the many aspects of human life, knowledge is an original perhaps least damaged by representation, since it is in its original form a representation of reality. Thus we have seen that transmission of knowledge via representations grossly at variance with the original form of the knowledge—as in spoken and written renderings of visual perceptions—nonetheless provides the basis for effective criticism and dissemination of the knowledge. Further, speech and writing are the only media hospitable to detailed conveyance of knowledge that is intrinsically abstract or nonsensory. In all of these instances of knowledge transmission, however, we are dealing with a movement of knowledge already formulated—that is, a conveyance of ideas which, although subject to pruning and development, have already gone through a process of individual creation and usually social gestation in a small group as well. (This is certainly the case in the transmission of knowledge via printed publication.) In other words, representation works, and is indeed both essential and sufficient, as a vehicle for the growth of knowledge as an already existent object or product. But representation of human producers of knowledge, in place of the living human producers of knowledge themselves, is quite a different thing: here the representation must function not as a product of life but as life itself, and the result is something less than life. To return to the sexual metaphor, we can represent with highly effective consequence most if not all aspects of the sexual act, but for the act to prove fruitful in a procreative sense—to result in the production of life—living human beings must be present, or at least living cells of humans (or at very least the DNA codes of cells, which are not only information or representations of life but life itself). Thus, the use of electronic connections and their carried representations not as conveyors of knowledge but as proxies for the nervous systems of intellectual communities that produce or incubate knowledge falls short of the mark in a crucial way. The difference between conveyance and production of knowledge is admittedly often very thin, and the lack of full in-person presence may serve to improve the generation process in several ways (such as the reduction of in-person anxiety), but certain shortcomings of communities of representations seem persistent nonetheless.

Just how much of a toll these shortcomings will in fact exact on the growth of knowledge remains very much open. Indeed, the benefits obtained by the instant, asynchronous, permanent mingling of geographically dispersed minds in electronic global communities may well offset any loss due to the absence of in-person presence. When we recall that the first unifying climax of the Scientific Revolution spurred by the printing press occurred in the work of Isaac Newton more than 250 years after the introduction of the press in Europe, and when we consider that computer conferencing is now not even two decades old, we can see how minor the current accomplishments of computer conferencing and related communal media may be in comparison to what lies in store.

But representations are not the only technological means for attainment of global intellectual communities, although they are by far the easiest. If the global circle of proxies meets its limit not in lack of universality but in lack of in–person presence, we then may ask what technologies can do to extend in–person communities to the universal dimensions of electronic circles. We turn now to the transportation of complete, in-person human beings across great distances, and examine the prospects of this nonrepresentational mode for the creation and sustenance of worldwide intellectual communities.

FLESH AND THE GROWTH OF KNOWLEDGE

To remark that the world is shrinking is boringly commonplace, for the world has been shrinking for quite some time. The informational world of represented human relations began rapidly shrinking as early as 150 years ago with the invention of the telegraph and its speed-of-light transmission of products of cognition. The physical world of real-life access by in-person human beings has been shrinking a good deal longer. The trip of Magellan's fleet around the world from 1519 to 1523 was a signal accomplishment, and the development of steam, rail, automobile, air, and finally space transport in our own age has continued to reduce the effective size of our planet. The leap into space is of the utmost importance to our life, our knowledge, and our future as a species, but the pertinence of these and smaller triumphs of transportation to the creation of real-life global intellectual communities can best be assessed by the application of a simpler, mundane standard: does the space vehicle,[18] jet plane, or other conveyance allow people to commute as freely across the globe as they do across their neighborhood?

As in all matters technological, the immediate answers to this question are not only technological but economic, and for that matter political and social as well. A magnetic underground tube capable of moving passengers from New York to Los Angeles in 12 minutes is thought to be technologically feasible, for example, but the price tag is about 250 billion dollars,[19] far beyond the capacity of any company to sustain even with massive government funding. More practically, what contribution can supersonic transport between America and Europe make to global intellectual circles when the cost of such transport is well above the ability of most individuals or scholarly institutions to repeatedly afford? We need to acknowledge such economic realities, but having acknowledged them, we also need to be aware that the specific configuration of today's economic constraints is not relevant to the question of ultimate epistemological potentials, or of how far in principle technology can extend and amplify our cognitive capacities. (After all, the price of supersonic and even faster transport will likely sooner or later fall.)[20] In considering the contribution of transportation to the growth of knowledge via the possible creation of in-person global Academies and Vienna Circles, we therefore will try to cut through ubiquitous but fluctuating economic and similar intervening factors, and examine more fundamental relationships of transportation (movement of people in their entirety) and physical intellectual communities.[21] Such an examination in effect asks: given the economic and social feasibility of bringing any number of people from any place in the world into physical proximity for cognitive purposes, what would be the result for the growth of knowledge?

In one respect, the global mixing of minds on a nonrepresentational physical basis is already underway. W. W. Bartley III writes about the intellectual ferment of Karl Popper's seminar at the London School of Economics in the 1960s, attended by students and scholars from Israel, Ireland, America, and of course England, as well as by emigrés from central Europe including Popper himself.[22] Nor is such a hybrid environment exceptional or even new in structure. Centers of learning throughout history have attracted seekers of knowledge from all parts of the known world; in this regard, the main difference between today's universities and the schools of ancient Alexandria is the size of the intellectual net cast.

Such universal universities, however, are not really communities of scholars functioning worldwide in transit and in situ, but are rather glorified intellectual localities enriched with the infusion of geographically disparate minds. Apropos of McLuhan's observation that electronic environments have centers everywhere and margins nowhere,[23] the true global physical community can have no center or locale other than the globe itself.

Is such a world unequal to our current technologies of jet planes and bullet trains? Not at all. I myself, not ten years beyond my doctorate and hardly a world-renowned scholar, have attended intellectual events in London, New York, and Los Angeles, each constituted of scholars from 15 or more countries worldwide. That such affairs are still occasions or events demonstrates that we have not yet achieved physical global intellectual communities wherein one can travel from New York to London for heated discussion over tea as easily as one now travels from New Jersey to New York for heated discussion over coffee.[24] Nevertheless, the fact that such events are becoming increasingly commonplace demonstrates that attainment of the in-person global community may not be too far off.

The question remains: will such a community be of unmitigated benefit to the growth of knowledge? Nothing in an evolutionary world is an unqualified blessing. A concern frequently raised about the ease of physical mingling engendered by transportation technology is that this ease threatens our privacy and deprives our intellect of a certain saving distance, a space from others, necessary for rational contemplation. Historian-turned-critic of technology Lewis Mumford has made this point with increasing fervor for more than 50 years,[25] and acts on it to the extent of living in a remote, small town in upstate New York, far from airports and without automobile or television. Less extreme analysts such as Richard Sennett and Joshua Meyrowitz suggest that modern technologies—especially electronic media that bring the public into the formerly private home—are transforming the nature of privacy[26] so as to make it, if not explicitly inimical to leisurely reflection, certainly very much different from the type of quietude we popularly associate with rational thinkers from the Age of Pericles through the Enlightenment. (Kant lived his entire life in the vicinity of his native Königsberg, though after publication of *Critique of Pure Reason* he received numerous philosophers and dignitaries; Descartes on the other hand was something of a world traveller in the Europe of his day.) Related to these concerns about the loss of privacy is the view that too many cooks spoil the soup—that the very abundance of minds that can be communally brought to bear on a problem in our age of rapid transportation and instant communication may swamp the intellectual process and render the problem untreatable. This is indeed the rationale in favor of a representative as opposed to a referendum-type democratic government.[27]

A "nursery" stage in the development of ideas, or a time when newly created knowledge is not exposed to unrelenting criticism, seems to have a place in the knowledge process. (As we have seen in Chapter 5, an early protected stage is compatible even with the Popperian ideal of opening our ideas to the widest and strongest criticism available.[28]) However, for such a stage to be jeopardized by transportation and communication technology—for the concerns of Mumford and the others to hold merit—the technological intrusion would have to lead to the following consequences: first, the privacy lost or compromised by the technological incursion would have to entail cognitive as well as social aspects of the person, that is, the thinking as well as the social being would have to be impaired (although the two are interrelated, only Mumford really makes the equation between lack of privacy and the impairment of rational thought); and second, the loss of privacy would have to be technologically and socially uncontrollable or irremediable (lack of effective control is implied in most of the above concerns). In fact, neither consequence seems to be the case.

The existential philosopher Wolfgang Schirmacher, for example, argues that the very technological manipulability of privacy shows that this "privacy" is not part of the deeper human self which creates technology and, though affected by technology, continues as a human element throughout history.[29] We may add that the situation of our cognitive faculties in a Kantian/evolutionary framework places these too in a part of ourselves that is transcultural (at least until such time as we create new brains via gene splicing and artificial intelligence). On the second criterion—the lack of control or remedy—McLuhan's observation that North Americans go inside to be public (for access to television and telephone) and outside to be private (in automobiles)[30] reminds us that while some technologies unintentionally diminish aspects of our solitude, others surprisingly provide new occasions for privacy.

I would take this last recognition even further, and point out that often the very technologies which endanger our privacy also offer means to insure and improve it. The same car with which I contribute to and fight choking traffic in Manhattan takes my family and me to summers of peace and privacy on Cape Cod. Automobiles, buses, trains, and planes do much the same for many people in this crowded and shrunken world of ours, which yet has innumerable out-of-the-way nooks and crannies for refuge. Moreover, even if we imagine a Mumfordian nightmare world in which technology has rendered every last corner of the Earth accessible to millions, sectors of privacy could still be maintained by custom or law. Such social restraints have already been operating for many years in use of electronic media like telephones which, despite their round-the-clock vulnerability or capacity to ring 24 hours a day, disturb us in the middle of the night exceedingly rarely.

A further important measure of control comes from our ability to deliberately construct technologies that regulate or neutralize the loss of privacy of existing devices. The telephone answering machine gives us protection from unwanted calls when custom may fail, and continues a tradition of remedial media—technologies to offset problems caused by other technologies. Remedial media for the preservation of privacy hark back at least as far as the invention of the window shade to keep out Peeping Toms brought into being by the window.[31]

To the extent that we can control the degree and quantity of mingling fostered by the ease and speed of transportation technologies, we can expect such technologies to make their contribution along with telescopes and microscopes; data processing of computers; speech, writing, and printing; photographic and electronic media as communicators, vehicles of self-discovery, and creators of representational communities; and the cumulative weight of knowledge physically embodied in all technologies; to the continuation and expansion of the golden Age of Knowledge we have been building since the Renaissance. With the exception of some computers, swift global transportation technologies are the newest devices in this series (jet planes for civilian transport are but three decades old), and current economic conditions render these less readily usable than computers, since travelling around the world is more expensive than telecommunicating. The transportative–cognitive consequences of jet travel are thus only now beginning to be felt (in the case of space craft, not yet felt at all). However, the differences between artificial and human intelligence discussed in the previous chapter, and between representational and in-person communities considered here, suggest that the increase of physical human presence on the cognitive scene made possible by affordable air and eventual orbital travel will result in vivid dividends in the growth of knowledge (a primary reason, incidentally, that I favor human presence in space exploration[32]).

In comparison to the infinity of what we can know, any increase in our knowledge, however monumental, still leaves the sum total of what we know infinitesimal. Nonetheless, as we come to understand even a little of the knowledge process—how we know what little we know—we quite naturally wonder what meaning this cognitive process and the knowledge it produces have in the greater scheme of things. Humanity and this globe seem but a speck in the universe, albeit a speck with remarkable self-transcendent qualities. In our final chapter, we will consider the role of human knowledge and its technological constituents in the cosmic realm.

BEYOND PLANET

Transportation across and around the globe takes us to the verge of transportation beyond the globe: the leap from the Earth that brings us closer to our origins and future in the cosmos in a physical sense. Although the technological impact on the cosmos may be seen to have begun, from one perspective, on the day the first organic compound reacted on its environment on Earth, another reasonable candidate for birthdate would be the day the first humanly made machine was hurled into space. (Perhaps the former was the date of conception and the latter the date of birth?) On that latter occasion the embodied idea, the material difference, truly left its womb or nursery on Earth to make a difference, however small, in the universe around us; as was so clearly depicted in Kubrick's *2001*, on that day the tool of human affairs became a needle in the stars. We are, after all, minutiae in a vastly larger realm, and the launching of Sputnik and other first space probes announced to this likely non–listening realm—or perhaps confirmed an announcement already made by radio broadcasts—the presence of an antientropic spiraling community of life, intelligence, and technology on this planet which may be without parallel in the universe. The global community of Earth, itself not yet fully formed, thus has already begun spilling into the cosmos. We accordingly conclude with some thoughts about the ultimate lodging of the technological cognitive evolution we have considered in this book.

NOTES

1. See "Communities of Representations" later in this chapter for a detailed description of computer teleconferencing via telephone lines, and associated notes for sources and references. Carl Townsend's *Electronic Mail and Beyond,* Belmont, Calif.: Wadsworth, 1984, provides a thorough introduction.

2. My wife Tina Vozick calls my attention to a variety of media designed to accompany rather than substitute for humans in the communication process (blackboards and posters would be examples). Nonetheless, these media are peripheral to the primary purpose of writing, which, as Freud put it, is "in its origin the voice of an absent person" (*Civilization and Its Discontents,* p. 38). See also chapter 3 note 19 for Mallarmé on the relationship of writing and speech.

3. See Felipe Korzenny, "A Theory of Electronic Propinquity," *Communications Research,* January 1978, pp. 3–23, for a summary of nonverbal gestures gleaned from a series of filmed conversations, and arguments on the impoverishment of communication that results when these visual cues are absent.

4. Giambattista Vico (1668–1744) objected to Descartes' separation of science and history, and attempted to develop a philosophy of science as a cultural (not merely mathematical-abstract) activity. See Giorgio Tagliacozzi, ed., *Vico: Past and Present,* Atlantic Highlands, N.J.: Humanities, 1981, and in particular Eckward Kessler's "Vico's Attempt Towards a Humanistic Foundation of Science," pp. 73–88 in that volume.

John Stuart Mill and later the Vienna Circle (see, for example, Rudolf Carnap's *The Logical Structure of the World,* trans. R. George, Berkeley: University of California Press, 1967 [1928]) strove for a philosophy of science as a predominantly (often purely) logical, transcultural phenomenon. The Vienna Circle urged this in conscious contrast to the Marxist view of economic relations providing a conditioning and even determining "infrastructure" for science, and to the Freudian perspective of science—like all human activities—as an expression of an irrational unconscious.

Popper's transcultural view of science and his critiques of Marxism and Freudianism are in line with the Vienna Circle's goals. But he expressly departs from the Circle by emphasizing the origins of knowledge in areas other than and in addition to sensory experience (Popper's intellectual forebear is Kant, as opposed to the Lockean-British empiricist roots of the Vienna Circle), and by stressing the falsificationist as opposed to verificationist-probabilistic nature of scientific progress.

Most importantly, Popper's evolutionary epistemology places human knowledge in a transcultural context not via abstraction into mathematics or pure sense data (as in Descartes and the Vienna Circle), but via connection to a naturally selective process of interaction with the environment which is fundamental to all living organisms. In this way, Popper in effect subsumes the cultural concerns of Marxism into a more far-reaching biological model—culture (economics and so forth) is integrated, not denied.

See Popper's "Who Killed Logical Positivism" in Schilpp, ed., *The Philosophy of Karl Popper*, pp. 69–71, and Victor Kraft's "Popper and the Vienna Circle" in the same volume (pp. 185–204) for discussions of Popper and the Circle. See Popper's "Normal Science and Its Dangers" in I. Lakatos and A. Musgrave, eds., *Criticism and the Growth of Knowledge*, Cambridge: Cambridge University Press, 1970, pp. 51–58, for Popper's disagreement with the leading cultural relativist in philosophy of science of our time (Thomas Kuhn, *The Structure of Scientific Revolutions*, 1962/1969). See also note 5 below.

5. See Gerard Radnitzky, "Popper as a Turning Point in the Philosophy of Science" in P. Levinson, ed., *In Pursuit of Truth*, pp. 68–69, for elaboration of why cultural relativism is nonevolutionary even though its daily meat is change.

6. The reshaping of reality via the partially cultural (and partially biological) activity of technology defines the degree to which reality is not culturally transcendent.

However, the laws of reality or nature—more specifically the aspects of these humanly perceived laws contributed by nonhuman reality and nature—still seem to have the upper hand or greater resiliency in the technology-reality relationship. Thus, we may locally contravene the unplanned aspect of the laws of natural selection via gene-splicing technology, but this contravention can take place only by appeal to the patterns of molecular biology which govern the behavior of genes; and the outcome—the organic characteristic or organism produced—will ultimately survive or not in accordance with the naturally selective vicissitudes of the universe at large.

See "Can Technology Change Natural Laws" in my "What Technology Can Teach Philosophy." See also chapter 7, note 23; and see also chapter 5, note 14 for Kantorovich's view (compatible with Dewey) of discovery of the deeper laws of nature via disturbance of surface phenomena through technology.

7. Stephen Jay Gould, "The Evolutionary Biology of Constraint," *Daedalus*, Spring 1980, pp. 39–52.

8. Resistance from the scientific establishment is frequently the most serious social impediment to growth of knowledge, and has been explored in Kuhn's *Structure of Scientific Revolutions*, chapter 12, and Stephen Toulmin's *Human Understanding*, New York: Oxford University Press, 1972. The so-called "strong programme" in the sociology of knowledge emanating from the University of Edinburgh has studied the proneness of knowledge (especially scientific knowledge) to become a creature or captive of dominant economic and political structures. Standard works in this field are Barry Barnes, *Interests and the Growth of Knowledge*, London: Routledge and Kegan Paul, 1977, and *Scientific Knowledge and Sociological Theory*, London: Routledge and Kegan Paul, 1974; and David Bloor, *Knowledge and Social Imagery*, London: Routledge and Kegan Paul, 1976. See I. C. Jarvie's critique of these and similar works in "Popper on the Difference Between the Natural and the Social Sciences" in P. Levinson, ed., *In Pursuit of Truth*, pp. 83–107. See also William Broad and Nicholas Wade, *Betrayers of the Truth*, New York: Simon & Schuster, 1983 for a popular account of the distortion of science via social factors ranging from pride to simple greed, and Loren Graham's *Science and Philosophy in the Soviet Union*, New York: Knopf, 1972 for a detailed assessment of the growth of various fields of knowledge in a Marxist, totalitarian environment. See also Walter Sullivan, "Soviet Scientists Often Thwarted," *New York Times*, October 7, 1986, pp. C 1,7 for discussion of recent political difficulties in Soviet research.

A problem with many of these social investigations of knowledge (especially Barnes' and

Bloor's) is that they tend to blur the distinction between investigation of external reality as warped (or perhaps, on occasion, pushed in the right direction) by noncognitive social factors, and investigation of external reality as nothing more than a reflection of social factors (or cognition as intrinsically social, with nonsocial biological and logical facets viewed as fictions or ideologies). Kuhn sets the tone for this by viewing resistance to scientific progress and revolution not as an obstruction of knowledge progressing closer to the truth, but as a resistance to the shift of one paradigm to another paradigm that may or may not be closer to the truth, whatever "truth" itself may be. See also note 8 in chapter 1 of the present volume.

9. Aharon Kantorovich, "The Collective A Priori in Science," *Nature and System*, 1983, pp. 77–96.

10. A modem is an electronic device that translates computer-generated printed characters into a form that can be transmitted through ordinary telephone wires, much as the conventional telephone transforms the human voice into an electronic energy pattern suitable for transmission. (The modem's job is actually easier than that of the conventional phone, since computer-generated text already exists in electronic code.) Commercially available modems usually transmit at speeds ranging from 300 to 2400 bits per second (eight bits equal one alphabetic or numerical character), and may be inside the personal computer, or mounted externally via wire connections.

See Alfred Glossbrenner, *The Complete Handbook of Personal Computer Communications*, New York: St. Martin's, 1985, ch. 3, and Townsend, *Electronic Mail*, ch. 3 for further details on modems and computer communication via telephone lines.

11. See the March 1986 *IEEE Transactions on Professional Communication's Special Issue on Computer Conferencing*, edited by Valarie Arms, with articles by Andrew Feenberg, Elaine Kerr, myself, and others, for further descriptions of the methods and outcomes of computer conferences. Hiltz and Turoff's *The Network Nation*, Reading, Mass.: Addison-Wesley, 1978, and Hiltz's *On-Line Communities*, Norwood, N.J.: Ablex, 1983 offer detailed sociological analyses of computer conferencing. See also Hiltz's forthcoming report on "The Virtual Classroom," and my edited *Electronic Chronicles*, Greenwich, Conn.: JAI Press, forthcoming.

12. Electronic mail (one-to-one or one-to-many) and computer conferencing (many-to-many) have been used at least since the 1960s by the U.S. Army's "ARPANET" and similar government systems. EIES is the oldest and most sophisticated central computer conferencing system available to the professional world and sectors of the general educated public. See Glossbrenner, *Handbook,* ch. 15 for brief descriptions of other current computer conferencing systems, including Notepad, ConferII, and Participate, all of which have been used for a variety of scholarly and educational activities.

I am a member of the on-line faculty of the Western Behavioral Sciences Institute, and have conducted three seminars with their group via computer conferencing since 1984: "Information Technologies as Vehicles of Evolution" in September 1984; "The New Global Village" in August 1985; and "Space: Humanizing the Universe" in September 1986. (The last is also the title of a public "Electure" I conducted with more than 400 participants on Participate's computer conferencing system on The Source in April 1984.)

I'm founder and director of Connected Education (my wife Tina Vozick helped me with the founding and is associate director), and have taught on-line courses in "Artificial Intelligence and Real Life," "Computer Conferencing in Business and Education," "Issues in International Telecommunications" (together with faculty located in Washington, DC, London, England, and Tokyo, Japan), and "Ethics in the Technological World." Other courses offered in the Connect Ed Program include "Applications in Telecommunications," "Telelaw," "The Microchip Economy," "Technological Forecasting," "Electronic Publishing," "Desktop Publishing," and "On-Line Writer's Workshop."

Connected Education is also planning an on–line Ph.D. in Philosophy of Technology with Polytechnic University of New York.

National Technological University and Telelearning are examples of educational programs that make partial use of computer conferencing, relying in addition on computer-assisted instruction (sending computer disks through the mails), and on audio and video teleconferencing. See Robert Cowan, *Teleconferencing: Maximizing Human Potential,* Reston, Va.: Reston, 1984, for descriptions of these and related forms of teleconferencing.

For an early perceptive prediction of computer conferencing, see Vannevar Bush's "As We May Think," *Atlantic Monthly,* July 1945, pp. 101–108, where he envisions a "Memex" device that writes, reads, files, and communicates with a screen and keyboard that fits in a desk-size cabinet.

13. See "Computer Clubs Growing as Hobbyists Share Data," *New York Times,* July 24, 1986, pp. C1, 10, for accounts of some of the social aspects of computer conferencing. "People meet, shop, play, and learn," the article says.

People also have been known to fall in love and have a variety of sexual experiences via exchange of written messages in computer conferencing. See my "Impact of Personal Information Technologies" for some general data, and Lindsay van Gelder, "The Strange Case of the Electronic Lover," *Ms,* October 1985, for one notorious case.

14. See Chuck Taylor, "Lower Costs, Better Sound via Audio Teleconferencing," *Washington Business Journal,* September 23, 1985, p. 13, for description of some of the activities of the Darome Connection, one of the leading audio teleconferencing facilitators. Although extensive training can be conducted via telephone conferencing classes, the primary use is among business people who already belong to a structured community (for example, a corporation)—thus the audio conference serves to buttress or augment an already existing in-person community, rather than provide the basis for a wholly or primarily representational group.

15. See note 3 above.

16. See Paul Levy's *G. E. Moore and the Cambridge Apostles,* New York: Holt, Rhinehart, Winston, 1979, for details on the Apostles' initiation rites.

17. See Sherry Turkle, *The Second Self,* New York: Simon and Schuster, 1984, ch. 6, and Steven Levy, *Hackers,* Garden City, N.Y.: Doubleday, 1984, for accounts of the boy geniuses whose passion and labor with computers provided the spiritual basis and much of the literal architecture and programs of the personal computer revolution, but whose public images have since come to be identified with those in the community who pride themselves on illegally entering private computer records. For example, the protagonist in the movie "War Games" (1983) unexpectedly gains access to a Defense Department computer while trying to break into another computer system and nearly initiates World War III. His understanding of computers, however, saves the day in the end.

18. See Stephen W. Kothals-Atles, "The Aerospace Plane: The Technological Feasibility and Policy Implications," Report #15, Program in Science and Technology for International Security, MIT, July 1986, for a detailed assessment of various military and commercial possibilities for hypersonic air transport (twice the speed of current Concorde SST), often referred to as the "Orient Express." Kothals-Atles concludes that such a vehicle is "probably unsuitable as hypersonic civilian transport"—but this assessment is based on current projections of operating expenses, which may or may not hold in the aftermath of unpredictable technological breakthroughs in the future. Kothals-Atles also concludes that current estimates of three billion dollars needed for development of such aircraft are probably low.

Civilian use of the space shuttle in the near future remains in doubt in the wake of the Challenger explosion in early 1986. Ronald Reagan's decision to shift commercial launchings of the shuttle into the private sector (David E. Sanger, "President's Order on Space Shuttle

Called Era's End," *New York Times,* August 17, 1986, pp. 1, 32), however, may stimulate eventual use of the shuttle for workaday (including academic) transport.

19. Reported on "Eyewitness News," WABC-TV, February 13, 1978, 6:00 to 7:00 PM. Henry Kolm explains in "An Electromagnetic 'Slingshot' for Space Propulsion," *Technology Review,* June 1977, pp. 61–62, that if such "vehicles were run inside an evacuated [vacuum] tunnel, speeds would be essentially unlimited."

20. In 1986, airlines operating from the East Coast of the United States offered flights to many parts of the United States for less than $100 and to some parts of the world for under $200 round-trip. These are reductions of several hundred percent from average airfares in the 1970s.

21. A search for transportation-cognitive relationships that cut deeper than fluctuating costs of transportation runs counter to Marx's view of economics as the deepest infrastructure (and cognitive activities as a dependent surface structure). See chapter 4, "Marx on Mind in Material" for a discussion of Marx's epistemology of the stomach, and a critique of this view from the standpoint of the evolutionary epistemology of the brain and tool advocated in this book.

22. W. W. Bartley, III, "A Popperian Harvest," in P. Levinson, ed., *In Pursuit of Truth,* pp. 249–256.

23. See, for example, Marshall McLuhan, "The Brain and the Media: The 'Western' Hemisphere," *Journal of Communication,* Autumn 1978, pp. 54–60. See also Murray Schafer, "McLuhan and Acoustic Space," *The Antigonish Review,* Summer–Fall 1985, pp. 105–113, and his discussion of McLuhan's manuscript "Changing Concepts of Space in an Electronic Age."

24. Fully integrated globalization would likely eradicate such differences in caffeinated style, resulting in coffee and tea being equally available and well-brewed in both London and New York. This would on the one hand make the world less interesting to travelers, but on the other hand make the world more accommodating as a whole to differences in tastes, as people could consume either beverage at its best in two places rather than only in one place or the other.

The challenge for global intellectual communities is how to bring diverse cognitive styles together without compromising their diversity—how to give different minds access to each other without blurring their differences. See the text immediately following in this chapter and the notes below for discussion and citation of various views on technology and social homogenization.

25. Thus, Mumford contends "that the maintenance of distance both in time and space was one of the conditions of rational judgement and cooperative intercourse, as against unreflective responses and snap judgements" (*Technics and Civilization,* New York: Harcourt Brace, 1934, p. 295).

26. Richard Sennett, *The Fall of Public Man,* New York: Knopf, 1977; Joshua Meyrowitz, *No Sense of Place,* New York: Oxford University Press, 1985. Meyrowitz focuses on electronic media (principally television) and Sennett on broader segments of technology (for example, industrialization and urbanization) as the source of this mixing of public and private spheres.

27. Personal telecommunicating computers in homes and places of business obsolesce the physical (but not necessarily the social-political) reason for representative government, allowing citizens to vote directly on bills and issues. Problems arise in that citizens with the technological wherewithal to vote are not necessarily informed on the object of the vote, and, in addition, representative democracy entails a combination of following the dictates of the electorate and exercising creative leadership which the electorate can later either support or reject.

Donald B. Straus explores some of these problems in his forthcoming *Computers and the Democratic Process* (in preparation).

28. See chapter 5, note 36.

29. Wolfgang Schirmacher, "Privacy as an Ethical Problem in the Computer Society," in C. Mitcham and A. Huning, eds., *Philosophy and Technology II,* Boston: Reidel, 1986, pp. 257–268.

30. Marshall McLuhan, "Inside on the Outside, or the Spaced-Out American," *Journal of Communication,* Autumn 1976, pp. 46–53. Cellular telephones available in automobiles may soon begin to invalidate this example, but the principle of solutions or balances to technological problems arising unexpectedly in other technologies remains.

31. See my "Media Evolution and Rationality as Checks on Media Determinism" in S. Thomas, ed., *Studies in Mass Communication and Technology, Volume 1,* Norwood, N.J.: Ablex, 1984, pp. 231–237, for further discussion of media-remedial media. The example of the Peeping Tom and the window shade occurred to me after reading Edward Wachtel's "The Influence of the Window on Western Art and Vision," *The Structurist,* 1977/1978, pp. 4–10.

32. See my "Cosmos Helps Those Who Help Themselves: Historical Patterns of Technological Development, and their Applicability to the Human Development of Space," paper presented at Space Agenda: Context and Opportunity Conference at M.I.T., Cambridge, Massachusetts, on April 5, 1986, and to be published in C. Mitcham, ed. *Research in Philosophy and Technology,* vol. 9, 1988. See also my "Space: Humanizing the Universe," paper presented at the Western Behavioral Sciences Institute, La Jolla, California, on July 15, 1986.

Chapter 9

Technology and the Evolution of the Cosmos

In a universe constantly evolving, occasions when the rules of this evolution change are of profound importance. And if in observing these rarest of occasions we notice some pattern in their occurrence, then a change in this pattern of rule changes is of profounder importance still.

In the course of this volume we have made reference to several evolutionary breakpoints, including the changes from inanimate matter to life and from life to human intelligence. In each of these cases, the second or emergent phenomenon operates according to rules that radically exceed (yet subsume rather than obliterate) the properties of the first type of existence. Thus, the matter of life must obey the laws of chemistry and physics, even though it copies and extends itself in a way that inanimate matter never can; and the human being is thus far as mortal as any living organism, and subject to the laws of chemistry and physics as well, even as we use our unique intelligence to inquire into the rules of life and matter, perhaps to change them in some humanly desirable way. We have also seen that the rational component of our intelligence, emerging from a nonrational matrix, is still very much prone to the lure of nonrationality in the form of irrationality, even as this emergence gives us the option of rational action. The status of technology in this yet unfolding scheme of things is the topic of this chapter.

We can better understand technology in the context of these more-from-less occurrences by considering the conditions that must precede or set the stage for such departures. Klaus Krippendorf, writing of the relationship of paradox and information theory,[1] has provided a model that may be applicable to the emergence of new rule systems in the universe. Krippendorf suggests that a paradox may be seen as a problem that requires an infinite (that is, in principle an ever unobtainable) amount of information to resolve in its originating situation. For example, an attempt to make sense of the seemingly simple statement, "This statement is a lie," sets off a quest for resolution in one of two possible conclusions, "The statement in question is true," or "The statement in question is false." Upon arrival at either pole, however, the hopeful supplicant is directed to the opposite number—that is, given to understand that the solution lies in information somewhere over the next hill—where the seeker is promptly returned to the point of origin, to be redirected back to the opposite pole, and so forth and so on, ad infinitum.[2] No amount of following such instructions and journeying back and forth can provide information sufficient to comprehend the statement; the only way out of this ping-pong existence is the realization that the statement is undecipherable in a universe of just truth and falsity—that is, resolution requires information radically beyond that found in the either/or universe of truth and falsity of which the troublesome statement is presumed to be a part. The crisis generated by the undecipherable statement thus propels us to recognize a new class of statements, paradoxes, which are neither true nor false yet have characteristics of both. Here in this new realm of paradox we at last are able to make some peace with the difficult statement: although we can never understand it in terms of truth and falsity, this very recognition is itself a form of understanding, and provides a sort of stasis that we can live with.

For what unresolvable crises of existence, what conflicts among rules governing the evolution of the universe, might technology provide such stasis?

A human universe without technology would be unremittingly paradoxical: via human intelligence, the universe can think, deliberate, and attempt to plan its future, that is, operate in a manner sharply at variance with the apparently blind, unplanned procedure of the universe from which human intelligence came; but without technology to implement such thoughts and plans, the human brain can make no real difference in the universe, and would accomplish as much as a chicken with its head cut off running to and fro between "This statement is a lie is true," "This statement is a lie is false." Of course as we have seen, there is no such animal as human intelligence without technology—nor indeed life itself without some prototechnological manipulation of the environment—but we can plainly observe that the less technology humans have had available, the less they have been able to surmount the futility of unimplementable knowledge or of viewable yet unreachable goals. Even today, our advanced technological state is not up to the task of giving us immortality, freedom from disease, unimpeded movement through the world, or the realization of many other ancient dreams. In some instances, such as the technological engendering of various diseases (for example, lung cancer from exposure to asbestos) technology may progress in a two-steps-forward/one-step-backward manner. All in all, however, the import of technology is clear: through these adventures in materialization, the universe attains a new plane, a new era in its history, in which rationality, aspiration, will, dreams, and the host of human ghostly undertakings find increasingly influential expression.

The technologically implemented plan departs not only from previous types of existence, but also from the rules that governed their emergence, since all forms of existence prior to human technology—even those as drastically divergent as life and nonlife, or human intelligence (which actually emerged coequally with rather than prior to technology) and nonintelligent life—share a common mode of conception characterized by an utter lack of planned parenthood. The best laid technological plans, of course, have no more guarantee of success than the unintended outcomes of DNA combinations, and this endemic uncertainty of the universe in which technologically endorsed forethought makes its entrance means that the new technological order thus far has been more a challenge than a sequel to the universe's established, unthinking mode of doing business. However, the very challenge and initiation of planned material, regardless of its outcome, is revolutionary on a meta or rules level. Just how far this outcome may go—whether it can and will ever overthrow the iron rule of uncertainty which has governed the universe's conduct heretofore—must be a central theme in a cosmology of technology. Entailed in this issue too is the question of whether such a shift should be encouraged to occur— whether fallible planning is preferable to the mercy of uncertain winds which nonetheless have brought us quite far.[3]

We will consider the nature and prospects of the technological agenda in the cosmos by examining its two prime, interrelated components: rationality and duration. The first, the agency of planning, speculation, and foresight, is what makes technology different from any other known material in the universe; as we look at how this human wellspring seeks to inform technology and thereby the universe, we also will consider the extent to which the technological implementation of rationality may pervert the rational factor. The second component—duration—is what makes the technological difference count, and discussion of the materialization and relative permanence of mental products in technology will bring us face-to-face with the mind-and-matter issue one more time. Crucial to these considerations is the role of knowledge, which in being both embodied in and created by technology constitutes an inextricable loop with technology that works for the growth of both.

THE BURDEN OF RATIONAL TECHNOLOGY

The question of technology as an instrument of deliberate design in the universe amounts to a question of the extent to which technology can be rationally directed. If technology were exclusively an expression of human irrational will, then the outcome of such expression could in no sense be construed as a plan or design, for the consequence of blind will is no different on the issue of foresight than the consequence of natural selection. But is any human activity exclusively irrational—that is, irrational to the point of zero rational content? Rational reconstructions (like Freud's) of supposedly irrational systems, or attempts such as Hegel's to impute an underlying cunning or logic to seemingly nonrational, even impersonal, vast historical systems, suggest that even these irrational systems have an intrinsic rational component (as indeed "post"-rational systems certainly would—see Chapter 2). And if an apparently irrational system turns out to fulfill some rational agenda, then its expression in technology surely would constitute some sort of plan, albeit perhaps unconscious, on the part of its human initiators.[4] Furthermore, the requirement of rationality need not be met in all or even most technological expressions in order for the cosmos to be rationally directed via technology; an occasional or even isolated rational technological thread awash in a sea of irrationality or nonrationality could in principle still be the source of a deliberately constructed cosmos. Thus, to entertain the possibility of a purposely created universe through technology we have only to counter the view that a rationally directed technology is for some reason never the case.

Despite the seeming extremity of such a view, it is not easily countered. To begin with, strong arguments against the possibility of a rationally directed technology abound in the many arguments against the efficacy and even the existence of a rational faculty independent of technology. As we saw in Chapter 2, these arguments are impressive, and ought not to be treated too lightly. Nonetheless, on the basis of the evolutionary rejoinders developed in that chapter, we would seem on good ground in concluding that the rational factor is real and effective in human affairs, at least in the hypothetical absence of possible technological distortion.

But technology is in fact rarely if ever absent from attempts to implement reason, and the issue of what technology does to the human rational faculty is probably the most frequent objection to the feasibility of a rationally directed technology raised by those who do not deny the operation of rationality on nontechnological grounds. The strongest form of this argument is that the expression of rationality in technology is inherently incompatible with or destructive of the rational impulse: a self-defeating oxymoron. Elaborations and variations espoused at times by Jaspers,

Ellul, Marcuse, and many others, and accepted in many elements of the popular culture, hold that technology functions autonomously or semi-autonomously of its human creators, and this social autonomy dilutes, perverts, or otherwise compromises the rational spark. A favorite variant (for example, Ellul's) has the technological system acquiring an antihuman design and momentum of its own; Marcuse argues more specifically that rational technology becomes a humanly wielded totalitarian technology that extinguishes the rational, inventive force from which it arose.[5]

We encountered related arguments, in less refined form, in the claims discussed in Chapter 6 that television is destructive of the mind. The bottom-line response to the more sophisticated scenarios of Ellul and Marcuse, however, is much the same: the very fact that we are aware of and can attempt to assess the dimensions of this technologically induced irrationality demonstrates that this irrationality is not as pervasively crippling as its critics suppose it to be.

Moreover, remedial technologies such as the window shade and the telephone answering machine (mentioned in Chapter 8), deliberately designed to address a technologically engendered problem (in these cases, lack of privacy), demonstrate that the rational awareness of technological danger need not be limited to awareness only, and in fact can be channeled or embodied in a technology rationally directed toward coping with the problem. The evolution of television provides an especially vivid example. For years a television hallmark, often considered one of its definitive characteristics, was the evanescence or instantly disappearing quality of its pictures. Unlike the words on printed pages, which admit to repeated and leisurely inspection, the images of television were gone for good the instant after they were seen. Critics such as Mumford cited this drawback to great effect, even to the point of suggesting that television was in the process of creating a society without memory, akin to the brain-damaged individual who lacks all sense of retrospection and anticipated time.[6] The danger of a memory-less society of course was never real—sales of books have continued to increase throughout the tenure of television[7]—and as early as the 1960s the widespread broadcasting of video recordings rather than live performances on television (and the video recording of live performances for posterity) undercut the charge of the intrinsic fleetingness of the medium. By the beginning of the 1970s, students were using portable videotape recorders to preserve lectures given on campus by critics such as Mumford about the memory-destroying quality of television. But the decisive development in giving television viewers an actionable memory and sense of anticipation was the introduction of home videotape recorders in the mid-1970s; these are now estimated to sit next to televisions in 30 percent of the homes in the world with television.[8] Via the VCR (video cassette recorder), the viewer can record any program on television for repeated viewing and perusal—indeed, most VCRs can be easily set to record programs on television for a period of two weeks subsequent to the setting. The VCR thus serves as a striking counterexample to theories about the irresistibility and irrationality of technology: here our presumably wounded rationality identified a problem in a prevalent, overwhelmingly popular technology, and then went on to devise a technological addendum that has redressed this problem. Could this have been accomplished by a rationality irreparably mangled by technique, or in a technological system effectively beyond our rational control?

Such remedial developments indicate not that technology does not engender many unintended and sometimes dangerous consequences, but that such consequences are not in principle utterly or even effectively extinguishing or crippling of our rational faculty. That this rationality will prevail in any given situation or indeed even ultimately is by no means assured. The buildup of nuclear weapons starkly demonstrates how insanely dangerous a technological consequence (in this case, not completely unintended) can be; the question of whether the nuclear arsenal will be as susceptible as television to a rationally directed technological solution remains unacceptably unanswered. When I look in the eyes of my children I feel deeply ashamed and angered that our species should ever have let things come to this. But I know that the blame lies not in the technological impulse but in the savagery of the human soul—this faculty that we have for meting out death and despair, that apparently predates all but the most primitive technologies (see, for example, a report of the sadistic massacre of 1400 men, women, and children in pre-Columbian America[9]—a massacre conducted by the epitome of Rousseau's natural technologically-unperverted man). And I know that were it not for the intervention of another sort of technology, the medical kind, many children would not be alive at all today. But most of all I would argue as a purely practical matter that the only hope we have of disarming the nuclear demon lies in a nurturing of whatever technologically expressible rationality we have left at our disposal in this massively self-destructive age. If, as many claim, it is our rationality which has already been totally disarmed by technological systems gone utterly beyond our control, then all is lost. However if, as I would argue, our capacity for rationality not only survives but flourishes in powerful mixtures with technology even in these rampantly irrational times, then I think it possible and even likely that through this rational-technological vehicle we will leave the menace of a seething nuclear hell behind.[10]

Our survival of the nuclear age has direct bearing on the question of the ultimate permanence of technological change, because in an exquisite irony we have developed the capacity to blow our planet to smithereens at precisely the same time that we have developed the ability to leave our planet and extend our technological existence to the cosmos at large. A cardinal characteristic of technology from the standpoint of the growth of knowledge is the fixation or permanentization of knowledge in technological embodiments—knowledge that otherwise would be totally dependent on the vagaries of individual memories and lifespans. The huge storehouse of knowledge that the technologies of Earth have already become, however, would likely be ruined beyond repair in an all-out nuclear war, with the result that the heretofore permanent flame of technology, the amalgam of human wisdom throughout the ages, would be extinguished on Earth and precluded from ever illuminating the universe around us. Nuclear holocaust would be not only a personal and planetary but a cosmic tragedy, for it would deprive the technological revolution of any possible cosmic significance.

All the accumulated wonder of humankind, all the cognitive-technological successes and more than we have attempted to explicate in this volume, would come to naught in a nuclear rubble. Even in the absence of nuclear war, the promise of our technological endeavor could well suffer the same lonely, inconsequential fate without a transportation technology to take our endeavor to the universe. An Earth and a human population remade by appropriate technology might well endure forever; but minus some vehicle to connect the Earth to the cosmos, the human technological spark would be as unknown and insignificant to the cosmos as if it were snuffed out tomorrow in a nuclear rain. Moreover, the Earth is at ultimate risk from the nova of our sun and perhaps sooner from comet showers;[11] thus an Earth-bound civilization, regardless of its technological sophistication, would likely vanish without cosmic import. (Endurance of an Earth-limited technology would allow its possible discovery by some hypothetical alien intelligence at some future date, thereby hooking our intelligence into the universe in this passive—from the human perspective—way. In the event of our planet's destruction by natural or technological means, remnants of our civilization—depending on the extent of the destruction—could similarly be available to some future alien inspection. Then again, were the cosmos already inhabited by advanced alien space travel technologies—or generative of these on a regular basis—our gifts of intelligence and technology would presumably not spell a fundamental difference in the technological order.)

But we have indeed already touched the stars, albeit only with the very edges of our fingertips, and this reality of space travel means that the cost to the cosmos of nuclear or other self-inflicted annihilation would be commensurately real, a sudden and possibly final scotching of an ancient evolution which has already begun spilling out into the boundless universe. Beginning with our first tentative gropings into space—and continuing, one hopes, with the eventual development of multigenerational, cryogenic, near speed-of-light or perhaps faster-than-light travel[12]—our technological progeny and material imaginings ripple across all eternity and infinity. This is what a petty nuclear squabble would lose for us, and for the cosmos.

Essential as human attainment of the stars is for us and the cosmos, however, it will be no bed of roses. Human commerce with the galaxies will carry not only our vision and problems on Earth, but also the unmeasured aspects of our pathology not a consequence of our existence till now in the womb of our planet. Further, despite the technological equalization of physically minuscule humans and the vast, horizonless universe, human civilizations in the stars will undoubtedly be subject to difficulties due in large part to the very enormity of these new surroundings. Thus a successful movement into space will if anything heighten the need for a continuingly active, unrestricted rationality. And a technological shift of our sphere from Earth to the universe beyond entails an equivalent increase in ethical responsibility: should we set off a nuclear or equally destructive event on this larger scale, it could take not only the Earth but parts of the universe down as well.

Advocacy of technological solutions to our current problems on Earth is thus not, as many critics claim, advocacy of a technological escapism. It is rather the urging of an evolutionary maturity, the encouragement of a course of action grounded in the observation that we are intrinsically a species that not only seeks to describe and understand but also remedy and surmount our problems by taking them in technological hand. The logic of this lesson is that the ultimate residence of our species—the environment in which we must play out and create to some degree our destiny—is no more the Earth than it is the country or neighborhood or building in which we were born or now happen to live. It is all of these, and more. We end our inquiry into the products of the human mind with several observations on what the extension of human production to the cosmos may mean for this final human environment.

TURNING THE UNIVERSE INSIDE OUT

On the day our first space ships pierced the clouds around the Earth, a thinking universe might have shuddered with delight at the prospect of its long, tedious eons of unawareness being near an end. But as far as we know, we are the thinking part of the universe, and the new era at hand is the hand of human thought, of cosmic material molded to the unpredictable, multiple dictates of human mentality.

Awareness and self-awareness, intention, deliberation, dreams, fallible rationality, and unclear foresight—these are the hope of a universe awakening from something deeper than slumber. But human hope remains only a fairy tale until embodied in the human steel of technology and its properties of physical extension, durability, and permanence. The rational, directed component of technology is the hope of the universe; the nuts-and-bolts material of technology is what makes the hope real.

Mind and matter, matter and mind—how these two have danced across the history of philosophy. How the human intellect hungers, seemingly in vain, to get at just that moment when the material brain begets an immaterial idea. But the thrust of this volume is that that moment is just half the story, or perhaps not the story at all. For we have seen that matter and mind interact not just at the one spectacular instant of an idea's conception, but at the equally breathtaking instant of an idea's embodiment in a material technology. And we have also seen that the conceiver of immaterial ideas is best described not as an immaterial mind or mentality, but as the living, thinking material of the brain—atoms, molecules, proteins, and larger structures coaxed to generation of ideas by conversance with ideas already generated by other brains.

What existence does a transmaterial idea have independent of this material sandwich of brain on the one side, technology on the other? Does an idea ever exist at all without some material substrate—either of the human brain; communication (representation) via speech, writing, photochemical, or electronic media; or embodiment in any technology? The uniqueness of the circumstances of authorship of ideas in the brain in comparison to all other physical and living processes amply warrants our designation of this mental material as extraordinary material: mind. Yet we must not allow the uniqueness and criticality of this distinction to obscure an equally significant sameness—an essential materiality that runs from inert molecules to bubbling peptides of life through the neurons that hum in the brain and refashion the inert matter of the cosmos as technology. Mind serves as the crucial value-adding handler in this long process—the giver of grace, to put the matter spiritually—in which the permanentization or fixation of ideas in technology is less a conversion of immaterial into material than a transfer of properties from a relatively unstable (though generative) material base (the brain) to a relatively stable (though nongenerative) material container: technology. The revolution of technology is thus a revolution of spirit—a triumph of mind over matter—only insofar as spirit and mind are understood to be terms that justly call attention to the unique generational capabilities of neuronal matter situated in a human environment.

The extension and relative endurance of ideas in technology comes into a cosmos still very much subject to entropy and vicissitudes far crueler than even the worst that human rationality has on occasion dished out (natural plagues and volcanos have taken far more human lives than our most brutal wars). To what degree can even the best of human intentions embodied in the most enduring of technologies counter this universal force? Even Norbert Wiener, who pioneered an understanding of life and technology as coequal antientropic developments, was prone in the end to view the mightiest of human-technological accomplishments as antientropic islands in an overwhelmingly entropic sea[13]—as just so much heroic running up an infinite down escalator. However, Buckminster Fuller, no less a pioneer in the explication of human technology in an entropic universe than Wiener, saw that the very existence and evolution of life, intelligence, and technology in the first place would be unlikely in a universe ruled pervasively and ultimately by entropy, and thus he concluded that the living-technological revolution might well go deeper than entropy—and eventually find a way through human ingenuity to reverse the very motor of the downward cycle.[14]

The key to an optimistic interpretation lies in recognizing the unusual nature of the stability that the technological embodiment of fleeting ideas brings to the cosmos. Ordinarily an increase in stability—as in the backward slide from life to nonlife—results in an increase in sameness that feeds entropy. The dynamic process of life counters this tendency by incessant creation of diversities of structures; but in the absence of technology, these structures are highly vulnerable and (in the case of human ideas) evanescent. Technology combines the best of stability and diversity by permanentizing the increase in organization and design, of complexity and differentiation, that results from human thinking. Herbert Spencer anticipated the significance of this development more than 100 years ago, in his description of progressive evolution as a change from disorganized sameness to organized differences.[15]

An utterly successful technological reformation of the cosmos thus would remake all the universe's material into organized material, designed to be more hospitable to life and humanity. As we have already seen in Chapter 7, however, the restructuring of natural material into mentally organized material via technology is as yet in the most preliminary of stages: thus far we have been able to create from natural material only artificial nonliving material, perhaps the very beginnings of artificial life, and artificial intelligence not truly at all. Critics of artificiality often miss this preliminary character of the current artificial state, and point to discrepancies between technology and natural systems as if this clash were an inherent and inevitable consequence of the technological order.[16] Incompatibilities between technological and living systems, however, flow not from any essential quality of technology, but rather from the primitive, undeveloped condition of the offending chemical or machine. Thus an automobile that pollutes the air does so not because all technologies of transportation must inevitably despoil the environment, but because at this early stage of technological development the only economically effective transportation vehicles that humans know how to construct are those that have the unhealthy side-effect of polluting the air. Again, regressive economic and social conditions can severely retard the development of life-compatible technologies and therein disfigure this evolutionary process—the ascendancy of gas-burning rather than electrical automobiles at the turn of the century due to the financial power of oil companies would be a striking example here—but the recent development of pollution-reducing devices for cars, as well as the ultimate prospects of solar energy, suggest that eventually humans will construct vehicles that move us from one place to another rapidly and efficiently without air pollution.[17]

Complete success for a technological transformation of the cosmos—a universe remade entirely in accordance with human design—is extremely unlikely, since even the best planned technologies are inescapably touched by unforeseen consequences. This means that rationally directed technology, even at its finest, replaces the blind process by which the universe heretofore generated variations, or the raw material upon which natural selection operates, but does not replace the unpredictable selection process itself, that is, the winnowing of variations or raw material based on how they in fact perform in the environment. Human technology consciously makes suggestions to the cosmos—a radical enough departure from the previous mode of generation—but does not and cannot yet dictate or control how such suggestions are acted upon. In the new evolutionary schema, human technology proposes, but the cosmos still disposes.

The environment that determines the success or failure of technological offerings is itself only a composite of all modes and aspects of existence, and to the extent that humans keep pumping materialized ideas into the cosmos, we may gradually gain some influence over this ultimate gatekeeper. The universe is boundless, however, and even a rapidly increasing portion of infinity still leaves a remainder that is infinite. Thus, though the future may have a human face, it will come into focus exceedingly slowly.

THE FACE IN OUR FUTURE

As far as we know, ours is the only game in the cosmos. We have no evidence and no compelling reason to think that we are not the first, last, and only hope of the universe. Even had we cause to think otherwise, we surely lose nothing in assuming the mantle of uniqueness, in view of all we could bring to the universe and all we would betray of our own past and existence should we bungle the job.

Our past, the crawl from matter to life to intelligent-technological life and all that we have done with intelligence and technology has been long and difficult. Long, but by no means fruitless. For we at the close of the twentieth century possess powers exceeded only by those of the universe from which we came, and to which we now verge on returning via knowledge embodied and emboldened in technology.

This is an imperialism of intelligence of which I speak. Colonization of the cosmos by products of the human intellect. Architecture of the galaxies to make them more comfortable for humans, a plumbing of the stars for all that they may give us. I am utterly and unashamedly a human chauvinist, because we are the only meaning I know.

As Kant so keenly saw, we cannot know what the cosmos would be without us. We give the stars their coordinates, their meaning, our meaning. Yet for most of our past we supplied this meaning rather passively and selfishly, in ideas that usually died with their creators or the small groups who heard or read them. A few ideas got away, escaped the prison of temporality, by being picked up by a communicated representation, a technological embodiment, or both. Gradually the weight of these durable ideas began adding up and achieving a certain momentum of knowledge— a slowly but steadily growing self-increasing loop of ideas, representations, technological embodiments, technologies designed to discover knowledge, more ideas, ideas about building better technologies to communicate and discover knowledge, and so on. Telescopes and microscopes were invented to increase our sensory data, computers to help process this data, media of all kinds to disseminate and criticize the knowledge produced, transportation machines that had the unintended effect of enlarging the sources and scope of intellectual circles, and all the while technologies of every sort, whatever their purpose, quietly added to the growing weight of knowledge by testing and preserving ideas by virtue of their mere existence. These are the ships we would take to the stars.

Many would argue that with the pitifully, fallibly little that we understand of the cosmos or indeed of anything else, how dare we presume this imperial undertaking? How dare we take it upon ourselves to alter the lines of previous evolution through gene-splicing, to construct machines that would think like our soul, and worst of all, to dare endeavor to impose our tiny bit of wisdom and embodiments on a virgin cosmos?

My answer would be that more-from-less, growth-from-ignorance, has always been the way of the cosmos and the upshot of evolution. The cosmos is neither virgin nor unsoiled, for we came from it. The logic of more from less, of increasing our small pool of understanding, of improving upon the paradox of our existence, bids us now to return and grow and supervise and learn. This road—the road of knowledge become technology become knowledge—is really the only road open.

NOTES

1. Klaus Krippendorf, "Paradox and Information," paper presented at the Fifth International Conference on Culture and Communication, Temple University, Philadelphia, Pennsylvania, March 26, 1983.
2. If one assumes or concludes that the statement "This statement is a lie" is true, this means that the statement is indeed a lie, which means it is not true. If one then concludes that the statement "This statement is a lie" is not true, this means that the statement is not a lie, which means that the statement is true . . . which means that the statement "This statement is a lie" is indeed a lie, which means it is not true . . . and so on.
3. The value of endemic uncertainty in the development of biological organisms and human knowledge is explored in depth in chapter 3 of this book.
4. Moreover the logic would not necessarily have to remain unknown to its human executors—discovery of unconscious patterns is indeed the central goal of the Freudian program.
5. Karl Jaspers and Jacques Ellul see loss of reason as part of a general human impalement on the spikes of technological systems. "The technical-life order . . . did at the outset preserve real worlds of human creatures, by furnishing them with commodities," Jaspers writes in *Man and the Modern Age*, trans. E. Paul & C. Paul, London: Routledge & Kegan Paul, 1931/1951, pp. 44–45. "But when at length the time arrived when nothing in the individual's immediate and real environing world was any longer made, shaped, or fashioned by that individual for his own purposes . . . when the dwelling place was machine-made, when the environment had become despiritualized . . . then man was, as it were, bereft of his world. Cast adrift in this way, lacking all sense of historical continuity with past or future, man cannot remain man. The universalization of the life-order threatens to reduce the life of the real man in a real world to mere functioning."
Cf. Ellul's *The Technological Society*, trans. J. Wilkinson, New York: Knopf, 1954/1964, p. 321: "Man was made to do his daily work with his muscles; but see him now, like a fly on flypaper, seated for eight hours, motionless at a desk. . . . The human being was made to breathe the good air of nature. . . . He was created for a living environment, but he dwells in a lunar world of stone, cement, asphalt, glass, cast iron, and steel. . . . See him now, enclosed by . . . rules and architectural necessities . . ."
Jaspers' view of technological humans alienated from a sense of past and future anticipates Mumford's claim that electronic media "reduce all human experience into that of the present generation and the passing moment" (*The Pentagon of Power*, New York: Harcourt Brace Jovanovich, 1970, p. 294). Jaspers' mourning of the passing of remoteness—"Thanks to the technical conquest of time and space by the daily press, modern travel, the cinema, wireless, etc., a universalization of access has become possible. No longer is anything remote, mysterious, wonderful" (p. 48)—similarly presages Mumford's. See note 25 in chapter 8 of the present volume for Mumford's critique of immediacy; see also notes 9 and 36 in chapter 6 for Dewey's misgivings about the unmediated, which predate Jaspers'. See also Jaspers' arguments that technological "Children become like grown-ups as soon as they possibly can,

and join in grown-up conversations . . ." and analogously "when the old pretend to be young, of course the young have no reverence for their elders" (p. 50); compare with Meyrowitz's recent assessment of the ways that television blurs distinctions between adults and children (*No Sense of Place,* 1985, ch. 13).

Heidegger's view of the impact of technology on rationality is both more and less pessimistic than Jaspers', Ellul's, or Mumford's. He argues that "subjective" rationality—or the human practice of attempting to define and understand the world in human terms—concludes with the crystalization of human will upon the universe in modern, massive technology. Once realized in technology, the Kantian project of passively understanding the world via human constructs is no longer necessary or practicable. A new type of critical thinking—one that attempts to recognize existence as it is, rather than as an object of human domination—is, however, possible and advisable. Heidegger thus sees technology as intrinsically destructive of traditional rationality, but not in a way that prevents the emergence of a new mode of critical thinking more suitable for the technological world. See chapter 4, note 20 for references on Heidegger.

Marcuse's critique of technologically perverted rationality, written from the stance of a spiritual revolutionary, is less pessimistic still. Thus, although Marcuse argues throughout *One Dimensional Man,* Boston: Beacon, 1964, that "the techniques of industrialization . . . prejudge the possibilities of Reason and Freedom. . . . Technological rationality reveals its political character as it becomes the great vehicle of better domination, creating a truly totalitarian universe in which society and nature, mind and body are kept in a state of permanent mobilization for the defense of the universe" (p. 18), he also is able to say in the same volume that "paperbacks . . . and long-playing records are truly a blessing" (p. 65) in their attempt to undermine the upperclass monopolization of art and culture. In other words, the "totalitarian" consequence of technology is for Marcuse something less than total, at least insofar as allowing the implementation of given technologies that may subvert or offer opposition to the dominating trend. Such a view of possible value for modern technology runs counter to Ellul, for example, who chastises Mumford for praising the printing press (*The Technological Society,* p. 95).

These and similar assessments of the derationalizing, dehumanizing consequences of modern technology have found their way into much of our popular thinking and culture, such as movies ranging from Charlie Chaplin's *Modern Times* (1936) to George Lucas' uncharacteristically bleak *THX-1138* (1970). The roots of these assessments go back at least as far as Marx—for example "In the factory we have a lifeless mechanism independent of the workman, who becomes its mere living appendage," *Capital,* 4th German edition, 1st American edition, ed. Frederick Engels, rev. E. Untermann, trans. S. Moore and E. Aveling, New York: Modern Library, 1906 (1867), pp. 461–462—although Marx's critique was more concerned with the capitalist misuse of technology than with technology in general.

See Marx Wartofsky, "The Critique of Impure Reason II: Sin, Science, and Society," *Science, Technology, & Human Values,* Fall 1980, pp. 5–23, for a discussion and summary of the works of Althusser, Habermas, and numerous other neo– and non–Marxists on the topic of the technological constriction of reason. Some of these views straddle the theses that, first, technology totally co-opts reason (generally held by Ellul, for example) and, second, technology injures reason, but leaves us enough reason to defend against and/or attempt to repair this injury. See the text in this chapter and note 10 for discussion of the latter thesis.

6. "To be aware of only immediate stimuli and immediate sensations is a medical indication of brain injury," Mumford concludes in a passage on electronic media and their emphasis of the immediate present in *The Pentagon of Power,* p. 298.

7. Gross sales of books exceeded three billion dollars for the first time in the early 1970s. See Edward Jay Whetmore, *Media America,* Belmont, Calif.: Wadsworth, 1979, ch. 2, for a

discussion of books and reading in the age of television. See also note 51 in chapter 6 of the present volume.

8. *Channels 1986 Field Guide*, p. 74.

9. Boyce Rensberger, "Mass Grave Tells Savage Story of Indian Strife," *The New York Times*, May 27, 1980, pp. C1, 3.

10. The success of our ability to overcome the irrationality of technological domination of others hinges to a large extent on where in the technological-human relationship the problem of irrational technological domination emerges. The conventional view is that technology begins as a mode of dominating nature for human ends such as production of food and shelter (a rational, positive application of technology) and then gets perverted into a tool for the domination of humans by humans. Marx Wartofsky makes this point in "Is Science Rational?" in Truitt and Solomons, eds., *Science, Technology and Freedom*, Boston: Houghton-Mifflin, 1972, pp. 202–210, where he distinguishes between "liberating" versus "repressive" reason (liberation from bondage to natural circumstances and old problems, repression via imposition of one's will upon another), and suggests that the seeds of the second are in the first ("Mastery over nature does not turn a 'free' nature into an 'unfree' nature. Nature is neither free nor unfree. . . . But I can suppress another human being . . . to my will," p. 205). Jaspers makes a comparable observation about technological enhancement of human life at the outset (see note 5 above), and Marcuse says that "growing mastery of nature," which at first frees people from mindless work to realize their human potential, comes to fulfill "human needs only as a byproduct: increasing cultural wealth and knowledge [provides] the material for progressive destruction" (*Eros and Civilization*, New York: Vintage, 1955/1962, pp. 79–80).

Against this Platonic view of technology and rationality (Platonic in the sense of positing a de-evolution or accelerating degradation in the relationship), we have the generally neglected thinking of Francis Bacon about technology and domination. Acutely aware of the enormity of the human instinct to dominate, Bacon described a progression of three types of ambition or domination in human affairs: "The first is of those who desire to extend their own power in their native country; which kind is vulgar and degenerate. The second is of those who labor to extend the power of their country and its dominion over men. This certainly has more dignity, though not less covetousness. But if a man endeavor to establish and extend the power and dominion of the human race itself over the universe, his ambition (if ambition it can be called) is without doubt both a more wholesome thing and a more noble than the other two. Now the empire of man over things depends wholly on the arts and sciences. For we cannot command nature except by obeying her," *Novum Organum*, 1620, Book I, section 129, reproduced in Hugh Dick, ed., *Francis Bacon: Selected Writings*, New York: Modern Library, 1955, p. 539.

Here, then, we find a response to the predominating view that the nuclear buildup and similar destructive applications of technology are the logical culmination of the technological-rational union: in Bacon's schema, aggressive tribalism and nationality are an expression of human existence prior to, and ultimately supplanted by, human mastery of nature via technology. The difference between the Indian slaughter of Indians 500 years ago (note 9 above) and the slaughter of other peoples in many parts of the globe today is not only that today's slaughters are magnified by the assistance of weapon technologies, but that today's slaughters are publicly known via communications media—a state of affairs that allows for their rational condemnation which creates pressure for their cessation. Further, the application of numerous technologies whose effects are just beginning to be felt (including medical and agricultural technologies that make life more livable around the planet), as well as the eventual possibility of leaving this planet for a less crowded civilization in the cosmos, may reduce the material and psychological motives for national (and personal) aggression. In this context, the nuclear threat may be seen not as an apocalyptic climax of technological development, but as a last gasp of an essentially pre– or primitive technological impulse that is already being co-opted by our technological mastery of nature. The problem—still a very real and potentially calamitous one—then becomes one of how to increase our control of nature to the point where the nuclear threat becomes defused or is no longer a threat. (Such a perspective is supportive in principle of the goal of the current United States Strategic Defense Initiative—popularly called the Star Wars defense because of its plan to use lasers based in satellites around the Earth—to render nuclear weapons impotent and obsolete. At the same, even the peaceful use of nuclear fission technologies is an anomaly or exception in this schema, because this type of human mastery of nature can jeopardize the lives of many members of our species—as witness the accidents at Three Mile Island in 1979 and Chernobyl in 1986. I thus support SDI research and oppose nuclear fission for energy production.)

See Benjamin Farrington, *Francis Bacon: Philosopher of Industrial Science,* London: Lawrence and Wishart, 1951, for more on Bacon's philosophy of technology; see chapter 8 note 6 in the present volume for a discussion of the degree to which technology must "obey" natural laws.

11. The explosion of our sun is usually expected to occur in about five billion years; comet bombardments may pose risks that are somewhat more imminent. See R. C. Molander, "Only 14 Million Years Until Doomsday," *The New York Times,* May 12, 1984, p. 23.

12. See Karl Popper, *Quantum Mechanics and the Schism in Physics,* London: Hutchinson's, 1982, p. 25, for an interpretation of recent quantum mechanic test results that supports the possibility of faster-than-light travel.

The background is as follows: First, special relativity theory (1905) expressly prohibits faster-than-light travel, and general relativity theory (ca 1916) predicts gross distortions (for example, increase to infinite mass) that occur at faster-than-light speeds. Second, Eddington's eclipse observations of 1919 corroborated Einstein's relativity approach and resulted (among many other consequences) in the generally held view that faster-than-light travel is impossible. Third, in the mean time, investigation by Bohr, Heisenberg, and others of subatomic levels produced the "Copenhagen Interpretation" of quantum mechanics, a cluster of interpretations which among other conclusions held that observation of particle A in a "field" (or relationship) with particle B influences the behavior of B. Fourth, Einstein opposed this subjectivist interpretation, and proposed (along with Podolsky and Rosen) an experiment (1935) to test the subjectivist claim:

1. particles A and B collide and move away from each other at light year distances;
2. measure A and observe some aspect of B (momentum, position), measure A again and observe the same aspect of B, repeat for various aspects;
3. if the subjectivist claim is correct—if observation of one part of a field influences the behavior of another—then B's behavior should vary in accordance with measurement of A's;
4. if Einstein's "realist" interpretation of quantum mechanics is correct (that is, changes in B due to measurement of A are the result of "hidden" real variables in small environments: for example, waves that travel from A to B at sublight speeds), then the E-P-R experiment should show no correlation between measurement of A and behavior of B.

Fifth, E-P-R remained a thought experiment, until mathematically refined by J. S. Bell, amended to measure "spin" (rather than position and momentum) by David Bohm, and actually performed by John F. Clauser, Abner Shimony, and others in the 1970s. Sixth, the results corroborated the subjectivist Copenhagen Interpretation: B was indeed influenced by measurements of A. Seventh, Popper suggests that a realist (hidden variable) interpretation could still apply, if we were willing to forego the special relativity theory limitation on faster-than-light travel: "The consequences [of the experiment], if correct, would go against Einstein's interpretation of special relativity and in favor of Lorentz's interpretation and its formulae and of Newton's 'absolute space.'"

The results of these experiments would have posed an exquisite dilemma for Einstein, who would have been obliged to give up either his realist interpretation of quantum mechanics or the special relativity proscription on faster-than-light travel.

Other possible avenues of faster-than-light travel are suggested by mathematical projections of "tachyons," hypothetical particles travelling at superluminary speed. See E. Recami, ed., *Tachyons, Monopoles, and Related Topics*, New York: North Holland, 1978, for a highly technical but instructive summary.

13. Norbert Wiener, *The Human Use of Human Beings*, New York: Avon, 1950/1957, p. 66.

14. See R. Buckminster Fuller, *Utopia or Oblivion*, New York: Bantam, 1969, ch. 8, for his view of human-technological existence as "doing increasingly more with increasingly less." The ultimate triumph of the living-intelligent-technological counterentropic enterprise is also held possible by Freeman Dyson, who writes that "life may succeed against all odds in molding the universe to its own purpose . . . the design of the inanimate universe may not be as detached from the potentialities of life and intelligence as scientists of the twentieth century have tended to suppose" ("Energy in the Universe," *Scientific American*, September 1971, pp. 50–59). Prigogine approvingly reproduces this quote in *Order from Chaos*, p. 117.

15. "Evolution is definable as a change from an incoherent homogeneity to a coherent heterogeneity" (Herbert Spencer, *First Principles*, New York: Appleton, 1864/1896, p. 371).

16. Ellul is one of the most unremitting in this regard, arguing in *The Technological Society* that the very technological intention (even prior to the deployment of a particular tool) is totalitarian, and that any attempt to remedy the situation via technological means is thus inherently self-defeating. (See note 5 above.) However, see also Berta Sichel, "New Hope for a Technological Society: An Interview with Jacques Ellul," *Et Cetera*, Summer 1983, pp. 192–206, where Ellul calls attention to the preface of *The Technological Society* in which he writes that "if man changes politically, if he thinks differently, if he changes the goals of his life, he will or might evolve differently." Ellul adds in the 1983 interview that "with new developments in computer science . . . we have new alternatives . . ."

17. Removal of exhaust pollution will have no impact on the hazards to health of automobiles reflected in the millions of people who die each year in automobile accidents. This threat to life can and has been mitigated by remedial or preventative technologies such as seatbelts, and by social custom and law including the 55-mile-per-hour speed limit and stern punishment of drunk drivers. The National Safety Council estimated in 1986 that 40,000 lives had been saved since the enactment of the 55–mile-per–hour speed limit for automobile traffic in the United States in 1973.

The loss or injury of even small numbers of humans in automobile accidents is deplorable, but needs to be evaluated in such contexts as: how many lives have been saved by automotive travel (for example, ambulances rushing heart attack victims to hospitals), how does automobile use improve the quality of some people's lives (such as the ability to live in the country and work in the city), what were the fatalities from previous or alternative forms of transportation (for example, horse manure was a source of deadly tetanus disease at the turn of the century), and what sort of risks should people be allowed to take voluntarily in a technological society.

Bibliography

Abelson, P. H. & A. L. Hammond (1977) *Electronics: The Continuing Revolution,* Washington, D.C.: AAAS.
Agassi, Joseph (1968) *The Continuing Revolution,* New York: McGraw-Hill.
Alland, Alexander Jr. (1977) *The Artistic Animal,* New York: Anchor.
Andersson, Gunnar (1982) "Naive and Critical Falsificationism," in *In Pursuit of Truth,* ed. P. Levinson, Atlantic Highlands, N.J.: Humanities.
Apple, R. W. Jr. (1978) "Two Britons Devise a Computer That Can Communicate in Chinese," *New York Times,* January 25, p. 2.
Aristotle *De Anima.*
Aristotle *Physica.*
Asimov, Isaac (1953) *The Caves of Steel,* Garden City, N.Y.: Doubleday.
Asimov, Isaac (1956) *The Naked Sun,* Garden City, N.Y.: Doubleday.
Asimov, Isaac (1985) *Robots and Empire,* Garden City, N.Y.: Doubleday.
Asimov, Isaac (1983) *The Robots of Dawn,* Garden City, N.Y.: Doubleday.
Ayala, Francisco J. & Theodosius Dobzhansky (eds.) (1974) *Studies in the Philosophy of Biology,* Berkeley: University of California Press.
Bacon, Francis (1620) *Novum Organum,* Book I, section 129 in *Francis Bacon: Selected Writings,* ed. H. Dick, New York: Modern Library, 1955.
Baker, William O. et al. (1977) "Computers and Research," in *Electronics: The Continuing Revolution,* ed. P. H. Abelson & A. L. Hammond, Washington, D.C.: AAAS, pp. 56–61.
Baldwin, James Mark (1898/1907) *The Story of the Mind,* New York: Appleton.
Barnes, Barry (1977) *Interests and the Growth of Knowledge,* London: Routledge & Kegan Paul.
Barnes, Barry (1974) *Scientific Knowledge and Sociological Theory,* London: Routledge & Kegan Paul.
Barnouw, Eric (1975) *Tube of Plenty,* New York: Oxford University Press.
Bartley, W. W. III (1976) "Biology and Evolutionary Epistemology," *Philosophia,* 6 (3–4), September–December, pp. 463–494.
Bartley, W. W. III (1977) Letter to the Editors, *The New York Review of Books,* October 27, p. 45.
Bartley, W. W. III (1982) "Philosophy of Biology versus Philosophy of Physics," *Fundamenta Scientiae,* 3 (1), pp. 55–78.

Bartley, W. W. III (1982) "A Popperian Harvest," in *In Pursuit of Truth*, ed. P. Levinson, Atlantic Highlands, N.J.: Humanities, pp. 249–289.
Bartley, W. W. III (1982) "Rationality, Criticism, and Logic," *Philosophia*, 11 (1–2), February, pp. 121–221.
Bartley, W. W. III (1984 [1962]) *The Retreat to Commitment*, La Salle, Ill.: Open Court.
Barzun, Jacques (1941) *Darwin, Marx, Wagner*, Boston: Little, Brown.
Bateson, Gregory (1979) *Mind and Nature*, New York: Bantam.
Baumer, Franklin L. (1977) *Modern European Thought*, New York: Macmillan.
Baur, Stuart (1975) "Kneedeep in the Cosmic Overwhelm with Carl Sagan," *New York*, September 1, p. 30.
Bazin, André (1967 [1958–1965]) *What Is Cinema?*, trans. and ed. H. Gray, Berkeley: University of California Press.
Bell, Daniel (1975) "Technology, Nature, and Society," in *The Frontiers of Knowledge*, Garden City, N.Y.: Doubleday, pp. 27–78.
Bergson, Henri (1911) *Creative Evolution*, trans. A. Mitchell, New York: Holt.
Berlin, Isaiah (1939/1963) *Karl Marx*, New York: Oxford University Press.
Blake, William (ca 1800) "Auguries of Innocence," in *The Portable Blake*, ed. A. Kazin, New York: Viking, 1946, pp. 150–154.
Bloor, David (1976) *Knowledge and Social Imagery*, London: Routledge & Kegan Paul.
Bock, W. J. (1959) "Preadaptation and Multiple Evolutionary Pathways," *Evolution*, 13, June, pp. 194–211.
Boden, Margaret (1977) *Artificial Intelligence and Natural Man*, New York: Basic.
Bohr, Niels (1958) *Atomic Physics and Human Knowledge*, New York: Wiley.
Bolter, J. David (1984) *Turing's Man*, Chapel Hill, N.C.: University of North Carolina Press.
Brand, Stewart (ed.) (1985) *Whole Earth Software Catalog for 1986*, Garden City, N.Y.: Doubleday.
Briskman, Larry (1985) "Articulating Our Ignorance: Hopeful Scepticism and the Meno Paradox," *Et Cetera*, 42 (3), Fall, pp. 201–227.
Broad, William & Nicholas Wade (1983) *Betrayers of the Truth*, New York: Simon & Schuster.
Broughton, J. M. & D. J. Freeman-Moir (eds.) (1982) *The Cognitive Developmental Psychology of James Mark Baldwin*, Norwood, N.J.: Ablex.
Brown, Ben (1985) "Bracing for the Aftershock," *Channels 1986 Field Guide*, pp. 22, 24.
Brumbaugh, Robert S. (1966) *Ancient Greek Gadgets and Machines*, New York: Crowell.
Buchler, Justus (ed.) (1955) *Philosophical Writings of Peirce*, New York: Dover.
Bunge, Mario (1980) *The Mind-Body Problem*, New York: Pergamon.
Burbank, Luther (1921) *How Plants Are Trained to Work for Man*, New York: Collier.
Burbank, Luther (1939) *Partner of Nature*, New York: Appleton-Century.
Burlingame, Roger (1959) "The Hardware of Culture," *Technology and Culture*, 1 (1), pp. 11–28.
Bush, Vannevar (1945) "As We May Think," *Atlantic Monthly*, July, pp. 101–108.
Butler, Samuel (1879/1911) *Evolution: Old and New*, London: Fifield.
Butler, Samuel (1865) "Lucubratio Ebria," *The Press*, Christchurch, New Zealand, July 29, in *The Notebooks of Samuel Butler*, ed. H. F. Jones, New York: AMS, 1968, pp. 35–40.
Campbell, Donald T. (1960) "Blind Variation and Selective Retention in Creative Thought as in Other Knowledge Processes," *Psychological Review*, 67, pp. 380–400.
Campbell, Donald T. (1974) "Evolutionary Epistemology," in *The Philosophy of Karl Popper*, ed. P. Schilpp, La Salle, Ill.: Open Court, pp. 413–463.
Campbell, Donald T. (1982) "Evolutionary Epistemology: Partial Supplementary Bibliography," unpublished.

Campbell, Donald T. (1974) "Unjustified Variation and Selective Retention in Scientific Discovery," in *Studies in the Philosophy of Biology*, ed. F. J. Ayala & T. Dobzhansky, Berkeley: University of California Press, pp. 139–161.
Campbell, Jeremy (1982) *Grammatical Man*, New York: Simon and Schuster.
Capra, Fritjof (1976) *The Tao of Physics*, New York: Bantam.
Carnap, Rudolph (1967 [1928]) *The Logical Structure of the World*, trans. R. George, Berkeley: University of California Press.
Carpenter, Edmund & Marshall McLuhan (eds.) (1960) *Explorations in Communication*, Boston: Beacon.
Cathcart, Robert & Gary Gumpert (eds.) (1979/1986) *Inter/Media: Interpersonal Communications in a Media World*, New York: Oxford University Press.
Charnot, Jean (1939) *Art from the Mayans to Disney*, New York: Sheed and Ward.
Chase, Stuart (1929) *Men and Machines*, New York: Macmillan.
Chester, G. et al. (1971) *Television and Radio*, 4th ed., Englewood Cliffs, N.J.: Prentice-Hall.
Chin, P. & B. Martin (1986) "Solving the Chinese Puzzle," *Newsweek*, August 18, p. 43.
Chomsky, Noam (1972) *Language and Mind*, 2nd ed., New York: Harcourt Brace Jovanovich.
Chomsky, Noam (1975) *Reflections on Language*, New York: Pantheon.
Cleveland, Harlan (1985) *The Knowledge Executive*, New York: Dutton.
Cohen, I. B. (1981) "Newton's Discovery of Gravity," *Scientific American*, March, pp. 167–179.
Cohen, Robert S. et al. (1976) *PSA 1974*, Boston: Reidel.
Coleridge, Samuel Taylor (1907 [1817]) *Biographia Literaria*, ed. J. Shawcross, London: Oxford University Press.
"Computer Clubs Growing as Hobbyists Share Data," (1986) *New York Times*, July 24, pp. C1, 10.
Cowan, Robert (1984) *Teleconferencing: Maximizing Human Potential*, Reston, Va.: Reston.
Curtis, James M. (1978) *Culture as Polyphony*, Columbia, Mo.: University of Missouri Press.
Dalton, Richard (1984) "Electronic Notebooks," *Whole Earth Software Review*, Summer, pp. 87–95.
Darwin, Charles (1968 [1859]) *The Origin of Species*, Hammondsworth, UK: Penguin.
Darwin, Charles (1868) *The Variation of Animals and Plants Under Domestication*, London: Murray.
Darwin, Charles (1984 [1862]) *The Various Contrivances By Which Orchids Are Fertilised By Insects*, Chicago: University of Chicago Press.
Davies, Nigel (1979) *Voyages to the New World*, New York: Morrow.
Davis, D. D. (1949) "Comparative Anatomy and the Evolution of Vertebrates," in *Genetics, Paleontology, and Evolution*, ed. G. L. Jepsen et al., Princeton: Princeton University Press, pp. 64–89.
de Haan, David (1977) *Antique Household Gadgets and Appliances*, Woodbury, N.Y.: Barron's.
Dessauer, Friedrich (1927) *Philosophie der Technik; Das Problem der Realisierung*, Bonn: Cohen.
Dessauer, Friedrich (1956) *Streit um die Technik*, Frankfurt: Knecht.
Dewey, John (1925) *Experience and Nature*, Chicago: Open Court.
Dick, Hugh (ed.) (1955) *Francis Bacon: Selected Writings*, New York: Modern Library.
Dorhn, A. (1875) "Der Ursprung der Wirtbelthiere und das Princip des Functionswechsels," in E. S. Russell, *Form and Function*, Chicago: University of Chicago Press, 1982 [1916], pp. 274–278.
Drake, Stillman (ed.) (1957) *Discoveries and Opinions of Galileo*, New York: Doubleday.

Dretske, Fred (1986) "Minds, Machines, and Meaning," in *Philosophy and Technology II*, ed. C. Mitcham & A. Huning, Boston: Reidel, pp. 97–109.
Dreyfus, Hubert L. (1972/1979) *What Computers Can't Do*, New York: Harper Colophon.
Dreyfus, Hubert L. & Stuart E. Dreyfus (1986) "From Socrates to Expert Systems," in *Philosophy and Technology II*, ed. C. Mitcham & A. Huning, Boston: Reidel, pp. 111–130.
Dumbach, Annette E. & Jud Newborn (1986) *Shattering the German Night*, Boston: Little, Brown.
Durbin, Paul (ed.) (forthcoming) *Philosophy and Technology IV*, Boston: Reidel.
Durbin, Paul (ed.) (1979, 1982, 1985) *Research in Philosophy and Technology*, vols. 2, 5, 8, Greenwich, Conn.: JAI Press.
Dyson, Freeman (1971) "Energy in the Universe," *Scientific American*, September, pp. 50–59.
Edman, I. (ed.) (1942) *The Philosophy of Santayana*, New York: Modern Library.
Edmondson, William (ed.) (1985) *The Age of Access: The Posthumous Papers of Colin Cherry*, Dover, N.H.: Croom Helm.
Eigen, Manfred & Peter Schuster (1979) *The Hypercycle: A Principle of Natural Self-Organization*, New York: Springer.
Eirich, Frederick (1984) Private correspondence to Paul Levinson, December 20.
Eisenstein, Elizabeth (1979) *The Printing Press As An Agent of Change*, New York: Cambridge University Press.
Eisenstein, Sergei (1957 [1946]) *Film Form and The Film Sense*, ed. and trans. J. Leyda, New York: Meridian.
Eldridge, Niles & Joel Cracraft (1980) *Phylogenic Patterns and the Evolutionary Process*, New York: Columbia University Press.
Ellis, Don Carlos & Laura Thornborough (1923) *Motion Pictures in Education*, New York: Crowell.
Ellul, Jacques (1964 [1954]) *The Technological Society*, trans. J. Wilkinson, New York: Vintage.
Emerson, Ralph Waldo (1870) *Society and Solitude*, Boston: Osgood.
Engels, Frederick (1882) *Dialectics of Nature*, excerpted in *Technology and Human Affairs*, ed. L. Hickman & A. al-Hibri, St. Louis, Mo.: Mosby, 1981, pp. 215–220.
Enterline, James R. (1972) *Viking America*, Garden City, N.Y.: Doubleday.
Ericson, R. (ed.) (1979) *Proceedings of the Silver Anniversary Meeting of the Society for General Systems Research*, New York: Springer.
Evans, R. I. (ed.) (1975) *Konrad Lorenz: The Man and His Ideas*, New York: Harcourt Brace Jovanovich.
Farrington, Benjamin (1951) *Francis Bacon: Philosopher of Industrial Science*, London: Lawrence and Wishart.
Feigenbaum, Edward & Pamela McCorduck (1983) *The Fifth Generation: Artificial Intelligence and Japan's Computer Challenge to the World*, Reading, Mass.: Addison-Wesley.
Feuer, Lewis S. (ed.) (1959) *Marx & Engels: Basic Writings on Politics and Philosophy*, New York: Anchor.
Feyerabend, Paul (1975) *Against Method*, London: New Left Books.
Feyerabend, Paul (1978) "The Gong Show—Popperian Style," in *Progress and Rationality in Science*, ed. G. Radnitzky & G. Andersson, Boston: Reidel, pp. 387–392.
Freeman, Eugene & Henryk Skolimowski (1974) "The Search for Objectivity in Peirce and Popper," in *The Philosophy of Karl Popper*, ed. P. Schilpp, LaSalle, Ill.: Open Court, pp. 464–519.
Freud, Sigmund (1930) *Civilization and Its Discontents*, trans. J. Riviere, New York: Cape and Smith.

Freud, Sigmund (1939/1967) *Moses and Monotheism*, trans. K. Jones, New York: Vintage.
Fuller, R. Buckminster (1938) *Nine Chains to the Moon*, Carbondale, Ill.: Southern Illinois University Press.
Fuller, R. Buckminster (1969) *Utopia or Oblivion*, New York: Bantam.
Gernsheim, Helmut & Alison Gernsheim (1956/1968) *L. J. M. Daguerre*, New York: Dover.
Giedion, Siegfried (1969 [1948]) *Mechanization Takes Command*, New York: Norton.
Giedion, Siegfried (1960) "Space Conception in Prehistoric Art," in *Explorations in Communication*, ed. E. Carpenter & M. McLuhan, Boston: Beacon, pp. 71–89.
Gleick, James (1983) "Exploring the Labyrinth of the Mind," *The New York Times Magazine*, August 21, pp. 23 ff.
Gleick, James (1986) "Less Drastic Theory Emerges on Freezing after Nuclear War," *New York Times*, June 12, pp. 1, 20.
Glossbrenner, Alfred (1985) *The Complete Handbook of Personal Computer Communications*, New York: St. Martin's.
Gödel, Kurt (1962) *On Formally Undecidable Propositions*, New York: Basic.
Gödel, Kurt (1931) "Über Formal Unentscheidel Satze der *Principia Mathematica* und Verwaandter System I" in K. Gödel, *On Formally Undecidable Propositions*, New York: Basic, 1962.
Goldstine, Herman (1972) *The Computer: From Pascal to von Neumann*, Princeton: Princeton University Press.
Gombrich, E. H. (1960) *Art and Illusion*, Princeton: Princeton University Press.
Gombrich, E. H. (1979) *The Sense of Order*, Ithaca, N.Y.: Cornell University Press.
Gould, Stephen J. (1982) "Darwinism and the Expansion of Evolutionary Theory," *Science*, 216, April 23, pp. 380–387.
Gould, Stephen J. (1980) "The Evolutionary Biology of Constraint," *Daedalus*, 109, Spring, pp. 39–52.
Gould, Stephen J. (1981) *The Mismeasure of Man*, New York: Norton.
Gould, Stephen J. (1977) "The Problem of Perfection," *Natural History*, 86, January, pp. 32–35.
Gould, Stephen J. & Niles Eldridge (1977) "Punctuated Equilibria," *Paleobiology*, 3 (2), pp. 115–151.
Gould, Stephen J. & Elizabeth Vbra (1982) "Exaption: A Missing Term in the Science of Form," *Paleobiology*, 8 (1) pp. 4–15.
Graham, Loren (1972) *Science and Philosophy in the Soviet Union*, New York: Knopf.
Gray, Asa (1876) *Darwiniana*, New York: Appleton.
Gray, Asa (1861) *Natural Selection not Inconsistent with Natural Theology*, London: Truebner.
Gruber, Howard (1982) "Piaget's Mission," *Social Research*, 49 (1), Spring, pp. 239–264.
Haas, Warren J. (1982) "Computing in Documentation and Scholarly Research," *Science*, 215, February 12, pp. 857–861.
Haeckel, Ernst (1900) *The Riddle of the Universe*, trans. J. McCabe, New York: Harper.
Hall, Edward T. (1959) *The Silent Language*, New York: Fawcett.
Hardy, Alistair (1965) *The Living Stream*, London: Collins.
Hartline, Beverly K. (1980) "Lobster-Eye X-Ray Telescope Envisioned," *Science*, 207, January 4, p. 47.
Hartog, Marcus (1913) *Problems of Life and Reproduction*, New York: Putnam's.
Havelock, Eric (1982) *The Literate Revolution in Greece and Its Cultural Consequences*, Princeton: Princeton University Press.
Havelock, Eric (1963) *Preface to Plato*, Cambridge, Mass.: Harvard University Press.
Hawthorne, Nathaniel (1962 [1851]) *House of the Seven Gables*, New York: Collier.
Hayek, Friedrich (1952) *The Sensory Order*, Chicago: University of Chicago Press.

Heelan, Patrick (1983) *Space-Perception and the Philosophy of Science*, Berkeley: University of California Press.
Heidegger, Martin (1973) *The End of Philosophy*, trans. J. Stambaugh, New York: Harper & Row.
Heidegger, Martin (1977) *The Question Concerning Technology*, trans. W. Lovitt, New York: Harper & Row.
Hendrickson, Robert (1981) *The Literary Life and Other Curiosities*, New York: Viking.
Hickman, Larry (1981) "Filosofia Hoy: El Filosofo y Techne," paper presented at conference "Para Que Filosofia Hoy," University of Navarra, Spain.
Hickman, Larry (1984) "Making and Doing in a Democracy: Dewey's Experience of Technology," paper presented at American Philosophical Association Meeting, New York City, December.
Hickman, Larry (1983) "Philosophy, Techne and the Body," paper presented at Fifth International Conference on Culture and Communication, Temple University, Philadelphia, March 26.
Hickman, Larry (ed.) (1985) *Technology, Philosophy, and Human Affairs*, College Station, Tex.: Ibis Press.
Hickman, Larry & Azizah al–Hibri (eds.) (1981) *Technology and Human Affairs*, St. Louis, Mo.: Mosby.
Hiltz, Roxanne (1983) *On-Line Communities*, Norwood, N.J.: Ablex.
Hiltz, Roxanne (forthcoming) "The Virtual Classroom," report.
Hiltz, Roxanne & Murray Turoff (1978) *The Network Nation*, Reading, Mass.: Addison-Wesley.
Hofstadter, Douglas R. (1982) "Artificial Intelligence: Subcognition as Computation," November, unpublished paper.
Hofstadter, Douglas R. (1979) *Gödel, Escher, Bach*, New York: Basic.
Hofstadter, Douglas R. (1985) *Metamagical Themas*, New York: Basic.
Hofstadter, Douglas R. (1982) "Self-Referential Sentences: A Follow-Up," *Scientific American*, January, pp. 16ff.
Hofstadter, Douglas R. & D. C. Dennett (eds.) (1981) *The Mind's I*, New York: Basic.
Hogan, John V. L. (1923) *The Outline of Radio*, Boston: Little, Brown.
Hood, William F. (1982) "Dewey and Technology," in *Research in Philosophy and Technology*, vol. 5, ed. P. Durbin, Greenwich, Conn.: JAI Press, pp. 189–207.
Hughes, R. I. G. (1981) "Quantum Logic," *Scientific American*, October, pp. 202–213.
Innis, Harold (1951) *The Bias of Communication*, Toronto: University of Toronto Press.
Innis, Harold (1950) *Empire and Communications*, Toronto: University of Toronto Press.
Izzillo, Theresa & Jeffrey Wolf (1985) "Banking on a Windfall," *Channels 1986 Field Guide*, pp. 36, 38.
Jackman, Jarrell & Carla Borden (eds.) (1983) *The Muses Flee Hitler*, Washington, D.C.: Smithsonian Institute Press.
Jantsch, Erich (1980) "Ethics and Evolution," in *The Responsibility of the Academic Community in the Search for Absolute Values*, vol. 1, New York: International Cultural Foundation Press, pp. 373–384.
Jantsch, Erich (1981) "Unifying Principles of Evolution," in *The Evolutionary Vision*, ed. E. Jantsch, Boulder, Colo.: Westview.
Jantsch, Erich (ed.) (1981) *The Evolutionary Vision*, Boulder, Colo.: Westview.
Jarvie, I. C. (1982) "Popper on the Difference Between the Natural and the Social Sciences," in *In Pursuit of Truth*, ed. P. Levinson, Atlantic Highlands, N.J.: Humanities, pp. 83–107.
Jaspers, Karl (1931/1951) *Man in the Modern Age*, trans. E. Paul & C. Paul, London: Routledge & Kegan Paul.

Jaynes, Julian (1976) *The Origin of Consciousness in the Breakdown of the Bicameral Mind*, Boston: Houghton-Mifflin.
Jensen, Arthur (1980) *Bias in Mental Testing*, London: Methuen.
Jepsen, G. L. et al. (eds.) (1949) *Genetics, Paleontology, and Evolution*, Princeton: Princeton University Press.
Jones, Gwyn (1984) *A History of the Vikings*, New York: Oxford University Press.
Jones, H. F. (ed.) (1968) *The Notebooks of Samuel Butler*, New York: AMS.
Kant, Immanuel (1873 [1788]) *Critique of Practical Reason*, trans. T. K. Abbott, as *Kant's Theory of Ethics*, London: Longman's.
Kant, Immanuel (1934 [1787]) *Critique of Pure Reason*, trans. J. M. D. Meiklejohn, London: Dent.
Kantorovich, Aharon (1983) "The Collective A Priori in Science," *Nature and System*, 5, pp. 77–96.
Kantorovich, Aharon (1982) "Quarks: An Active Look at Matter," *Fundamenta Scientiae*, III, pp. 297–319.
Kapp, Ernst (1877) *Grundlinien einer Philosophie der Technik*, Braunschweig: Westermann, republished as *Einfuehring in die Technik–Philosophie von Ernst Kapp*, ed. H. Sass, Dusseldorf: Stern–Verlag, 1978.
Kaufmann, Walter (ed.) (1954) *The Portable Nietzsche*, New York: Viking.
Kazin, Alfred (ed.) (1946) *The Portable Blake*, New York: Viking.
Kessler, Eckward (1981) "Vico's Attempt Towards a Humanistic Foundation of Science," in *Vico: Past and Present*, ed. G. Tagliacozzi, Atlantic Highlands, N.J.: Humanities.
Kimura, Motoo (1979) "The Neutral Theory of Molecular Evolution," *Scientific American*, November, pp. 98–126.
Klein, Stanley (1980) "Computers That Draw Pictures," *New York Times*, July 6, Section 3, pp. 1, 5.
Koestler, Arthur (1964) *The Act of Creation*, New York: Macmillan.
Kolm, Henry (1977) "An Electromagnetic 'Slingshot' for Space Propulsion," *Technology Review*, 79 (7), June, pp. 60–66.
Korzenny, Felipe (1978) "A Theory of Electronic Propinquity," *Communications Research*, 5, January, pp. 3–23.
Kothals-Atles, Stephen W. (1986) "The Aerospace Plane: The Technological Feasibility and Policy Implications," report #15, Program in Science and Technology For International Security, MIT, July.
Kraft, Victor (1974) "Popper and the Vienna Circle," in *The Philosophy of Karl Popper*, ed. P. Schilpp, LaSalle, Ill.: Open Court, pp. 185–204.
Krippendorf, Klaus (1983) "Paradox and Information," paper presented at 5th International Conference on Culture and Communication, Temple University, Philadelphia, March 26.
Kuhn, Thomas (1978) *The Essential Tension*, Chicago: University of Chicago Press.
Kuhn, Thomas (1962/1970) *The Structure of Scientific Revolutions*, Chicago: University of Chicago Press.
Kuhns, William (1971) *The Post-Industrial Prophets*, New York: Harper Colophon.
Lachenbruch, David (1985) "From Gizmo to Household Word," *Channels 1986 Field Guide*, pp. 74–75.
Lakatos, Imre (1968) "Criticism and the Methodology of Scientific Research Programmes," *Proceedings of the Aristotelian Society*, 69, pp. 149–186.
Lakatos, Imre (1970) "Falsification and the Methodology of Scientific Research Programmes," in *Criticism and the Growth of Knowledge*, ed. I. Lakatos & A. Musgrave, Cambridge, UK: Cambridge University Press, pp. 91–196.
Lakatos, Imre & Alan Musgrave (eds.) (1970) *Criticism and the Growth of Knowledge*, Cambridge, UK: Cambridge University Press.

Land, Michael F. (1978) "Animal Eyes With Mirror Optics," *Scientific American*, December, pp. 126–134.
Levinson, Paul (forthcoming) "Cosmos Helps Those Who Help Themselves," in *Research in Philosophy and Technology*, vol. 9, ed. C. Mitcham, Greenwich, Conn.: JAI Press.
Levinson, Paul (1982) "Evolutionary Epistemology Without Limits," *Knowledge*, 3 (4) June, pp. 465–502.
Levinson, Paul (1986) "Guns, Knives, and Pillows: On the Normative Neutrality of Technological Applications," paper presented at the New York Colloquium on Philosophy and Technology: Phenomenology and Technology, Polytechnic University, Brooklyn, NY, October 3.
Levinson, Paul (1979) "Human Replay: A Theory of the Evolution of Media," Ph.D. diss., New York University.
Levinson, Paul (forthcoming) "Impact of Personal Information Technologies on American Education, Interpersonal Relationships, and Business, 1985–2010," in *Philosophy and Technology IV*, ed. P. Durbin, Boston: Reidel.
Levinson, Paul (1986) "Information Technologies as Vehicles of Evolution," in *Philosophy and Technology II*, ed. C. Mitcham & A. Huning, Boston: Reidel, pp. 29–47.
Levinson, Paul (1981) "McLuhan and Rationality," *Journal of Communication*, 31 (3) Summer, pp. 179–188.
Levinson, Paul (1986) "Marshall McLuhan and Computer Conferencing," *IEEE Transactions on Professional Communication*, PC-29 (1), March, pp. 9–11.
Levinson, Paul (1984) "Media Evolution and Rationality as Checks on Media Determinism," in *Studies in Mass Communication and Technology*, ed. S. Thomas, Norwood, N.J.: Ablex, pp. 231–237.
Levinson, Paul (1976) Review of N. Chomsky's *Reflections on Language*, Media Ecology Review, 4 (4), May, pp. 24–26.
Levinson, Paul (1979/1980) Review of J. Mander's *Four Arguments for the Elimination of Television*, The Structurist 19/20, pp. 107–114.
Levinson, Paul (1986) "Space: Humanizing the Universe," paper presented at Western Behavioral Sciences Institute, La Jolla, Calif., July 15.
Levinson, Paul (1985) "Technology as the Cutting Edge of Cosmic Evolution," in *Research in Philosophy and Technology*, vol. 8, ed. P. Durbin, Greenwich, Conn.: JAI Press, pp. 161–176.
Levinson, Paul (1977) "Toy, Mirror, and Art: The Metamorphosis of Technological Culture," in *Technology, Philosophy, and Human Affairs*, ed. L. Hickman, College Station, Tex.: Ibis Press, 1985, pp. 162–175.
Levinson, Paul (1982) "What Technology Can Teach Philosophy," in *In Pursuit of Truth*, ed. P. Levinson, Atlantic Highlands, N.J.: Humanities, pp. 157–175.
Levinson, Paul (ed.) (forthcoming) *Electronic Chronicles: A Compendium of the Connected Education On–Line Intellectual Community*, vol. 1, Greenwich, Conn.: JAI Press.
Levinson, Paul (ed.) (1982) *In Pursuit of Truth: Essays on the Philosophy of Karl Popper*, Atlantic Highlands, N.J.: Humanities.
Levy, Paul (1979) *G. E. Moore and the Cambridge Apostles*, New York: Holt, Rhinehart, Winston.
Levy, Steven (1984) *Hackers*, Garden City, N.Y.: Doubleday.
Lewin, Roger (1982) "Adaptation Can Be a Problem for Evolutionists," *Science*, 216, June 11, pp. 1212–1213.
Lewin, Roger (1981) "Lamarck Will Not Lie Down," *Science*, 213, July 17, pp. 316–321.
Lewontin, R. C. (1982) "Organism and Environment," in *Learning, Development, and Culture*, ed. H. Plotkin, New York: Wiley, pp. 151–170.

Löfgren, Lars (1979) "Goals for Human Planning," in *Proceedings of the Silver Anniversary Meeting of the Society for General Systems Research*, ed. R. Ericson, New York: Springer, pp. 460–467.
Löfgren, Lars (1981) "Knowledge of Evolution and Evolution of Knowledge," in *The Evolutionary Vision*, ed. E. Jantsch, Boulder, Colo.: Westview, pp. 129–152.
Löfgren, Lars (1981) "Life as an Autolinguistic Phenomenon," in *Autopoeisis*, ed. M. Zeleny, New York: Oxford University Press, pp. 236–249.
Lorenz, Konrad (1973/1977) *Behind the Mirror*, trans. R. Taylor, New York: Harcourt Brace Jovanovich.
Lorenz, Konrad (1941) "Kant's Doctrine of the A Priori in the Light of Contemporary Biology," in *Konrad Lorenz: The Man and His Ideas*, ed. R. I. Evans, New York: Harcourt Brace Jovanovich, 1975, pp. 181–217.
Lum, Man–Kong (1984) "A Study of the Impact of the Telephone on Human Sexuality," MA thesis, The New School for Social Research, May.
Lyell, Charles (1871) *Geological Evidences for the Antiquity of Man*, Philadelphia: Lippincott.
McKeever, William (1910) "The Moving Picture: A Primary School for Criminals," *Good Housekeeping*, August, pp. 184–186.
McLuhan, Marshall (1978) "The Brain and the Media: The 'Western' Hemisphere," *Journal of Communication*, 28 (4), Autumn, pp. 54–60.
McLuhan, Marshall (1962) *The Gutenberg Galaxy*, New York: Mentor.
McLuhan, Marshall (1976) "Inside on the Outside, or the Spaced-Out American," *Journal of Communication*, 26 (4), Autumn, pp. 46–53.
McLuhan, Marshall (1977) "Laws of the Media," preface by P. Levinson, *Et Cetera*, 34 (2), June, pp. 173–179.
McLuhan, Marshall (1977) "The Rise and Fall of Nature," *Journal of Communication*, 27 (4), Autumn, pp. 80–81.
McLuhan, Marshall (1964) *Understanding Media*, New York: Mentor.
McLuhan, Marshall & Barrington Nevitt (1972) *Take Today: The Executive as Dropout*, New York: Harcourt Brace Jovanovich.
McNeill, William (1982) *The Pursuit of Power*, Chicago: University of Chicago Press.
McWilliams, Peter A. (1982) *The Word Processing Book*, Los Angeles: Prelude.
Maeroff, Gene (1979) "Reading Achievement of Children in Indiana Found as Good as in '44," *New York Times*, April 15, p. 10.
Mallarmé, Stéphane (1886) "Crise de Vers," in *Selected Poetry and Prose*, ed. and trans. M. A. Caws, New York: New Directions Books, 1982, pp. 75–76.
Mander, Jerry (1978) *Four Arguments for the Elimination of Television*, New York: Morrow.
Mander, Jerry (1985) "Six Grave Doubts About Computers," *The Whole Earth Review*, January, pp. 10–20.
Manuel, Frank E. (1968) *Portrait of Isaac Newton*, Cambridge, Mass.: Belknap.
Marcuse, Herbert (1955/1962) *Eros and Civilization*, New York: Vintage.
Marcuse, Herbert (1964) *One Dimensional Man*, Boston: Beacon.
Marshack, Alexander (1972) *The Roots of Civilization*, New York: McGraw-Hill.
Martin, Douglas (1984) "McLuhan Center Says A-Bomb May Be Good," *New York Times*, February 12, p. 20.
Marx, Karl (1906 [1867]) *Capital*, 4th German ed., 1st American ed., ed. F. Engels, rev. E. Untermann, trans. S. Moore & E. Aveling, New York: Modern Library.
Marx, Karl (1845) "Theses on Feuerbach" in *Marx & Engels*, ed. L. Feuer, New York: Anchor, 1959, pp. 243–245.
Mast, Gerald (1971) *A Short History of the Movies*, New York: Pegasus.

Mast, Gerald & Marshall Cohen (eds.) (1974) *Film Theory and Criticism*, New York: Oxford University Press.
Mayr, Ernst (1963) *Animal Species and Evolution*, Cambridge, Mass.: Harvard University Press.
Merleau-Ponty, Maurice (1962/1981) *Phenomenology of Perception*, trans. C. Smith, Atlantic Highlands, N.J.: Humanities.
Meyrowitz, Joshua (1985) *No Sense of Place*, New York: Oxford University Press.
Meyrowitz, Joshua (1979/1986) "Television and Interpersonal Behavior," in *Inter/Media*, ed. R. Cathcart & G. Gumpert, New York: Oxford University Press, pp. 253–272.
Milgram, Stanley (1977) "The Image-Freezing Machine," *Psychology Today*, 10, January, pp. 50–55 ff.
Miller, Jonathan (1985) "The Global Picture," *Channels 1986 Field Guide*, pp. 16, 18.
Miller, R. K. (1983) *Intelligent Robots*, Fort Lee, N.J.: Technical Insights.
Minsky, Marvin (1982) "Why People Think Computers Can't Think," *The AI Magazine*, Fall, pp. 3–15.
Mitcham, Carl (ed.) (forthcoming) *Research in Philosophy and Technology*, vol. 9, Greenwich, Conn.: JAI Press.
Mitcham, Carl & Alois Huning (eds.) (1986) *Philosophy and Technology II*, Boston: Reidel.
Mitcham, Carl & Robert Mackey (eds.) (1972/1983) *Philosophy and Technology*, New York: Free Press.
Modern Times (1936) motion picture released in the United States by Charlie Chaplin.
Molander, R. C. (1984) "Only 14 Million Years Until Doomsday," *New York Times*, May 12, p. 23.
Monaco, James (1977) *How To Read a Film*, New York: Oxford University Press.
Morison, Samuel Eliot (1942) *Admiral of the Ocean Sea*, Boston: Little, Brown.
Mumford, Lewis (1970) *The Pentagon of Power*, New York: Harcourt Brace Jovanovich.
Mumford, Lewis (1934) *Technics and Civilization*, New York: Harcourt Brace.
Münsterberg, Hugo (1970 [1916]) *The Film: A Psychological Study*, New York: Dover.
Munz, Peter (1985) "DNA, Falsificationism, and Dogmatism," *Et Cetera*, 42 (3), Fall, pp. 254–271.
Munz, Peter (1985) *Our Knowledge of the Growth of Knowledge*, London: Routledge & Kegan Paul.
Murray, Margaret (1949) *The Splendor That Was Egypt*, New York: Philosophical Library.
Muybridge, Eadweard (1969 [1887]) *Animal Locomotion*, New York: Da Cato Press.
Nagorski, Andrew et al. (1986) "Moscow Faces the New Age," *Newsweek*, August 18, pp. 20–22.
Nietzsche, Friedrich (1878) *Human, All-Too-Human*, in *The Portable Nietzsche*, ed. W. Kaufmann, New York: Viking, 1954, pp. 51–64.
North by Northwest (1959) motion picture released in the United States by MGM.
Ong, Walter (1977) *Interfaces of the Word*, Ithaca, N.Y.: Cornell University Press.
Ong, Walter (1982) *Orality and Literacy*, New York: Methuen.
Ong, Walter (1967) *The Presence of the Word*, Ithaca, N.Y.: Cornell University Press.
Pagels, Heinz (1983) "Fires in Space," *The New York Times Book Review*, August 21, pp. 9, 18.
"Paperless Office?" (1986) *Wall Street Journal*, February 27, p. 1.
Pascal, Blaise (1950 [1669]) *Pensées*, trans. H. F. Stewart, New York: Pantheon.
Pearson, Emily C. (1887) *From Cottage to Castle: The Boyhood, Youth, Manhood, Public and Private Career of Gutenberg*, Boston: Earle.
Peirce, Charles Sanders (1896–1899) "The Scientific Attitude and Fallibilism," in *Philosophical Writings of Peirce*, ed. J. Buchler, New York: Dover, 1955, pp. 42–59.

Phillips, David et al. (1954) *Introduction to Radio and Television*, New York: Ronald Press.
Piatelli-Palmarini, M. (ed.) (1980) *Language and Learning: The Debate Between Jean Piaget and Noam Chomsky*, Cambridge, Mass.: Harvard University Press.
Plato *Meno*.
Plato *Phaedrus*.
Plotkin, H. (ed.) (1982) *Learning, Development, and Culture*, New York: Wiley.
Popper, Karl R. (1962/1968) *Conjectures and Refutations*, London: Routledge & Kegan Paul.
Popper, Karl R. (1959) *The Logic of Scientific Discovery* (translation of *Logik der Forschung*), London: Hutchinson's.
Popper, Karl R. (1935) *Logik der Forschung*, Vienna: Springer.
Popper, Karl R. (1966 [1945]) *The Open Society and Its Enemies*, 5th ed., vols. 1 and 2, Princeton: Princeton University Press.
Popper, Karl R. (1970) "Normal Science and Its Dangers," in *Criticism and the Growth of Knowledge*, ed. I. Lakatos & A. Musgrave, Cambridge, U.K.: Cambridge University Press, pp. 51–58.
Popper, Karl R. (1972/1979) *Objective Knowledge*, New York: Oxford University Press.
Popper, Karl R. (1982) *Quantum Theory and the Schism in Physics*, ed. W. W. Bartley, III, London: Hutchinson's.
Popper, Karl R. (1974) "Replies to My Critics," in *The Philosophy of Karl Popper*, ed. P. Schilpp, La Salle, Ill.: Open Court, pp. 961–1197.
Popper, Karl R. (1974) "Who Killed Logical Positivism," in *The Philosophy of Karl Popper*, ed. P. Schilpp, La Salle, Ill.: Open Court, pp. 69–71.
Popper, Karl R. & John C. Eccles (1977) *The Self and Its Brain*, New York: Springer.
Postman, Neil (1985) *Amusing Ourselves to Death*, New York: Viking.
Postman, Neil (1982) *The Disappearance of Childhood*, New York: Delacorte.
Postman, Neil (1979) *Teaching As A Conserving Activity*, New York: Delacorte.
Prigogine, Ilya & Isabelle Stengers (1984) *Order Out of Chaos*, New York: Bantam.
Quest for Fire (1981) motion picture released in the United States by Twentieth Century-Fox.
Radnitzky, Gerard (1982) "Popper as a Turning Point in the Philosophy of Science," in *In Pursuit of Truth*, ed. P. Levinson, Atlantic Highlands, N.J.: Humanities, pp. 64–80.
Radnitzky, Gerard & Gunnar Andersson (eds.) (1978) *Progress and Rationality in Science*, Boston: Reidel.
Recami, E. (ed.) (1978) *Tachyons, Monopoles, and Related Topics*, New York: North Holland.
Rensberger, Boyce (1978) "The Oldest Works of Art," *The New York Times Magazine*, May 21, pp. 26–29 ff.
Rensberger, Boyce (1980) "Mass Grave Tells Savage Story of Indian Strife," *New York Times*, May 27, pp. C1, 3.
Restivo, Sal (1978) "Parallels and Paradoxes in Modern Physics and Eastern Mysticism, 1," *Social Studies of Science*, 8, pp. 143–181.
Restivo, Sal (1982) "Parallels and Paradoxes in Modern Physics and Eastern Mysticism, 2," *Social Studies of Science*, 12, pp. 37–71.
Rogers, M. (1987) "Mimicking the Human Mind," *Newsweek*, July 20, pp. 52–53.
Rohrlich, Fritz (1983) "Facing Quantum Mechanical Reality," *Science*, 221, September 23, pp. 1251–1255.
Ronan, Colin A. & Joseph Needham (1978) *The Shorter Science and Civilization in China: 1*, New York: Cambridge University Press.
Roszak, Theodore (1986) *The Cult of Information*, New York: Pantheon.
Russell, Bertrand (1914/1917) *Mysticism and Logic*, London: Allen & Unwin.
Russell, Bertrand (1920/1949) *The Practice and Theory of Bolshevism*, London: Allen & Unwin.

Russell, Bertrand (1931) *The Scientific Outlook,* New York: Norton.
Russell, Bertrand (1958) *The Will To Doubt,* New York: Philosophical Library.
Russell, E. S. (1982 [1916]) *Form and Function,* Chicago: University of Chicago Press.
Sanger, D. E. (1986) "President's Order on Space Shuttle Called Era's End," *New York Times,* August 17, pp. 1, 32.
Santayana, George (1923) *Scepticism and Animal Faith,* in *The Philosophy of Santayana,* ed. I. Edman, New York: Modern Library, 1942, pp. 376–450.
Schafer, Murray (1985) "McLuhan and Acoustic Space," *The Antigonish Review,* 62–63, Summer–Fall, pp. 105–113.
Schilpp, Paul (ed.) (1974) *The Philosophy of Karl Popper,* La Salle, Ill.: Open Court.
Schirmacher, Wolfgang (1986) "Privacy as an Ethical Issue in the Computer Society," in *Philosophy and Technology II,* ed. C. Mitcham & A. Huning, Boston: Reidel, pp. 257–268.
Schreiber, Flora (1953) "The Battle Against Print," *The Freeman,* April 20.
Schwegler, Albert (1856/1888) *History of Philosophy,* trans. J. Seelye, New York: Appleton.
Searle, John R. (1980) "Minds, Brains, and Programs," in *The Mind's I,* ed. D. R. Hofstadter & D. C. Dennett, New York: Basic, 1981, pp. 353–373.
Sennett, Richard (1977) *The Fall of Public Man,* New York: Knopf.
Shannon, Claude & Warren Weaver (1949) *The Mathematical Theory of Communication,* Urbana, Ill.: University of Illinois Press.
Sichel, Berta (1983) "New Hope for a Technological Society: An Interview with Jacques Ellul," *Et Cetera,* 40 (2), Summer, pp. 192–206.
Simpson, George Gaylord (1963) "Biology and the Nature of Science," *Science,* 139, January 11, pp. 81–88.
Simpson, George Gaylord (1953) *The Major Features of Evolution,* New York: Columbia University Press.
Skolimowski, Henryk (1980) "Evolutionary Illuminations," *Alternative Futures,* 3 (4), Fall, pp. 3–34.
Skolimowski, Henryk (1976) "Evolutionary Rationality," in *PSA 1974,* ed. R. S. Cohen et al., Boston: Reidel, pp. 191–213.
Sohn-Rethel, Alfred (1978) *Intellectual and Manual Labor,* Atlantic Highlands, N.J.: Humanities.
Spencer, Herbert (1864/1880) *First Principles,* 4th ed., New York: Appleton.
Stearn, Gerald E. (ed.) (1967) *McLuhan: Hot and Cool,* New York: Dial.
Stebbins, G. Ledyard (1950) *Variation and Evolution in Plants,* New York: Columbia University Press.
Stebbins, G. Ledyard & Francisco J. Ayala (1985) "The Evolution of Darwinism," *Scientific American,* July, pp. 72–82.
Stebbins, G. Ledyard & Francisco Ayala (1981) "Is a New Evolutionary Synthesis Necessary?," *Science,* 213, August 28, pp. 967–971.
Steele, E. J. (1979) *Somatic Selection and Adaptive Evolution,* Toronto: Williams & Wallace.
Stent, Guenther (1975) "Limits to the Scientific Understanding of Man," *Science,* 187, March 21, pp. 1052–1057.
Straus, Donald B. (1984) "Artificial Intelligence in Maintenance," *Applied Artificial Intelligence Report,* November, pp. 11, 18, 20.
Straus, Donald B. (in preparation) *Computers and the Democratic Process.*
Straus, Donald B. (1984) "Just to Jump In," *Information Technologies,* Conference 349, comment 1637, Electronic Information Exchange System teleconference, September 19.
Stroke, George W. (1969) *An Introduction to Coherent Optics and Holography,* 2nd ed., New York: Academic Press.

Sullivan, Walter (1986) "Soviet Scientists Often Thwarted," *New York Times*, October 7, pp. C1, 7.
Suspicion (1941) motion picture released in the United States by R.K.O.
Tagliacozzi, Giorgio (ed.) (1981) *Vico: Past and Present*, Atlantic Highlands, N.J.: Humanities.
Tarski, Alfred (1956) *Logic, Semantics, Metamathematics*, trans. J. H. Woodger, New York: Oxford University Press.
Taylor, Chuck (1985) "Lower Costs, Better Sound Via Audio Teleconferencing," *Washington Business Journal*, September 23, p. 13.
The Egyptian (1954) motion picture released in the United States by 20th Century-Fox.
The Frontiers of Knowledge: The Frank Nelson Doubleday Lectures (1975) Garden City, N.Y.: Doubleday.
The Responsibility of the Academic Community in the Search for Absolute Values, vol. 1 (1980) New York: International Cultural Foundation Press.
Thomas, Hugh (1979) *A History of the World*, New York: Harper & Row.
Thomas, Sari (ed.) (1984) *Studies in Mass Communication and Technology*, Norwood, N.J.: Ablex.
THX-1138 (1970) motion picture released in the United States by Warner Brothers.
Toulmin, Stephen (1972) *Human Understanding*, New York: Oxford University Press.
Townsend, Carl (1984) *Beyond Electronic Mail*, Belmont, Calif.: Wadsworth.
Treisman, Michel (1977) "Motion Sickness: An Evolutionary Hypothesis?," *Science*, 197, July 29, pp. 493–495.
Truffaut, François (1967) *Hitchcock*, New York: Touchstone.
Truitt, Willis H. & T. W. G. Solomons (eds.) (1972) *Science, Technology and Freedom*, Boston: Houghton-Mifflin.
Turkle, Sherry (1984) *The Second Self*, New York: Simon & Schuster.
2001: A Space Odyssey (1968) motion picture released in the United States by MGM.
Utsumi, Takeshi (1985) *China Connection*, Conference 783, comment 57, Electronic Information Exchange System teleconference, December 6.
van Dam, Andries (1984) "Computer Software for Graphics," *Scientific American*, September, pp. 122–137.
van Gelder, Lindsay (1985) "The Strange Case of the Electronic Lover," *Ms*, October, pp. 94ff.
Van Loon, Hendrik (1928) *The Story of Invention*, New York: World.
Vollmer, Gerhard (1982) "Kant and Evolutionary Epistemology," Proceedings of the 7th International Wittgenstein Symposium, Kirchberg am Wechsel, Austria, August.
Vonèche, Jacques (1982) "Evolution, Development, and the Growth of Knowledge," in *The Cognitive Developmental Psychology of James Mark Baldwin*, ed. J. M. Broughton & D. J. Freeman-Moir, Norwood, N.J.: Ablex, pp. 51–79.
Wachtel, Edward (1983) "The First Picture Show: Cinematic Aspects of Cave Art," paper presented at Fifth International Conference on Culture and Communication, Temple University, Philadelphia, March 25.
Wachtel, Edward (1977/1978) "The Influence of the Window on Western Art and Vision," *The Structurist*, 17/18, pp. 4–10.
Waechtershaeuser, Guenter (1984) "Light and Life: On the Nutritional Origins of Perception and Reason," paper presented at 150th Annual Meeting of the American Association for the Advancement of Science, New York City, May 27.
Walters, Ray (1983) "The Coming of the Computer," *The New York Times Book Review*, July 24, pp. 12–13.
War Games (1983) motion picture released in the United States by MGM/United Artists.

Wartofsky, Marx (1980) "The Critique of Impure Reason II: Sin, Science, and Society," *Science, Technology, and Human Values,* 6 (33), Fall, pp. 5–23.
Wartofsky, Marx (1972) "Is Science Rational?," in *Science, Technology and Freedom,* ed. W. H. Truitt & T. W. G. Solomons, Boston: Houghton-Mifflin, pp. 202–210.
Weizenbaum, Joseph (1976) *Computer Power and Human Reason,* San Francisco: Freeman.
Whetmore, Edward J. (1979) *Media America,* Belmont, Calif.: Wadsworth.
Whitehead, Alfred North & Bertrand Russell (1910, 1912, 1913) *Principia Mathematica,* vols. I, II, III, Cambridge, U.K.: Cambridge University Press.
Wiener, Norbert (1948/1961) *Cybernetics,* New York: MIT and John Wiley.
Wiener, Norbert (1950/1967) *The Human Use of Human Beings,* New York: Avon.
Wilder, Raymond L. (1981) *Mathematics as a Cultural System,* New York: Pergamon.
Williams, Archibald (1910) *The Wonders of Mechanical Ingenuity,* Philadelphia: Lippincott.
Wittgenstein, Ludwig (1972) *Philosophical Investigations,* trans. G. E. M. Anscombe, Oxford: Blackwell.
Wyatt, J. W. et al. (1979) "The Status of Hand-Held Calculator Use in School," *Phi Delta Kappan,* November, pp. 217–218.
Zeleny, M. (ed.) (1981) *Autopoiesis,* New York: Oxford University Press.
Zimmerli, Walther (1986) "Who Is To Blame for Data Pollution," in *Philosophy and Technology II,* ed. C. Mitcham & A. Huning, Boston: Reidel, pp. 291–305.
Zimmerman, Michael (1979) "Technological Culture and the End of Philosophy," in *Research in Philosophy and Technology,* vol. 2, ed. P. Durbin, Greenwich, Conn.: JAI Press, pp. 137–145.
Zuckerman, Joseph (1986) "Try This with a Computer!," *Connect Ed Cafe,* Connected Education Conference 1320, comment 539, Electronic Information Exchange System teleconference, June 23.

Biographical Sketch of the Author

Paul Levinson is president of Connected Education, Inc., a not-for-profit organization that offers a program of courses via computer teleconferencing for graduate and undergraduate academic credit in conjunction with the New School for Social Research. Connect Ed also consults with a wide variety of businesses and academic institutions, including the Polytechnic University of New York.

Dr. Levinson is Associate Professor of Communications at Fairleigh Dickinson University, a member of the faculty of the M.A. in Media Studies Program at The New School for Social Research, and on the faculty of the Western Behavioral Sciences Institute in La Jolla, California. He has taught at Fordham University, Hofstra University, and St. John's University. In 1987–1988 he will be Visiting Professor at the Philosophy and Technology Studies Center, Polytechnic University, Brooklyn, New York.

He is editor of *In Pursuit of Truth: Essays on the Philosophy of Karl Popper* (Humanities Press, 1982), and is author of more than 40 articles on the philosophy and history of technology and communications media. He holds a Ph.D. in Media Theory from New York University, an M.A. in Media Studies from the New School, and a B.A. in Journalism from New York University. He is special consultant for PBS-TV's "Ways of Knowing" series.

He lives with his wife and children in New York City.

Index

a priori knowledge (*see* knowledge-innate)
Abelson, P. H., 114
abstraction (*see also* communications media; language; writing)
   as content of communication, 123ff.
   as means of communication, 119ff., 150
   economic, 84
Academy Awards, 117–118
adaptation, evolutionary, 3–5, 11, 13, 20–21, 46, 47ff., 99, 166–167, 191 (*see also* evolutionary epistemology; natural selection; preadaptation)
Agassi, Joseph, 154
Age of Discovery, xv, 133, 154
Age of Faith, 133–134
Age of Invention, 66
Age of Reason, 45
agriculture, 101, 138
air flight, 217, 218 (*see also* aviation)

air pollution, 231, 237
al-Hibri, A., 85
Alexandria (Egypt), 93, 115, 210
Alland, Alexander Jr., 60
alphabet (*see* writing-alphabetic)
Althuser, L., 234
altruism and evolution, 27, 40
America, pre-Columbian, 226 (*see also* Vikings)
amoebas, 20, 22
anagenesis, 61
Andersson, Gunnar, 41, 60, 110
animism, 82
"anthropotropic" media, 111, 141, 142, 158, 183ff., 194
Apple Computers, 115
Apple, R. W., 152
Aristarchus, 89
Aristotle, 29, 66, 95, 99, 104, 110, 200
Arms, Valarie, 159, 216
art, 139, 152
   and computers, 155 (*see also* music)
   Platonic criticism of, 110

artificial intelligence, xv, 81, 86, 114, 148, 162, 172ff., 189–192
   and humor, 174–175
   and intentionality, 176–177
   and living systems, 177–182
   and "noise-plus", 177 (*see also* noise)
   auxiliary vs. autonomous, 190
   champions and critics of, 190–192, 193
   "top-down" vs. "bottom-up", 191
artificial life, 179ff.
artificial selection, 87, 179, 193, 195 (*see also* gene splicing)
Asimov, Isaac, 194
astronomy, 89–90
AT&T, 159
Athens (Greece), 197
Attenborough, David, 144
authenticity and artificial intelligence, 175
autocatalysis, 27, 153
automobile, 126, 212, 231, 237
Aveling, Edward, 84
aviation, 92–93, 213, 217–218
Ayala, Francisco J., 13, 85

Babbage, Charles, 187
Bacon, Francis, 67, 73, 81, 87, 110, 200, 235
Baker, W. O., 103, 114
Baldwin, James Mark, 56, 60–61
Barnes, Barry, 215
Barnouw, Eric, 159
Bartley, W. W. III, 16, 19, 23–24, 31, 37–38, 39, 41, 42, 59, 84, 161, 210, 218
Barzun, Jacques, 84
Bateson, Gregory, 82
Baumer, Franklin L., 84

Baur, Stuart, 195
Bazin, André, 118, 139, 149, 188
Behrens, Steve, 159
Bell, Alexander Graham, 87
Bell, Daniel, 84
Bergson, Henri, 118, 149
Berlin, Isaiah, 84
Bible, 132–133
biosphere, 36
Blake, William, 152
Bloomsbury Group, 197, 199
Bloor, David, 215, 216
Bock, W. J., 59
Boden, Margaret, 162, 194
Bohm, David, 236
Bohr, Niels, 14, 42–43, 236
Bolter, J. David, 187
books, xiv–xv, 10, 81, 92, 105–109, 116, 128ff., 137
Borden, Carla, 160
Bracken, Harry, 15
brain, xiii, 10–11, 59, 69, 148, 155, 161, 192–193 (*see also* artificial intelligence; mind)
   global, 146 (*see also* intellectual circles)
Brand, Stewart, 152, 155
Bricken, Uilliam M. Jr., 41
Briskman, Larry, 113
Broad, William, 215
Broughton, J. M., 60
Brown, Ben, 161
Brumbaugh, Robert S., 187
Buchler, J., 15
Buddhism, 14, 153
Bunge, Mario, xvii, 82, 83
Burbank, Luther, 179, 193
Burlingame, Roger, 187
Bush, Vannevar, 217
Butler, Samuel, 41, 86, 149

Cambridge Apostles, 206, 217

Campbell, Donald T., 2–3, 9, 13, 15, 16, 20–22, 38, 47, 52, 58, 59, 85, 99, 111, 120, 176, 190
Campbell, Jeremy, 41, 61, 85, 109
capitalism, 75, 152
Capra, Fritjof, 14
Carnap, Rudolf, 214
Carpenter, Edmund, 187
Cathcart, Robert, 188
cave paintings, 102, 187 (*see also* writing-pictographic)
cellular telephones, 219
certainty, 45–46, 57 (*see also* fallibilism; knowledge)
Challenger (Space Shuttle), 109, 113, 217 (*see also* space travel)
change, evolutionary vs. humanly created, 11–12, 34, 74, 185, 223, 231
Chaplin, Charlie, 234
Charnot, Jean, 187
Chase, Stuart, 149
chauvinism, protein, 180 (*see also* artificial intelligence)
Chernobyl, 113, 236
Cherry, Colin, 58, 162
Chester, G., 159
Chin, Paula, 152
China, 97, 111, 130–131, 150, 151, 152, 153, 159, 187
Chomsky, Noam, 6, 7, 15, 30, 41, 47, 58, 66, 83, 151, 183, 192
Christianity, 83, 132, 133, 145, 188
Churchill, Winston, 160
Clauser, John F., 236
Cleveland, Harlan, 116
cognition, 3, 6–7, 46ff., 84–85, 87, 90–91, 109, 146, 157–158 (*see also* evolutionary epistemology; mind; knowledge)
cultural impact on, 201–202, 215–216
structure of and external environment, 70–71
technological amplification of internal cognition (*see* computers; mathematics)
technological extension of sensory experience, xiv, 89–90, 96–97, 101, 111–112, 140–141
technological impact on, general, xiv, 1, 11–12, 89ff., 117ff., 197ff. (*see also* technology)
technological preservation of fruits of, 91–93 (*see also* communications media; permanence)
Cohen, I. Bernard, 113
Cohen, Marshall, 188
Cohen, Robert S., 42
Coleridge, Samuel Taylor, 38, 189
collectivism, 75
Columbus, Christopher, 116, 133, 154
communication (*see also* communications media)
and abstraction, 120–127
as extension of senses, 21, 97, 118, 149
as initiator of human technology, 80
brain-tongue vs. brain-hand, 155
electronic (*see* communications media; electronic texts)
primacy of in human life, 80, 83–84, 86, 115–116, 150, 174–175
Shannon-Weaver model of, 112, 121
vs. duplication, 120
vs. transportation, 97, 121–122,

145–146, 161, 199, 207–211
communications media, xiv–xv,
    91–92, 96, 108–109, 114,
    117ff. (*see also* books; electronic texts; television; etc.)
  abstract vs. replicative, 118, 122–123, 125–127, 139–141, 150
  "double" cognitive consequences of, 81
  extension action of, 21, 118, 149
  interactive vs. one-way (mass), 118, 142–143, 205
  permanence and portability, 118, 121, 129, 134
  speed, 118, 138, 142
communities, intellectual (*see* intellectual circles)
computer(s), xv, 40, 101ff., 126, 148, 165–166, 187 (*see also* art; icons; literacy; music)
  as amplifiers of primary cognition, 103, 105, 114, 155, 172, 177, 190
  as autonomous intelligences (*see* artificial intelligence)
  as facilitators of language (*see* electronic texts)
  conferencing, 104–105, 115, 136, 151, 157, 158–159, 160, 190, 198–199, 203ff., 216–217
  critiques of, 104–105, 114, 191, 194
  four services to growth of knowledge, 114–115
  graphics, 155–156
  "hackers", 217
  "laptop" (portable), 155
  mini, 105, 115
  modeling, 114, 152
  networks, 129
  personal, xvi, 16, 105, 115, 152, 155, 204, 207, 216, 217

  programming languages, 40, 138–139, 173
  vs. human infants and amoebas, 181 (*see also* artificial intelligence)
Comte, Auguste, 58, 83
Confucianism, 153
conjecture(s) and knowledge, 9, 34, 46, 53, 71, 85 (*see also* Popper)
Connected Education, Inc., 157, 205, 216–217
consciousness, 56
Copenhagen Interpretation, 236 (*see also* quantum mechanics)
Copernicus, 89
Cowan, Robert, 217
Cracraft, Joel, 12
creationism, 2, 13
criticism and knowledge, 23–24, 52–55, 60, 109, 133–134 (*see also* evolutionary epistemology; knowledge; falsification)
  internal vs. social, 107–108
Curtis, James M., 40, 149
cybernetics, 158, 183, 190, 194

Dalton, Richard, 155
Darwin, Charles, 8, 41, 49, 55, 58–59, 66, 71, 74, 84, 85, 87, 97, 134, 179, 193, 195, 203
Darwinian theory, 1–2, 4, 12–13, 31–33, 42, 75 (*see also* natural selection)
Davies, Nigel, 154
Davis, D. D., 59
deconstructionism, 174
de Forest, Lee, 159
de Haan, David, 156
de Kerckhove, Derrick, 61

Delaroche, Paul, 152
democracy, 211, 218–219
Dennett, D. C., 190
Descartes, René (Cartesian), 7, 15, 67–68, 82-83, 180, 200–201, 211, 214
design, argument from, 31-33
desktop publishing, 156, 216
Dessauer, Friedrich, 77, 85–86, 193
determinism (and indeterminism), 32, 42
Dewey, John, xiv, 61, 77, 81, 85, 86, 87, 92, 109, 111, 112, 116, 149–150, 157, 187, 193, 233
Dick, Hugh, 235
dinosaurs, 144
"disproportion of man" to cosmos (Pascal), 101
DNA, 10–11, 40, 41, 132–133, 201, 208, 223 (*see also* genetics)
Dobzhansky, Theodosius, 13
dogmatism, 5, 58
Dohrn, A., 59
Drake, Stillman, 110
Dretske, Fred, 177, 190–191
Dreyfus, Hubert, 177, 189–190, 191–192
Dreyfus, Stuart E., 189–190
dualism, 63ff., 82 (*see also* mind; mental)
Dumbach, Annette G., 161
Durbin, Paul, 85, 156
Dyson, Freeman, 237

Eccles, John C., 82
ecologies of rationality, 35
economics, impact of on technology, 145, 198, 209–210
Eddington, Arthur, 236
Edison, Thomas A., 87
Edman, I., 39

Edmondson, William, 58, 162
education
  rise of public and printing press, 134
  via computer conferencing, 204–205, 216–217
Egypt, ancient, 92, 93, 130–132, 151, 153
Eigen, Manfred, 27–30, 40, 43, 109
Einstein, Albert, 134, 236–237
Eirich, Frederick, 193–194
Eisenstein, Elizabeth, 114, 154
Eisenstein, Sergei, 188, 189
Eldridge, Niles, 12
electricity, xiv, 12, 68, 142, 156
Electronic Information Exchange System (EIES), 151, 157, 190, 205, 216
electronic texts
  mail, 104, 136, 216
  processing and transmission, xv, 60, 104, 135ff., 155, 156
  publishing, 60, 136, 156–157
  seminars (*see* computer-conferencing)
  vs. paper text storage, 136, 157
Ellis, Don Carlos, 158
Ellul, Jacques, 87, 158, 225, 233–234, 237
emergence (*see also* self-transcendence)
  of life from nonlife, 27, 43
  of rationality, 26–30
emergentist materialism and mentalism, 82
Emerson, Ralph Waldo, 42, 118, 149
emotion and rationality, 110, 189
empiricism, 4, 33, 35, 61, 67, 98, 110, 214
Engels, Frederick, 84, 85, 87, 234
Enterline, James R., 154

entropy, 61, 112, 113, 230 (*see also* noise)
environment, external, 5, 15, 56, 60–61, 66, 90 (*see also* reality)
epistemology, 2, 7, 33–34, 40, 42, 46–47, 65ff. (*see also* Kant; Popper; *etc.*)
  evolutionary (*see* evolutionary epistemology)
  fallibilist (*see* fallibilism)
  of the stomach, Marx's, 76, 218
  of the toothpick, 65, 72ff.
  techno-evolutionary (*see* evolutionary epistemology; technology)
  technological lessons for, 65–66, 71–72, 79–80, 81 (*see also* knowledge)
Ericson, R., 40
error, xiv, 9, 15–16, 50–56, 92–93, 156 (*see also* fallibilism; noise)
ethics, 99, 113, 124, 149, 216, 228, 235–236, 239
Euclidean conceptions, 111
Europe, 159–160
Evans, R. I., 15
evolution
  biological, 1–2, 10, 33, 42 (*see also* natural selection; preadaptation)
  cognitive, 2–4, 21–22, 42, 47, 64, 89ff., 232 (*see also* abstraction; cognition; knowledge; language)
  collective vs. individual, 6, 13, 75–76 (*see also* knowledge-individual)
  cosmic, xvi, 34, 178, 185, 221ff., 227–228, 230–231
  linguistic analogies to, 40 (*see also* language)
  molecular, 10, 13, 27
  of evolution, 35, 91, 221, 231
  of life from non-life (*see* emergence; self-transcendence)
  of technology (*see* anthropotropic; technology)
  social, 3, 74, 84
evolutionary epistemology, 2–4, 11, 16, 47, 54, 70, 83–84, 86, 93, 214, 218
evolutionary limitations hypothesis, 47–50, 90
exaption, 59 (*see also* preadaptation)
existentialism, 81
expert systems, 86, 189–190 (*see also* artificial intelligence)
extraterrestrials, 97, 194, 231–232

faith, 22–25, 39, 133 (*see also* metaphysics)
fallibilism, 2, 41, 45ff., 71, 77–78, 82, 110, 112, 200 (*see also* Popper)
falsification, 9, 166 (*see also* Popper)
Farrington, Benjamin, 236
faster-than-light travel, 236–237
Faust, 104
Feenberg, Andrew, 216
Feigenbaum, Edward, 189, 193
Fessenden, Reginald, 142, 159
Feuerbach, Ludwig, 16
Feyerabend, Paul, 25, 35, 40, 41, 110
Fichte, Johann, 69
film (*see* motion pictures)
fire, 179
Fontaine, Joan, 188
food (*see* "nutritional" theories of cognition)
Frankfurt School, 197
free will, 32–33, 146

Freeman, E., 15
Freeman-Moir, D. J., 60
Freud, Sigmund, xvi, 61, 118, 134, 148, 149, 153, 157, 161, 200, 203, 214, 224, 233
Fuller, R. Buckminster, 149, 230, 237
functionalism, 81 (*see also* Dewey)

Galileo, 95, 96, 110
gene splicing, 40, 87, 179, 215, 232
genetic engineering (*see* gene splicing)
genetics, 4, 6–7, 70 (*see also* natural selection)
Gernsheim, Helmut and Alison, 152
Giedion, Siegfried, 187, 194
Gleick, James, 114, 191
Glossbrenner, Alfred, 216
God, 3, 8, 27, 32–33, 41–42, 63, 74, 111, 191 (*see also* monotheism)
Gödel, Kurt, 5, 14
Goldstine, Herman, 187
Gombrich, E. H., 101, 113, 151
Gorbachev, Mikhail, 161
Gould, Stephen J., 12–13, 15, 49, 59, 85, 202, 215
gradualism (evolutionary theory), 1, 3
Graham, Loren, 215
grammar, generative (*see* Chomsky)
Grant, Cary, 188
Gray, Asa, 41–42
Gray, Hugh, 149
Greece, ancient, 115, 131
Gregory, R. L., 161
Griffith, D. W., 188
Gruber, Howard, 41
Gumpert, Gary, 188
Gutenberg, Johann, 133

Haas, Warren J., 156
Habermas, J., 234
Haeckel, Ernst, 70, 83
Hall, Edward T., 149
Hammond, A. L., 114
Hardy, Alistair, 61
Hartline, Beverly K., 195
Hartog, Marcus, 86
Havelock, Eric, 131, 153
Hawthorne, Nathaniel, 146, 161
Hayek, Friedrich, 192
Hebrews (*see* Judaism)
Heelan, Patrick, 111
Hegel, G. W. F., 37, 55, 74, 83, 156, 189, 224
Heidegger, Martin, 77, 85, 86, 148, 161, 162, 193, 234
Heisenberg, Werner, 14, 42, 43, 188, 236
Hendrickson, Robert, 157
Heron of Alexandria, 187
Hertz, Heinrich, 58
Hickman, Larry, 85, 158, 191, 192
hierarchy of vicariousness, 20–22, 99, 120 (*see also* Campbell, D. T.)
hieroglyphics, 130ff., 151 (*see also* writing)
Hiltz, Roxanne, 216
Hitchcock, Alfred, 168, 188
Hitler, Adolf, 7, 143, 149
Hofstadter, Douglas, 14, 28, 29, 41, 156, 177, 190, 191
Hogan, John V. L., 159
Hollywood, 117
hologram (brain function as), 59
holography, 59, 95–96, 118, 140, 145, 148–149, 184, 195, 207
hominids, 36, 114
Hood, Webster F., 85
Hooke, Robert, 114
Hoover, Herbert, 159
Hughes, R. I. G., 36, 43

humans, unique attributes of, 11–12, 56, 119, 138, 151, 221–222
Hume, David, 67–69
humor and artificial intelligence, 175, 184 (*see also* artificial intelligence)
Huning, Alois, 111, 113, 189, 190, 219
Huxley, Julian, 13, 61
hypercycle (Eigen), 29–30, 40
hypersonic transport, 217

IBM, 115
icons, 150, 152 (*see also* China; computers—personal)
idealism, philosophical, 4, 63ff. (*see also* subjectivism)
ideographic writing, 122, 150–151 (*see also* writing)
Ikhnaton (Amenhotep IV, Pharaoh), 131–132, 153
immaculate conception of photography, 139ff.
indeterminism (*see* determinism)
individuals vs. groups (*see* knowledge-individual vs. group; intellectual circles)
Industrial Revolution, 66
innatist theories of knowledge (*see* knowledge-innate)
Innis, Harold, 131, 133, 149, 151, 153, 154
"instant coffee" metaphor, 115, 121
intellectual circles, 197ff.
 and physical locale, 197–198, 205, 210
 and physical presence, 198–199, 208, 211, 213
 electronic facilitation of, 199, 203–209, 216–217 (*see also* computer-conferencing)
 global, 145, 205–206, 209–211, 212–213, 216, 218
 interactive vs. observational, 145, 205
 transportive facilitation of, 199, 209–211, 213, 218
intelligence, 6–7, 15, 86–87, 180–181 (*see also* mind; artificial intelligence)
intention and artificial intelligence, 176–177
interactionism, wholly material vs. wholly intellectual, 73, 76, 82, 84 (*see also* mental/material interaction)
intuition, 41
inventions, unintended applications of, 80, 87, 90ff. (*see also* technology)
irrationality (*see* faith; rationality)
irreversibility, 43 (*see also* Prigogine)
Ives, Herbert, 159
Izzillo, T., 161

Jackman, Jarrell, 160
Jantsch, Erich, 27, 40, 43, 61
Japan, 151, 159, 175, 205, 216
Jarvie, I. C., 215
Jaspers, Karl, 224, 233–234, 235
Jaynes, Julian, 151
Jefferson, Thomas, 7
Jensen, Arthur, 6, 15
Jepsen, G. L., 59
Jerusalem, 197
jet planes, 199, 209, 210
Jones, Gwyn, 154
Jones, H. F., 86, 149
Judaism, 131–132, 153
Jung, Carl, 202

justice, as abstract content of communication, 123–124 (*see also* ethics)

Kant, Immanuel (Kantian), xiv, xv, 6, 11, 8–9, 15, 30, 58, 66–78, 81, 83, 84, 86, 89, 96, 111, 123–125, 137, 150, 161, 162, 163, 167–168, 171, 183, 184, 189, 192, 193, 200–201, 202, 211, 212, 214, 232, 234
"thing-in-itself," 70, 112, 113
Kantorovich, Aharon, xvii, 111, 112, 202–203, 215, 216
Kapp, Ernst, 148, 149, 161
Kaufmann, Walter, 41
Kerr, Elaine, 216
Kessler, Eckward, 214
Kimura, Motoo, 13
Klein, Stanley, 155
knowledge
   absolute vs. uncertain, 8–9, 57, 70–71, 83 (*see also* conjecture; fallibilism)
   biological basis of (*see* evolutionary epistemology)
   definition of, 20, 57, 66–69, 82, 84
   embodiment vs. representation in technology, xiv–xv, 81, 192–193
   expression in material form (*see* technology)
   generation, criticism, dissemination of, 2, 4, 52–54, 91–92, 106–108, 109, 110, 144–145, 208 (*see also* communications media)
   growth of, 2–3, 10–12, 46–50, 55–56, 62 ff., 71, 89, 91, 103–107, 109
   individual vs. group production of, 60, 75, 107–108, 115–116, 146, 186
   innate, 4, 6–8, 11, 22, 66–68
   instrumental, 87
   limits on, 40, 47–48, 57, 63, 202, 215 (*see also* evolutionary limitations)
   monopolies of, 151
   "nursery" stage, 116, 211
   of the self, 47, 148, 163ff., 192
   "productive", 81–82, 87, 162
   result of interaction of innate cognition and environment (*see* Kant)
   social constituents of, 108–109, 200ff. (*see also* sociology of knowledge)
   technological continuum with, 83–84
   technological contribution to (*see* technology)
   unembodied, 79–80, 83, 85, 86, 92–93
   via disturbance of environment, 111–112
Koestler, Arthur, 87, 214
Kolm, Henry, 218
Koran, 132
Korzenny, Felipe, 214
Kothals-Atles, Stephen W., 217
Kraft, Victor, 215
Krippendorf, Klaus, 221–222, 233
Kubrick, Stanley, 213
Kuhn, Thomas, 14, 134, 154, 201, 215–216
Kuhns, William, 149

labor, Marxist view of, 74–76, 87
Lachenbruch, David, 161
Lakatos, Imre, 59, 60, 215
Lamarckian theory, 2, 8, 13, 51, 55, 60

Land, Michael F., 195
language (linguistic)
   analogies to biology, 28–29, 40–41
   and abstraction (*see* abstraction)
   as technology, 83–84
   origins and development of, 40, 58, 151 (*see also* Chomsky)
   spoken, 61, 83–84, 102, 105, 108, 114, 121ff.
   theory of truth, 5, 8
   vs. computer "languages", 40, 139, 173ff.
   written, 28, 61, 81, 95–96, 102, 104–106, 109–110 (*see also* writing)
Laplace, P. S., 32, 42
Lascaux (France), 102
Leeuwenhoek, Anton van, 98
Leibniz, G. W., 102, 187
Leonardo da Vinci, 187
Levinson, Paul (other works by), 14, 15, 38, 42, 58, 59, 82, 84, 85, 86, 87, 110, 111, 141, 149, 153, 156, 157, 158, 160, 161, 192, 194, 215, 216, 218, 219
Levy, Paul, 217
Levy, Steven, 217
Lewin, Roger, 13, 59
Lewontin, R. C., 15, 61
libraries, 93, 105, 106, 134, 137, 155
   unintended technological, xiv, 90ff. (*see also* permanence)
Libri, Giulio, 110
light, properties of, 35, 58, 85, 97–98, 112, 152, 236–237
literacy (*see also* writing)
   effect of computers on, 135ff., 151–152
   effect of printing press on, 134
   effect of television on, 141ff., 160, 234–235
living vs. non-living material, 26–27, 33, 45
Locke, John, 67, 110, 214
locomotion and cognition, 20, 80, 112 (*see also* hierarchy of vicariousness; transportation)
Löfgren, Lars, 40–41
logic, 17, 35–36, 39, 52, 66, 102, 110 (*see also* paradox; quantum mechanics; rationality)
London (England), 210, 216
London School of Economics, 237
Lorentz, H. A., 237
Lorenz, Konrad, 8, 15, 47, 57–58, 70, 83
Lucas, George, 234
Lum, Man-Kong, 159
Luther, Martin, 133
Lyell, Charles, 41–42

McCorduck, Pamela, 189, 193
McKeever, William, 114
McLuhan, Marshall, xv, 14, 42, 57, 61, 111, 114, 117–118, 129, 131, 137, 148, 149, 150, 151, 153, 154, 156, 157, 158, 160, 187, 188, 205, 210, 212, 218, 219
McNeill, William, 153
McWilliams, Peter A., 155
Mackey, Robert, 86
Maeroff, Gene, 160
Magellan, Ferdinand, 209
magnetic transportation, 209
Mallarmé, Stéphane, 46, 61, 214
Mander, Jerry, 160, 194
Manuel, Frank E., 114
Marco Polo, 97

Marconi, Guglielmo, 58, 159
Marcuse, Herbert, 225, 234–235
Marey, E.J., 158
Mars, 21
Marshack, Alexander, 95, 109, 127, 138
Martin, Bradley, 152
Martin, Douglas, 61
Marx, Karl (Marxism), xiv, 11, 16, 55, 56, 61, 66, 73–77, 81, 84, 85, 87, 171, 187, 188, 189, 193, 200–201, 214–215, 218, 234
Mast, Gerald, 158, 187, 188, 189
material, ideated vs. unideated, 64ff.
materialism (philosophical), 63ff., 71, 74–76, 82–84
mathematics, 101ff., 109, 113–114, 115, 193–194
Mayan civilization, 151
Mayr, Ernst, 13, 59
Mecca, 197
medicine, 4–5, 138, 150
Méliès, Georges, 188
mental/material interaction, 70–71, 76–79, 81–82, 86–87 (*see also* technology)
mentality, innate, (*see* knowledge-innate)
Merleau-Ponty, Maurice, 177, 191–192
metaphysics, 3, 11, 35, 50, 86
Meyrowitz, Joshua, 160, 181, 188, 211, 218, 234
microscope, xiv, 90, 96ff., 134, 141, 147
Milgram, Stanley, 187
Mill, John Stuart, 200, 214
Miller, Jonathan, 159, 161
Miller, R. K., 194
mind (*see also* brain; cognition)

and matter, 63ff.
body problem, 82
transempirical, xiii, 147, 161, 229
Minsky, Marvin, 189
mise-en-scène, 169ff., 188, 189
Mitcham, Carl, xvii, 58, 86, 111, 113, 153, 189, 190, 219
modem, 136, 156, 216
Molander, R. C., 236
Monaco, James, 189
monotheism, 131–133, 153
montage, 169ff., 189
Morgan, C. Lloyd, 60–61
Morison, Samuel E., 154
Morse Code, 122, 159
Morse, Samuel, 138
motion pictures, xv, 21, 81, 87, 92, 114, 117–118, 143, 158, 194
and Kantian epistemology, 164–165, 166ff.
motion sickness and evolution, 167
Mumford, Lewis, 149, 161, 211, 212, 218, 225, 233–234
Münsterberg, Hugo, 168–169, 172
Munz, Peter, 14, 39, 116
Murray, Margaret, 153
Musgrave, Alan, 60, 215
music and computers, 155
Muybridge, Eadweard, 158
mysticism (*see also* metaphysics), 4, 14, 202

Nagorski, Andrew, 161
narcissism, 184
national states, rise of, 133, 154
natural selection, xiii, 5, 15, 34–37, 41–42, 75, 84, 87, 193, 195
natural "vs." artificial, 185, 230–231
Nazis, 6, 160, 161
Needham, Joseph, 150

neural networks, 194 (*see also* artificial intelligence)
"never" dogmatism, 58, 190 (*see also* fallibilism)
Nevitt, B., 57, 188
New Jersey Institute of Technology, 151, 157, 190, 205
New School for Social Research, 175, 188, 205
New York Institute of Technology, 205
Newborn, Jud, 161
Newton, Isaac, 67, 70, 102–103, 114, 134, 208, 237
Nielsen ratings, 160
Nietzsche, Friedrich, 25, 35, 37, 40, 41, 77
nihilism, 1, 3, 67, 174
noise (distortion), 59, 83, 98–100, 112, 127–128, 150, 174, 176–177
nuclear fission power, 113, 236
nuclear weapons, 3, 30, 57, 61, 105, 114, 226–227, 235–236
nuclear winter, 114
number systems, 102, 108 (*see also* mathematics)
"nursery" stage of knowledge, 116, 211–212
"nutritional" theories of cognition, 38, 85, 111, 112

Occam's razor, 33
Ong, Walter, 131, 153, 157
open vs. closed systems, 46, 56–57, 76
optimism (and pessimism), 3, 37, 45ff., 51, 55–56, 64, 71, 226–227, 230, 237
Osborn, Fairfield, 60–61
ox-ribs, Pleistocene, 95, 96, 101, 102, 109, 127

Pagels, Heinz, 114
pancritical rationalism, 23–24, 31, 39
paperback books, 129, 234
paradigms (*see* Kuhn)
paradox(es), 47, 221–222, 233
  apparent quantum mechanical, 58 (*see also* light; quantum mechanics)
  liar, 222, 233
  Meno and technological equivalent of, 110, 111, 112, 113, 115
  of duplication, 120, 149–150
  of self as object of its own cognition, 124, 192
particle chambers, 111, 179
Pascal, Blaise, xiv, 101, 113, 140, 187
Pavlov, Ivan, 119
Pearson, Emily C., 154–155
Peirce, Charles Sanders, 9, 15, 45, 47, 58, 70–71, 112, 174–175
perception, sensory, 20–21, 67–69, 97, 99–101, 107, 110 (*see also* cognition)
permanence and human knowledge, 91, 223, 230 (*see also* knowledge; technology)
persistence of vision, 166ff., 187
pessimism (*see* optimism)
Pharaohs (*see* Egypt, ancient)
Phillips, David, 159
philosophy (*see listings under names of philosophers, e.g.,* Kant)
phonetic writing (*see* writing—alphabetic)
phonograph, 87, 126, 129
photography, 57, 96, 121, 123, 125, 152, 158, 165, 184, 194
physical presence and the genera-

tion of knowledge, 208–210
  (*see also* transportation)
physics, 14, 32, 35, 42–43 (*see also* quantum mechanics)
Piaget, Jean, xv, 7, 15, 30, 41, 66, 83, 111
Piatelli-Palmarini, M., 15
Picasso, Pablo, 152
planetary probes, 21, 97, 144, 214
Plato, 32, 37, 61, 73, 77, 82, 83, 95, 98, 110, 113, 143, 188, 193, 235
Plato's Academy, 207
Plotkin, H., 15, 61
pocket calculators, 103, 104–105, 114, 115
Podolsky, B., 236
poetry and rationality, 38
political
  issues and search for truth, 7, 15, 114 (*see also* ethics)
  power and radio, 143, 159–160
  science, 6
Polytechnic University (of New York), 149, 193, 217, 253
Popper, Karl, xiv, 2, 9, 11, 14, 15, 16, 19, 22–25, 31, 34, 37, 38, 39, 41, 43, 45, 47, 51, 53, 58, 59, 60, 66, 70–71, 72, 76ff., 81, 82, 85, 86, 90, 109, 112, 116, 156, 166, 174–175, 177–178, 200, 210, 211, 214–215, 218, 236–237
portability (of communication systems), 134–135, 144, 155
  (*see also* books)
Porter, Edwin S., 188
positivism, 83
Post, John, 39
post-rationality, 25, 35–37, 42, 224
Postman, Neil, 154, 160
preadaptation, 49–50, 53–55, 58–59, 101, 113, 124–125
preformation (Kant's discussion of), 83
prehistoric art, 102, 109, 139, 187
prerationality vs. irrationality, 25ff.
  (*see also* rationality)
Pribram, Karl, 59
Prigogine, Ilya, 40, 43, 237
Prince Charles, 117
printing press, 111, 114, 129, 132, 151, 154
privacy, 75, 211–212, 219
propaganda, 53, 110
Protestant Reformation, xv, 133
psychology, xvi, 6, 147 (*see also* cognition)
Ptolemy, 89
public education, rise of and print, 134
publication and criticism of knowledge, 54–55, 60 (*see also* knowledge)
Pullinger, D. J., 158
punctuated equilibria (evolutionary theory), 12–13
purposive vs. non-purposive processes, 12, 31 (*see also* change)
Pytka, Steve, 157

quantum mechanics, 35, 43, 47–48, 112, 188, 161
Quine, W. V. O., 39

radical fallibilism, 51
radio, 126, 142, 143, 144, 149, 159, 160, 194
  "ham," 142
Radio Shack, 155
Radnitzky, Gerard, 41, 60, 215
rationality (and irrationality), 3–4, 17ff., 189, 224 ff.

biological necessity of, 21–22
"dialogue" with irrationality, 38
evolutionary solutions to logical problem of, 26–30
pancritical, 23–24
prerational vs. irrational, 25–26
technological perversion of, 224–228, 233–234, 235–236
RCA, 159
Reagan, Ronald, 217–218
reality, external, 3–6, 14, 42, 54, 63ff., 86, 92, 110, 172, 201, 216 (*see also* environment)
Recami, E., 237
refrigerators, cognitive, 138
Reger, Max, 157
Reisman, David, 133, 154
relativism
   cultural, 4, 215
   epistemological, 52, 200ff.
relativity, Einstein's theories of, 111, 236–237
Renaissance, 213
Rensberger, Boyce, 109, 113, 235
Rensch, Bernhard, 61
restaurants, epistemological significance of, 197
Restivo, Sal, 14
robots, 194 (*see also* artificial intelligence)
Rogers, M., 194
Rohrlich, Fritz, 112
Roman Catholic Church, 133, 145
Roman Empire and Roman numerals, 102
Rome, 197
Ronan, Colin A., 150
Roosevelt, Franklin Delano, 160
Rorshach test and Kant's philosophy, 69
Rosen, N., 236
Roszak, Theodore, 114

Rousseau, Jean Jacques, 7, 226
Russell, Bertrand, 5, 13, 14, 15, 19, 22, 24, 37, 38, 39, 58, 192
Russell, E. S., 59

Sadat, Anwar, 117
Sagan, Carl, 144, 185, 195
Saint, Eva Marie, 188
Sanger, David E., 217–218
Santayana, George, 22, 39
Schafer, Murray, 218
Schilpp, Paul, 13, 15, 16, 38, 215
Schirmacher, Wolfgang, 212, 219
Schreiber, Flora, 160
Schuster, Peter, 40
Schwegler, Albert, 83
science, history of, 66ff., 97
Scientific Revolution, 134, 145, 208
SDI—Star Wars Defense, 236
Searle, John R., 190, 191
self-reference, 19ff., 124, 128, 161, 192, 268
self-transcendence, 19, 25, 27–29, 31, 34–36, 232–233
Sennett, Richard, 211, 218
sensory perception (*see* cognition; perception)
   extension of (*see* microscope; photography; telescope)
serendipity, 87 (*see also* technology—unintended)
sex, 159, 184, 186, 217
Shannon, Claude, 112, 121, 150
Shawcross, J., 38
Shimony, Abner, 236
Sichel, Berta, 237
sign language, 125
Simpson, George Gaylord, 3, 10, 13, 49, 59, 99
Skinner, B. F., 189
Skolimowski, Henryk, 15, 42, 163, 186

sociology of knowledge, 199–203, 215–216
Socrates, xvi, 95, 104, 109–110, 114, 137, 141, 147, 189
Sohn-Rethel, Alfred, 84
Solomons, T. W. G., 235
soul, 104
Soviet Union, 14, 189, 144, 161, 215
space
  acoustic, 153
  humanization, xvi, 213–214, 219, 227ff., 235
  travel, 83, 93, 97, 100, 109, 110, 113, 117, 137, 209, 213, 217–218, 227–228, 236
spectroscope, 58
speech (*see* language-spoken)
Spencer, Herbert, 58, 74, 84, 230, 237
Spitzer, Michael, 158
Sputnik, 214
Stalin, Joseph, 7
Star Wars Defense (*see* SDI)
Stavis, Gene, 188
Stearn, Gerald, 157
Stebbins, G. Ledyard, 13, 59, 85
Steele, E. J., 13
Stengers, Isabelle, 40
Stent, Guenther, 42, 47, 58, 192
Stradanus, Johannes, 132, 154
Straus, Donald B., 190, 219
Stroke, George, 148
subjectivism, 4, 82, 161 (*see also* mind-body problem; quantum mechanics)
  in art and photography, 139–141
Sullivan, Walter, 215

Tagliocozzi, G., 214
Tarski, Alfred, 6, 14
Taylor, Chuck, 217

technology (*see also* knowledge)
  accidents, 93, 109, 113
  amplification of primary cognition (*see* computers)
  and conquest of time, 138–140 (*see also* permanence; photography)
  and human transformation, 90, 98
  and natural laws, 193, 215 (*see also* Kant)
  and rationality (presumed perversion of), 37, 224–228, 233–234, 235
  "anthropotropic" evolution of, xiii, xiv, 12, 40, 42, 46, 56–57, 61, 63ff.
  as embodiment of knowledge (first order), 75–78, 80–81, 89, 92–94, 109
  epistemological union with, 94–96
  evolution of, 71, 74–76, 90ff., 230–233 (*see also* anthropotropic)
  experience extending, 100–101, 103, 147, 161 (*see also* telescope)
  meta-cognitive (second order), 81, 84–85, 94, 96–97, 106ff., 146–148
  philosophic avoidance of, 65–66
  pre-human, 65, 90, 91, 94, 109, 151, 193
  remedial, 212, 219, 225–226
  role of collateral, 187
  speech as, 61, 83–84, 104–105
  unintended application of, 80, 87, 147
telegraph, 118, 126, 138, 141, 142, 146, 159, 198, 209
teleology, 32ff. (*see also* change)

telepathy, 121
telephone, xv, 68, 87, 121–122, 125, 126, 141, 142, 159, 160, 198, 205, 212, 217, 219
teleportation, 146
telescope, xiv, 89–90, 96ff., 107, 110, 111, 113, 134, 141, 147, 154, 164, 185
television, 106, 108, 117–118, 126, 137, 142ff., 149, 159–161, 194, 205, 218, 225–226
  critiques of, 143, 160, 225
Thales, 147, 161
Third World, 159
Thomas, Hugh, 153
Thomas, Sari, 38, 219
Thornborough, Laura, 158
Three Mile Island, 111, 236
three worlds, Popper's philosophy of, 76–80, 177–178
  techno-materialist reformulation of, 79, 86, 91
tools, 77–78, 85, 87, 92, 95, 151
touch, perceptual primacy of, 20–21, 99
Toulmin, Stephen, 215
Townsend, Carl, 214, 216
toys, new technologies as, 158
transcatalytic polymers, 109 (*see also* Eigen)
transcendentalism, 42
transportation, technologies of, xv, 97, 145–146, 209ff., 218
Treisman, Michel, 188
Truffaut, François, 188
Truitt, W. H., 235
truth, 3–7, 11, 14, 45, 70–71, 78, 100, 216 (*see also* environment; reality)
Turkel, Sherry, 217
Turoff, Murray, 205, 216
typewriter, 136, 155

unintended consequences (*see* technology-unintended)
Union of South Africa, 160
universe, 33–34, 46, 57, 77, 79–80, 82, 89, 91, 100
Utsumi, Takeshi, 151

van Dam, Andries, 155
van Gelder, Lindsay, 217
Van Loon, Hendrik, 149
Vico, Giambattista, 162, 200, 214
video
  camera, 124, 142, 186
  cassette recorders, 225–226
  "samizdat", 144
videophone, 142, 145, 159, 206
Vienna Circle, 197, 199, 200, 210, 214–215
Vikings, 116, 133, 154
Virtual Classroom, 216
vision, 68, 89–90, 96–97, 99–100, 111, 112, 158
  as surrogate for locomotion, 20–21, 85, 112
  lobster's, 185, 195
  nutritional basis of, 85, 111
  technological extension of (*see* microscope; photography; telescope)
  transportive extension of, 97
Vollmer, Gerhard, 83
Vonèche, Jacques, 61
Vozick, Tina, xvii, 214, 216
Vrba, Elizabeth, 59

Wachtel, Edward, 187, 219
Wade, Nicholas, 215
Waechtershaeuser, Guenter, 85, 111, 112
Wallace, A. R., 107
Walters, Ray, 157
Wartofsky, Marx, 234–235

"Watchmaker" theory of creation, 32–33
weaponry, 95, 149, 152, 154 (*see also* nuclear weapons)
Weaver, Warren, 112, 120, 150
Weizenbaum, Joseph, 114, 191, 194
Western Behavioral Sciences Institute (WBSI), 203–205, 216, 219
Whetmore, Edward Jay, 234
Whitehead, Alfred North, 14
Wiegall, Arthur, 153
Wiener, Norbert, 112, 158, 162, 194, 230, 237
Wigner, Eugene, 42
Wilder, Raymond L., 113
will, human, 47, 77
Williams, Archibald, 187
Wittenstein, Ludwig, 14, 60, 200–201
Wolf, J., 161
word processing (*see* electronic texts)
world, external (*see* reality; environment)
writing
  alphabetic, xv, 111, 122, 125, 130ff., 151, 153
  McLuhan's critique of, 153
  pictographic, 102, 129–131, 150, 151
  Socratic/Platonic critique of, 95, 109–110
Wyatt, J. N., 114

Xerox, 156, 161

Zeleny, M., 40
Zimmerli, Walther, 113
Zimmerman, Michael, 85
Zuckerman, Joseph, xvii, 157
Zworykin, Vladimir, 159
zygote, 26

The author thanks Sylvia Engdahl, Frank Giannizzero, and Tina Vozick for help in the preparation of this Index.

Research Annuals and Monographs in Series in ECONOMICS

Research Annuals

Advances in Accounting
Edited by Bill N. Schwartz, *School of Business Administration, Temple University*

Advances in Accounting Information Systems
Edited by Gary Grudnitski, *Graduate School of Business, The University of Texas at Austin*

Advances in Applied Micro-Economics
Edited by V. Kerry Smith, *Department of Economics, Vanderbilt University*

Advances in Business Marketing
Edited by Arch G. Woodside, *A.B. Freeman School, Tulane University*

Advances in Distribution Channel Research
Edited by Gary L. Frazier, *University of Southern California*

Advances in Econometrics
Edited by George F. Rhodes, Jr., *Department of Economics, Colorado State University* and Thomas Fomby, *Department of Economics, Southern Methodist University*

Advances in Financial Planning and Forecasting
Edited by Cheng F. Lee, *Department of Finance, University of Illinois*

Advances in Futures and Options Research
Edited by Frank J. Fabozzi, Visiting Professor, *Sloan School of Management, Massachusetts Institute of Technology*

Advances in Health Economics and Health Services Research
Edited by Richard M. Scheffler, *School of Public Health, University of California*, Berkeley and Louis F. Rossiter, *Department of Health Administration, Medical College of Virginia, Virginia Commonwealth University*

Advances in Industrial and Labor Relations
Edited by David B. Lipsky, *New York State School of Industrial and Labor Relations, Cornell University*

Advances in International Accounting
Edited by Kenneth S. Most, *College of Business Administration, Florida International University*

Advances in International Marketing
Edited by S. Tamer Cavusgill, *Center for Business and Economic Research, Bradley University*

Advances in Marketing and Public Policy
Edited by Paul N. Bloom, *Department of Marketing, University of North Carolina*

Advances in Mathematical Programming and Financial Planning
Edited by Kenneth D. Lawrence, *Department of Industrial and Systems Engineering, Rutgers University*, John B. Guerard, Jr. *Department of Finance, Lehigh University* and Gary R. Reeves, *Department of Management Science, University of South Carolina*

Advances in Nonprofit Marketing
Edited by Russell W. Belk, *Department of Marketing, University of Utah*

Advances in Public Interest Accounting
Edited by Marilyn Neimark, *Baruch College, The City University of New York*

Advances in Statistical Analysis and Statistical Computing
Edited by Roberto S. Mariano, *Department of Economics, University of Pennsylvania*

Advances in Taxation
Edited by Sally M. Jones, *Department of Accounting, The University of Texas at Austin*

Advances in the Economic Analysis of Participatory and Labor Managed Firms
Edited by Derek C. Jones, *Department of Economics, Hamilton College* and Jan Svejnar, *Department of Economics, University of Pittsburgh*

Advances in the Economics of Energy and Resources
Edited by John R. Moroney, *Department of Economics, Texas A&M University*

Advances in the Study of Entrepreneurship, Innovation and Economic Growth
Edited by Gary Libecap, Director, *Karl Eller Center, University of Arizona*

Advances in Working Capital Management
Edited by Yong H. Kim, *Department of Finance, University of Cincinatti* and V. Srinivasan, *College of Business Administration, Northeastern University*

Perspectives on Local Public Finance and Public Policy
Edited by John M. Quigley, *Department of Economics and Graduate School of Public Policy, University of California, Berkeley*

Research in Accounting Regulation
Edited by Gary John Previts, *Department of Accounting, The Weatherhead School of Management, Case Western Reserve University*

Research in Consumer Behavior
Edited by Elizabeth C. Hirschman, *Department of Marketing, New York University* and Jagdish N. Sheth, *School of Business, University of Southern California*

Research in Domestic and International Agribusiness Management
Edited by Ray A. Goldberg, *Graduate School of Business Administration, Harvard University*

Research in Economic History
Edited by Paul Uselding, *Department of Economics, University of Illinois*

Research in Experimental Economics
Edited by Vernon L. Smith, *College of Business and Public Administration, University of Arizona*

Research in Finance
Edited by Andrew H. Chen, *Edwin L. Cox School of Business, Southern Methodist University*

Research in Governmental and Nonprofit Accounting
Edited by James L. Chan, *Office for Governmental Accounting Research and Education, University of Illinois at Chicago*

Research in Human Capital and Development
Edited by Ismail Sirgeldin, *Department of Population Dynamics and Political Economy, The Johns Hopkins University*

Research in International Business and Finance
Edited by H. Peter Gray, *Department of Economics, Rutgers University*

Research in International Business and International Relations
Edited by Anant R. Negandhi, *Department of Business Administration, University of Illinois*

Research in Labor Economics
Edited by Ronald G. Ehrenberg, *New York State of Industrial and Labor Relations, Cornell University*

Research in Law and Economics
Edited by Richard O. Zerbe, Jr., *Graduate School of Public Affairs, University of Washington*

Research in Marketing
Edited by Jagdish N. Sheth, *School of Business, University of Southern California*

Research in Political Economy
Edited by Paul Zarembka, *Department of Economics, State University of New York at Buffalo*

Research in Population Economics
Edited by T. Paul Schultz, *Department of Economics, Yale University*

Research in Public Sector Economics
Edited by P.M. Jackson, *Department of Economics, Leicester University*

Research in Real Estate
Edited by C.F. Sirmans, *Department of Finance, Louisiana State University*

Research in the History of Economic Thought and Methodology
Edited by Warren J. Samuels, *Department of Economics, Michigan State University*

Research in Transportation Economics
Edited by Andrew F. Daughty, *Department of Economics, The University of Iowa* and Clifford Winston, *The Brookings Institute*

Research in Urban Economics
Edited by Robert Ebel, Director, *Economics and Finance, Corporate Competitive Strategies, Northwestern Bell, Minneapolis*

Research on Technological Innovation, Management and Policy
Edited by Richard S. Rosenbloom, *Graduate School of Business Administration, Harvard University*

Monographs in Series and Treatises

Contemporary Studies in Applied Behavioral Science
Edited by Louis A. Zurcher, *School of Social Work, The University of Texas at Austin*

Contemporary Studies in Economic and Financial Analysis
Edited by Edward I. Altman and Ingo Walter, *Graduate School of Business Administration, New York University*

Contemporary Studies in Energy Analysis and Policy
Edited by Noel D. Uri, *Division of Antitrust, Bureau of Economics Federal Trade Commission*

Decision Research: A Series of Monographs
Edited by Howard Thomas, *Department of Business Administration, University of Illinois*

Handbook of Behavioral Economics
Edited by Benjamin Gilad and Stanley Kaish, *Department of Management Studies, Rutgers University, Newark*

Industrial Development and the Social Fabric
Edited by John P. McKay, *Department of History, University of Illinois*

Political Economy and Public Policy
Edited by William Breit, *Department of Economics, Trinity University* and Kenneth G. Elzinga, *Department of Economics, University of Virginia*

Please inquire for detailed subject catalog

JAI PRESS INC., 55 Old Post Road No. 2, P.O. Box 1678
Greenwich, Connecticut 06836
Telephone: 203-661-7602   Cable Address: JAIPUBL